T0257991

Handbook of Solar Radiation

Handbook of Solar Radiation

Edited by **Catherine Waltz**

New York

Published by Callisto Reference,
106 Park Avenue, Suite 200,
New York, NY 10016, USA
www.callistoreference.com

Handbook of Solar Radiation
Edited by Catherine Waltz

International Standard Book Number: 978-1-63239-413-2 (Hardback)

This book contains information obtained from authentic and highly regarded sources. Copyright for all individual chapters remain with the respective authors as indicated. A wide variety of references are listed. Permission and sources are indicated; for detailed attributions, please refer to the permissions page. Reasonable efforts have been made to publish reliable data and information, but the authors, editors and publisher cannot assume any responsibility for the validity of all materials or the consequences of their use.

The publisher's policy is to use permanent paper from mills that operate a sustainable forestry policy. Furthermore, the publisher ensures that the text paper and cover boards used have met acceptable environmental accreditation standards.

Trademark Notice: Registered trademark of products or corporate names are used only for explanation and identification without intent to infringe.

Printed in the United States of America.

Contents

Preface

This book is an outcome of the contributions made by eminent scientists and experts in the field of solar radiation from all over the world. It focuses on the basics of solar radiation. It also discusses ecological impacts of solar radiation. The book covers numerous topics on solar radiation such as measurements and analysis of solar radiation, and also discusses in detail agricultural application – bioeffect. It provides scientific understanding on solar radiation for reference to researchers and students.

Various studies have approached the subject by analyzing it with a single perspective, but the present book provides diverse methodologies and techniques to address this field. This book contains theories and applications needed for understanding the subject from different perspectives. The aim is to keep the readers informed about the progresses in the field; therefore, the contributions were carefully examined to compile novel researches by specialists from across the globe.

Indeed, the job of the editor is the most crucial and challenging in compiling all chapters into a single book. In the end, I would extend my sincere thanks to the chapter authors for their profound work. I am also thankful for the support provided by my family and colleagues during the compilation of this book.

Editor

Section 1

Introduction

Solar Radiation, a Friendly Renewable Energy Source

E. B. Babatunde
Covenant University, Ota, Ogun State,
Nigeria

1. Introduction

'let there be light and there was light' , Genesis 1:1. This quotation from the Holy Bible refers to the coming into being, the **"Sun"**; thus **"Energy"**, by the spoken words of God. The sun is a common feature in our sky; it is seen crossing the sky from one extreme horizon to the other every day, giving us light and heat. However, little did the world realize what a prodigious and free source of energy God has made available for mankind. Among the alternative renewable energy sources, solar power is a prime choice in developing affordable, discentralizable global power source that can be adopted for use in all climate zones around the world. This energy is free but the equipment to collect it and convert it to useable energy can be costly. Energy is radiated from the sun in all directions in space in the form of electromagnetic radiations (sun rays). The average amount of solar energy radiated to earth is about $1kW/m^2$, depending on the latitude and regional weather pattern of a location on the Earth's surface (Green, 2001).

1.1 The uncertainty of fossil fuel energy sources to meet world's energy demand

Before we go to the specifics of solar radiation and solar energy applications, we will discuss the inadequacy of the fossil fuels to meet the energy demands of the world now and in future and the potential dangers inherent in continue to use them.

The known conventional energy sources are: fossil fuels, which include coal,oil,natural gas and nuclear. Among the conventional energy sources, fossil fuels are the chief and the world' s current main sources of energy .

The fossil fuels are unfortunately depleting fast to a point where it is unlikely to be able to sustain the great rate of the world energy consumption within the next 200 years. It is in fact understood that about 80% of the world's oil reserves have been consumed by 1980 at the rate of the world energy consumption in 1975 (Meinel and Meinel 1975) The remaining reserves of coal in the world is estimated to last for about 25 years, while the life expectancy of the oil and gas reserves in the world is not positively known.

As of now oil remains the chief source of energy of the world. According to Eden (1983) the projected world total energy demand, if oil were only the source, is 130×10^6 barrels per day by the year 2000 whereas at that time the possible production of it is put at about 53×10^6

barrels per day. This would represent about 38.5% of demand. This indicates the incapability of oil to continue to meet the energy demand of the world.

As the world population increases and the economic standard of third world countries improves, there is an expectation of an unprecedented rise in the global energy demands. To allow the traditional energy sources, that is, fossil, nuclear, or hydro fuel to meet these increasing energy demands now and for too long in the future will be unwise and suicidal. The reasons for this strong opinion being:

- There is a strong international consensus on the threat of dangerous climate change due to pollutants emitted from fossil fuels powered engines. This threat is heightened by the rapidly increasing demand for fossil fuels, which in recent years propelled the price of crude oil above US$ 60 per barrel for the first time. This has demonstrated that production of "cheap" fossil fuels, which we may deplete by the middle of this century, can no longer cope with the demand. We therefore have to pay more to quickly bring about dangerous climate change and, if we survive that, wait for the highly probable energy crisis.
- The ecological impact of turning every river into a dam for hydroelectric power if possible, is scary and hard to imagine.

It has also been recognized that the heavy reliance on fossil fuel has had an adverse impact on the environment. For example, gasoline engines and steam-turbine power plants that burn coal or natural gas send substantial amount of sulphurdioxide (SO_2) and nitrogen oxides (NO_2) into the atmosphere. When these gases combine with atmospheric water vapor, they form sulphuric and nitric acids, giving rise to highly acidic precipitations which are very dangerous to plants and human beings. Further more, the combustion of fossil fuels also releases carbon dioxide into the atmosphere; the amount of this gas in the atmosphere has been observed to have steadily risen since the mid 1800, largely as a result of the growing consumption of coal, oil and natural gas. More and more scientists believe that the atmospheric built up of carbon dioxide (along with that of other industrial gases such as methane and chlorofluorocarbon) may induce a green house effect, causing the rising of the surface temperature of the earth by increasing the amount of heat trapped in the lower atmosphere. This condition could bring about climate changes with serious repercussions for natural and agricultural ecosystems.

Similarly, nuclear power generation as a source of alternative energy faces lots of social objections due to the possible radiation hazard that it may cause during production. Scientists cannot estimate the extent and gravity of destruction, both immediate and long term, that nuclear radiation hazard can cause when nuclear power reactor accident occurs such as the case of the Russian's Chernobyl nuclear power plant accident in 1987, and the recent nuclear energy plants accident(tsunamis) in Japan, which gravity and extent of damage to life and properties cannot now be estimated and for how long the damaging radiation will be absolutely controlled. By this many countries are signing off nuclear energy utilization.

Moreover the nuclear power material if inappropriately stored could end up in wrong hands and get turned into weapon of mass destruction that will make terrorism assume a much more dangerous dimension.

However, nuclear energy is hoped to be potentially capable of at least deferring the world energy starvation for a long time. In fact it may be capable of taking over the bulk of energy supply as the fossil fuels become exhausted.

2. The sun, origin of solar energy

Here we will not bother ourselves with detailed specifications of the Sun, but give us just some relevant data of it.

The Sun is one of the many billion of stars in the Milky Way Galaxy, the galaxy of our solar system in the universe. It is the closest star to our planet earth; its effect and importance to us on the earth results from its closeness.

The sun is learnt to be formed about 5000 million years ago(Okeke and Soon 2004,). It is a great ball of hot gases with diameter of about $1.4x$ 10^6km, which is about 109 times that of the earth, and it is about $1.5 x 10^8$km distant from the earth. It is the most important celestial object to us because it is the source which supplies the energy that allows life to flourish on earth.

The energy of the Sun is derived from a process similar to that of nuclear fusion in which hydrogen nuclei are believed to combine to form helium nucleus. The excess mass in the process is converted to energy in accordance with Einstein's theory i.e., **E= mc²**.

Thus, the Sun produces a vast amount of energy but only a tiny part of it reaches the earth. The energy comes from the nuclear fusion occurring at the core of the sun. The sun is a stable star, it thus promises to remain at the same magnitude of its properties and surface temperature for a long time. It is interesting to note that the Sun is not one of the hot stars, but one of the cooler stars. Cooler stars are yellow in colour and the Sun is yellow in colour. Yet its heat from 93million miles away is very effective in keeping us warm and sustains lives on our planet earth.

The Sun radiates about $3.86 x 10^{26}$ Joules of energy every second, a value which is more than the total energy man has ever used since creation. Although some of this energy is lost in the atmosphere, the amount reaching the earth's surface every second, if properly harnessed, is still probably enough energy to meet the world's energy demand (Maniel, 1974). Today it is a common knowledge that the Sun is the primary source of energy for all the processes taking place in the earth-atmosphere system. All lives on earth depend upon its radiant energy directly or indirectly to survive.

The Sun, therefore, is one of the popular emerging feasible sources of energy being looked into and sought by the world today for long–term, possible source of renewable and reliable energy. The Sun is available free for all land and mankind. It is free of politics. It only needs suitable devices to capture its rays and translate it into useful heat or work.

The amount of solar energy available for any land depends only on its location with respect to the Sun. If we examine the following expression for the solar energy available at the top of the atmosphere of any location from the sun,

$$H_o = 24/\pi \ I_{sc} \ Cos\varphi Cos\delta (Sin\omega_s - \pi/180)\omega_s Cos\omega_s \qquad (1)$$

two angles in this expression are related to the location of a site on the earth's surface with respect to the sun:

Φ, the latitude, and δ, the declination angle of the Sun.

The amount of solar energy received per unit area per second at the outer edge of the earth's atmosphere above a site is known as Extraterrestrial radiation, and is about 3.0×10^{26} Joules. The extraterrestrial radiation being received at the normal incidence (i.e. Sun – earth average distance) at the outer edge of the atmosphere of a site is known as the solar constant I_{sc} which is about $4921 kJm^{-2}h^{-1}$.

If the Sun emits energy as said above, in form of electromagnetic radiation given by

$$E = mc^2 \qquad (2)$$

where **m** is mass and **c** is velocity of light, the energy therefore, radiated by the sun, is equivalent to a mass loss by the sun every second and can be evaluated to be:

$$m = 3 \times 10^{26}/c^2 = 3.3 \times 10^9 \; kgs^{-1}$$

If the Sun thus loses mass at this rate, it can be estimated that the Sun may extinct in about 2×10^4 b years. Hence the energy of the Sun can be said to be in-exhaustible by the earth, i.e., the Sun is with us for some time to come.

However the amount of the energy reaching the earth's surface is about $1.00 \times 10^3 Wm^{-2}$ at noontime at the equator. The depletion of the Sun's energy as it passes through the atmosphere to the earth's surface, coupled with the seasonal, night and weather interruptions, constitutes the major impediment to the full realization of solar energy utilization. This notwithstanding, solar energy is proving by far the most attractive alternative source of energy for mankind.

Solar energy is pollution free, communitarian, conservational, decentralizable, adaptable, and the related devices to utilize it require very little or no maintenance, safe and cost effective. Solar energy utilization has come to stay as the possible future long–term energy resource. It can be argued that it is the only recurrent source, large enough to meet mankind demands of energy supply if properly harnessed. All other renewable energy sources depend directly or indirectly on the Sun for their existence.

3. Solar radiation fundamentals

3.1 Electromagnetic spectrum of the sun

The sun emits energy in form of electromagnetic waves which are propagated in space without any need of a material medium and with a speed, $c = 3 \times 10^8 \; ms^{-1}$. Electromagnetic radiation emitted by the Sun reaching out in waves extends from fractions of an Angstrom to hundreds of meters, from x – ray to radio waves.

An angstrom is a unit of length given by $1A = 10^{-8} \; cm = 10^{-4} \; \mu m$.

Electromagnetic radiations are usually divided into groups of wavelengths. The wavelength regions of principal importance to the earth and its atmosphere are the;

Ultraviolet (UV) – (0.3 – 0.4 µm) representing 1.2%
Visible (VIS) -   (0.4 - 0.74µm) representing 49%
Infrared (IR) -   (0.74 – 4. 0 µm) representing 49%

It was discovered that 99% of the Sun's radiant energy to the earth is contained in these wavelength regions, that is, between 0.3 and 4µm and comes mostly from the photosphere part of the sun.

4. Factors affecting the amount of solar radiation received on the earth surface

4.1 Astronomical factor

As said above, only a tiny portion of the energy of the sun reaches the earth's surface. The sun-earth distance constitutes one of the factors affecting the amount of solar energy available to the earth. The earth is known to be orbiting round the sun once in a year and at the same time rotates about its own axis once in a day. The two motions determine the amount of solar radiation received on the earth's surface at any time at any place. The path or the trajectory of the earth round the Sun is an elliptical orbit with the Sun located at one of the foci of the ellipse. The implication of this is that the distance of the earth from the sun is variant; hence the amount of radiation received on the earth surface varies. For example, the shortest distance of the Sun from the earth is called the perihelion, and is 0.993AU. (Astronomical unit of distance(AU)=1.496 ×10^8km). It takes place on December 21st.

On 4th of April and 5th of October the earth is just at 1AU from the sun, while on 4th of July, the earth is at its longest distance, 1.017AU from the sun; this position is called Aphelion. The path of the sun's rays thus varies with time of the day, season of the year, and position of the site on the earth's surface. It becomes shorter towards the noon time, it decreases towards the perihelion position and increases towards aphelion. Thus the variation in the sun-earth distance causes variation in the amount of solar radiation reaching the earth surface. The path of the sun's ray through the atmosphere is perhaps the most important factor in solar radiation depletion. It determines the amount of radiation loss through **scattering** and **absorption** in the atmosphere.

The eccentricity (E_0) of the elliptical orbit is expressed in terms of the sun-earth distance (r) and the average, r_0 of this distance over a year. It is given by

$$E_o = (r_0/r)^2 = 1+0.033 \cos(2\pi d_n/365) \tag{3}$$

where d_n is the Julian day number in the year. For example d_1=1 on January 1 and d_{365} =365 on December 31.

The elliptical motion of the earth round the sun gives rise to the seasons we experience on earth, and its rotation about its own axis determines the diurnal variation of the amount of radiation received. The amount of solar radiation received on a unit horizontal surface area per unit time at the top of the atmosphere is known as the **Extraterrestrial** radiation H_0, and is given by

$$H_o = 24/\pi \, I_{sc} \, E_o \cos \phi \cos \delta \, (\sin\omega_s-(\pi/180) \, \omega_s \cos \omega_s) \tag{4}$$

This equation gives the average daily value of extraterrestrial radiation, H_0 on a horizontal surface at the top of the atmosphere, while

$$I_0 = I_{sc} E_0 \cos \phi \cos \delta \ (\cos \omega_i - \cos \omega_s) \tag{5}$$

gives the average hourly value of the extraterrestrial radiation.
where ϕ is the latitude of the site,
δ is the declination angle of the sun
ω_i is the hour angle
ω_s is the sun set hour angle

The corresponding expressions for computing the extraterrestrial radiation on a tilted surface toward the equator at any latitude in the northern hemisphere are given by Igbal (1983). For the daily average, we have

$$H_{0\beta} = 24/\pi \ I_{sc} E_0 \ [(\pi/180) \ \omega'_s \sin\delta\sin(\phi-\beta) + \cos\delta\cos(\phi-\beta)\sin\omega'_s] \tag{6}$$

And for the hourly average, we have

$$I_{0\beta} = I_{sc} E_0 \ [\sin\delta\sin(\phi-\beta) + \cos\delta\cos(\phi-\beta)\cos\omega_i \tag{7}$$

where β is the angle of tilt toward equator

$$\omega'_s = \min\{\omega_s, \cos^{-1}[\tan\delta\tan(\phi-\beta)]\} \tag{8}$$

4.2 The atmospheric factor

The extraterrestrial radiation mentioned above is the maximum solar radiation available to us at the top of our atmosphere. The variable quantities affecting its amount at the ground surface are the astronomical factors mentioned above and the atmospheric factors.

Solar radiation however has to pass through the atmosphere to reach the ground surface, and since the atmosphere is not void, solar radiation in passing through it is subjected to various interactions leading to **absorption, scattering** and **reflection** of the radiation. These mechanisms result in depletion and extinction of the radiation, thus reducing the amount of solar radiation we receive at the ground surface of the earth. Several atmospheric radiation books describe and discuss these radiation depletion mechanisms.

5. Other radiation and atmospheric related parameters

The knowledge of radiation parameters, such as cloudiness index, clearness index, turbidity, albedo, transmittance, absorbance and reflectivity of the atmosphere through which the solar rays pass to the ground surface is very necessary for the utilization of solar energy. Also the knowledge of the meteorological parameters such as number of **sun shine hours** per day, **relative humidity, temperature, pressure, wind speed, rainfall** etc is desirable and important for accurate calculation of parameters of some solar energy devices. For example it is needed to know the average number of sun shine hours per day for accurate calculation of PV (photovoltaic) power needed in sizing solar power electrification for any location. In Nigeria, for example, we have an average of 4.5 hours of sunshine in a day. In detailed work, however, this value varies with geographical locations. Because of these, the measurement of solar radiation amount and its spectral distribution under all atmospheric conditions is undertaken at many radiation networks around the world (Babatunde and Aro, 1990).

The knowledge of the spectral distribution of solar radiation available is also important for development of semiconductor devices such as photo detectors, light emitting diodes, power diodes, photo cells, etc; it is also essential in the design of some special solar energy devices for the direct conversion of solar energy to electricity.

6. Solar radiation measurement and analysis

It is inevitable to know the potential of solar energy available on daily and monthly bases at the site for solar energy application, not only in amount but in quality, particularly its spectral composition. For this, the measurement of solar radiation energy and its spectral distribution under all atmospheric conditions is undertaken also at many radiation networks around the world.

Solar radiation energy arriving at the edge of the earth's atmosphere is carried or conveyed in electromagnetic spectrum, of wavelengths ranging from about 0.2μm to 4μm, as said above. These groups of wavelengths of the solar radiation are of principal importance to the earth and its atmosphere, especially for the calculation of absorption by gases, clouds and aerosols in the atmosphere and to calculate the spectral variation of the earth – atmosphere albedo, and also essential for photosynthesis, photobiology and photochemistry in the atmosphere.

6.1 Basic radiation measurements

The basic radiation fluxes being actively measured and studied in many radiation network stations globally include the sw-total (global) solar irradiance, sw-direct solar irradiance, sw-diffuse or sky irradiance. Other radiation fluxes measured are global and diffuse photosynthetic active radiation(PAR), ultraviolet total optical depth and the sun photometric measurement, and commonly measured radiation parameter is the sun shine hours. However the brief analysis here on radiation measurements is on the global (total) solar irradiance, H, direct solar irradiance, H_b, and diffuse sky irradiance ,H_d.

6.1.1 Global (total) solar irradiance

Global solar irradiance, H, which is the total sw-radiation flux, measured on a horizontal surface on the ground surface of the earth, comprising the direct sw- solar irradiance, H_b and diffuse sw- sky irradiance, H_d. In simple mathematics, the three fluxes are connected as in the following

$$H = H_b + H_d \qquad (9)$$

If all measurements were accurate, wherever two of these fluxes are measured, the third can easily be obtained, but this is not always so.

Global (total) solar radiation flux is the most easily and commonly measured of all the radiation fluxes in almost all the radiation network throughout the world. Measurement is done in the shortwave regions, 0.2 to 4.0μm wavelengths, which includes the photo synthetically Active Radiation (PAR).

The measurement is done to date, for example, at BSRN station, Physics Department University of Ilorin using Eppley Precision Spectral Pyranometer (PSP), serial number, SN17675F3 and 28866F3 with calibration constant of 8.2×10^{-6} V/ Wm^{-2} and well documented calibration history. Data quality is ensured by eliminating spurious errors that could arise from incidental and shading or partial un-shading of sensor by discarding all observations for which the insolation is less than 20Wm^{-2}. The data assembled on minute – by – minute basis was used to generate the hourly, daily and monthly averages.

6.1.2 Direct solar irradiance, H_b

The direct solar irradiance or solar beam $\mathbf{H_b}$, is the component of the total solar irradiance H, which comes directly from the top of the atmosphere, through the atmosphere, to the ground surface not deviated, nor scattered nor absorbed. The ratios of it to the total H i.e H_b/H and to the extraterrestrial radiation H_o, i.e H_b/H_o, are very important atmospheric radiation parameters in the radiative property of the atmosphere. H_b/H can be used to indicate the clearness of the atmosphere while H_b/H_o may be used to indicate the cleanness of the atmosphere and to determine the transmittance property of the atmosphere.

The direct solar irradiance is similarly measured like the global solar irradiance. It is measured using the Eppley solar tracker(NIP) with calibration constant $8.42 \times 10\text{-}6$V/ Wm^{-2}. Unfortunate the incessant power outage prevented the continuous functioning of this radiation sensor in many developing nations.Therefore the data of direct solar irradiance is here, as in many other stations, obtained by computation.

6.1.3 Diffuse sky irradiance, H_d

This radiation flux is also known as the sky radiation. It is short wave radiation, coming from the sky covering angular directions of 180^0 to the sensor. It is incident on the ground surface as a result of scattering and reflection by particles in the atmosphere. Its ratio to the total flux H, i.e H_d/H measures the cloudiness and turbidity of the sky and its ratio to the extraterrestrial radiation H_o i.e H_d/H_o is expected to measure the scattering co-efficient of the atmosphere.

This radiation flux is measured in same manner as those above. An Eppley Black and White Pyranometer model 8-48 with calibration constant 9.18×10^{-6} V/Wm^{-2}, with a shadow ring across the sensor, is used for the measurement. Unfortunately and inevitably the shadow ring may cut off some diffuse radiation, thus making the measurement to be inaccurate. This is why eqn.6 may not be valid or suitable to obtain the correct direct solar irradiance H_b.

7. Radiation fluxes formulae

As part of measurements, formulas for generating the different radiation fluxes: global (total) solar irradiance, H and its components, direct solar irradiance H_b, diffuse solar irradiance H_d, are developed to generate the required data of these radiation fluxes where they are required and are not regularly measured. Some of the expressions were developed in terms of other easily measured radiation and meteorological parameters. Numerous of these formulae exist, developed by many workers and published in relevant journals all over the world.

However many of them may not be applicable globally or valid at other geographical locations different from where they were generated(Page, 1964, Schulze,1976), while some of them may be applicable at geographical locations similar in latitude to where they were originated (Chuah et.al, 1981). Some of them are the Angstrom type (Angstrom,1924; Rietveld, 1978). Some are linear (Shears et.al, 1981 ; Glover and McCullouch 1958). Some are polynomials, some are parametric while some are indicial.

7.1 Total (global) solar radiation prediction formulae

Some prediction formulae for the radiation fluxes generated by the author include:

$$H/H_o = 0.329 + 0.315(s/S_m) \tag{10}$$

where:
H is the global (total) sw - solar irradiance been predicted.
H_o is the extraterrestrial at the top of the atmosphere of the site.
s/S_m is the fraction of sun shine hours at the site.

Eqn.10 is of the Angstrom type obtained by the author in 1995 at the BSRN station University of Ilorin (Babatunde,1995). Another is a multivariate one given by

$$H/H_o = 0.0189 + 0.2599(s/S_m) + 0.0027V + 0.0101T \tag{11}$$

where:
H_o and s/S_m are already defined in eqn 10.
V is the average visibility and T is the average ambient temperature at the location.

Eqns. 10 and 11 are formulae for estimating or generating global (total) solar radiation fluxes. Eqn.11 however is a multivariate expression. The magnitude of contributions by the meteorological variables in the expression to the amount of radiation obtainable at the location are indicated by their coefficients. The amount of global solar radiation predicted at the location depends, as can be observed from the equation, strongly on the variant, s/S_m, the number of sun shine hours, less on the ambient temperature T and much less on visibility V. The equation was developed by Babatunde and Aro (1996).

When tested, the value of global radiation flux predicted by eqn.10 was within 2.5% while that of eqn.11 was within 0.6%. Thus an equation developed in terms of multivariate metrological variables, although cumbersome, gives a better value of the radiation flux than the one in terms of one single variable. However for estimating values of the flux, H, for engineering purposes, the two equations are found to be adequate and reliable.

7.2 The diffuse radiation prediction formulas

Some formulas for computing the diffuse sky radiation were developed at various times and also in terms of related radiation and meteorological parameters by Babatunde (1995 ; 1999). Three of them, two of which are Angstrom type, are presented.

$$H_d/H = 0.4949 - 0.1148S_h \tag{12}$$

$$H_d/H = 0.945 - 0.971K_c \qquad (13)$$

$$H_d/H = 1 - K_c \qquad (14)$$

where H_d/H is known as the cloudiness index.
S_h is the fraction of sunshine hours.
K_c is the clearness index H/H_o.

When they were tested on the year 2000 radiation data, the values predicted by eqn.12 were within 18% while that of eqn.13 was within 11% and that of eqn.14 was within 19%. Therefore it can be said of these equations that they will adequately produce diffuse sky radiation data with reasonable accuracy. Eqn.13 is however the best of the three. It is of the Angstrom type, obtained as a result of experimental analysis and not as a result of regression analysis like others.

7.3 Direct radiation prediction formulas

Direct radiation component data is the most difficult to acquire because of the nature of the equipment for measuring it. Estimation of its values has therefore been relied upon to provide the data when needed.

The following formulas by the author for computing it were developed at various times (Babatunde, 1999; 2000)

$$H_b = H^2/H_o \qquad (15)$$

$$H_b/H = 0.308 + 0.424 \, H/H_o \qquad (16)$$

The two equations were developed in terms of the total radiation H and extraterrestrial radiation H_o. The two radiation fluxes, the predictors, are easily measured and computed respectively with very reasonable accuracy. Eqn.15, in particular, is a unique equation, developed purely from experimental results, Eqn.15 and eqn.16 will produce dependable values of the direct radiation data in all atmospheric conditions.

Some other equations developed for predicting H_b for specific atmospheric conditions are:

$$H_b/H = 0.341 + 0.571 \, K_c \qquad (17)$$

and

$$H_b/H = 0.247 + 0.415 \, K_c \qquad (18)$$

They have been tested and proven to be much more suitable for clear – sky conditions and cloudy – sky conditions respectively. They are equally as good as eqns. 15 and 16 above but only at the atmospheric conditions specified.

8. Solar energy applications

The major areas of application of solar energy are in the provision of low and high grade heat, direct conversion to electricity through Photovoltaic cells and indirect conversion to electricity through turbines.

Thus solar energy is utilizable through the principle of energy conversion from one form of energy to another. In this case, the thermal and electrical conversions of sun's energy make realizable, the various applications of solar energy. The various applications feasible and in practice are enumerated as follow.

8.1 Solar energy thermal conversion application

i. Production of hot water for domestic use.
ii. Cooling and Refrigeration.
iii. Solar passive drier in;
    a. Agriculture drying.
    b. Wood seasoning.
    c. Mushroom culturing or growing
    d. Production of pure water- distillation.

8.2 Solar electrical conversion application

i. Thermal to electricity conversion.
ii. Solar electric power systems (PV) Photovoltaic cell.
    a. Solar water pumping.
    b. Hydrogen Fuel.

There are some other types of solar electric power systems based on different technologies. Some of which are in practice and some are under development. Some of them are:

a. Crystalline silicon
b. Thin films
c. Concentrators
d. Thermo-photovoltaic
e. Organic solar cells

The first four are the major ones while the fifth one, under development, is a latest technology in solar energy conversion. It is related to thin film, and will be discussed latter in the chapter.

8.3 Thin films

Thin films will be developed to become a reliable and more efficient source for solar energy application. The principle of its applicability in the solar energy application is discussed under spectral selectivity properties of a surface in solar energy application. An organic solar cell is an example of such thin films. Solar electric thin films are lighter, more resilient, and easier to manufacture than crystalline silicon modules. The best developed thin film technology uses amorphous silicon in which the atoms are not arranged in any particular order as they would be in a crystal. An amorphous silicon film, only one micron thick, absorbs 90% of the useable solar radiation falling on it. Other thin film materials include cadmium, telluride and copper indium dieseline. Substantial cost savings are possible with this technology because thin films require little semiconductor materials. Thin films are also produced as large complete modules. They are manufactured by applying extremely thin layers of semi conductor materials unto a low –

cost backing such as glass or plastics. Electrical contacts, anti-reflective coatings, and protective layers are also applied directly to the backing materials. The films conform to the shape of the backing, a feature that allows them to be used in such innovation product as flexible solar electric roofing shingles.

8.4 Organic solar cells

This is a new solar energy electric conversion technology in which solar cell is currently being developed from various organic matters (dyes). They are sort of thin films discussed above. The crystallized silicon solar cells have being a standard technology in solar conversion devices for over fifty years. However they are still expensive, and relatively inefficient (they have achieved only 50% efficiency so far). Right now, various types of organic solar cells from dye materials are being studied and may soon replace the silicon solar cells, because they (organic solar cells) will be fabricated with greater efficiency, low cost processes, and they will be more versatile than silicon solar cells. Further still, they have added advantages of being thinner, lighter and more colourful than silicon solar cells.

9. Spectral selectivity surface applications

We now discuss a new specialized area of solar energy application, based on the spectral selectivity property of a surface. It is a new and special innovative concept in solar energy application.

It was discovered that optical properties of materials can be modified to select wavelengths of the solar spectrum to transmit, or absorb or reflect. On these principles the following applications are possibe:

- Selective absorbers,
- Heat mirrors,
- Reflective materials,
- Anti-reflective,
- Fluorescent concentrator,
- Holographic films,
- Cold mirrors,
- Radiative cooling,
- Optical switching,
- Transparent insulating materials,
- Solar control window.

Spectral selectivity of a surface is achieved by applying special coatings on substrates, which may be transparent or opaque, with the intention of modifying the optical properties of the surface, such that the surface selects wavelengths of the solar spectrum to transmit or absorb or reflect. These properties are: transmittance, absorbance, reflectance, emittance, absorption coefficient (α) extinction coefficient (k) refractive index (n) to mention a few, and upon which relevant applications are based. Surfaces of different material coatings will produce different values of these optical properties at different wavelengths of the solar spectrum.

Solar radiation is transverse oscillating electric and magnetic fields. The electromagnetic fields interact with the electric charges of the material of the surface on which solar radiation is incident. The interaction results in the modifications of the solar radiation at different parts of its spectrum. As a result, some parts of the radiation are absorbed, some are transmitted, and some are reflected back to space (Granquist, 1985; Lovern, et al, 1976). Thus, by spectral selectivity of a surface, it is meant surfaces whose values of absorptance, emittance, tramittance and reflectance of radiation and other related optical properties vary with wavelengths over the spectral region, $0.3 \leq \lambda \leq 3\mu m$ (Loven, et al. 1976; Maniel and Maniel, 1976).

For example, a spectral selective surface having high absorptance in the wavelength range $0.3\ \mu m \leq \lambda \leq 3\mu m$, and high reflectance at $3\ \mu m \leq \lambda \leq 100\ \mu m$ will appear black with regards to the short wavelengths range, $0.3\ \mu m \leq \lambda \leq 3\mu m$ and at the same time appear an excellent mirror in the thermal region, i.e. $3\ \mu m \leq \lambda \leq 100\ \mu m$. A device with these properties is called a "heat mirror".

We shall discuss briefly, for example, the principle of the following spectral selectivity applications of solar energy.

i.   Heat mirror
ii. Cold mirror
iii. Solar control coatings.

9.1 Heat mirror

A solar collector with a highly selective absorber in the short wavelengths range of solar radiation, that is, at $0.2 \leq \lambda \leq 3\mu m$, will reflect very highly the thermal radiation (IR) component of solar radiation. This implies that the device is black to this short wavelengths range because it absorbs them, and forms an excellent mirror in the thermal region because it reflects them. The device is called a "Heat mirror". Thus heat mirror is essentially a device that transmits or absorbs the short wavelengths radiation (UV – VIS) and reflects long thermal wavelengths (IR) of solar radiation. That is, it is a window to the short wavelengths and a mirror to the long wavelengths. Such a surface is therefore suitable for architectural windows in buildings, where low temperature or cooling effects is desired. This device therefore may be adaptable for passive cooling in a tropical climate region.

The heat mirror device is obtained by using a semiconductor–Metal Tandem. Thus, it can be called absorber-reflector Tandem. The semiconductor components are arranged to reflect the thermal radiation (IR), while the metal components absorb or transmit the UV – VIS radiation. A heat mirror device is also called a transmitting selective surface.

In the arrangement of the components, the reflective layer surface is arranged to cover the non-selective absorber base. In this way, the selective reflector reflects the thermal infrared radiation ($\lambda > 3\ \mu m$) and transmits the short wavelength range ($\lambda < 3\ \mu m$). The short wavelength radiation transmitted by the reflector is absorbed by the black absorber base. Some highly doped semi conductors such as InO_2, SnO_2 or the mixture of the two, Indium-Tin-Oxide (ITO), have been used successfully to produce the reflector component of the device (Seraphin, 1979). A heat mirror may therefore be used to separate heat radiation (IR) and light radiation (VIS) of the solar spectrum. The IR energy separated could be used for thermal purposes such as the thermo-photovoltaic.

9.2 Cold mirror coatings

Spectral splitting coatings can be used to divide solar spectrum into various broad band regions. By this, various regions of the solar spectrum can be separated for use for different purposes such as photovoltaic or photo thermal devices (Lambert, 1985).

A "cold mirror" device has opposite spectral response to that of the "heat mirror". That is, cold mirror films reflect highly (low transmittance) in the VIS region of solar spectrum and reflect poorly, but transmits highly in the IR region, thus splitting the spectrum into short wavelengths and long wavelengths. The high energy waves i.e. the short wavelengths are used for photovoltaic generation while the low energy waves, the long wavelengths (IR) are used for photo thermal heating. This device can be used in "green house" with special arrangements of baffles on the roofs. The device will reflect the photosynthetic active radiation (PAR), $0.35 \leq \lambda \leq 0.75$ µm waves into the green house while transmitting the IR into the air channels which can be redeployed to maintain a suitable warm temperature in the green house. ZnS/MgF_s and $T_i O_2/ S_i O_2$ have been used to achieve these coatings.

9.3 Solar control coating

Solar control coating is a design intended to reduce the incoming heat radiation through windows of a building by reflecting off the heat radiation (IR). To achieve comfortable indoor temperatures, that is, to achieve cooling in a building, solar control coating surfaces that are transparent at $0.4 \leq \lambda \leq 0.7$ µm and reflecting at $0.7 \leq \lambda \leq 3$ µm may be used for the material of the windows in the building. By this, the infrared part of the solar radiation is reflected back, which is possible through the use of solar control windows. A 50% reduction in the internal heating of a building without noticeable reduction in the lightning of the interior of the building had been achieved. The use of such windows may achieve the same objective of a controversial air conditioner in a building. Solar control coating are particularly applicable in hot climate countries such as Nigeria.

In solar control and energy conserving windows, low transmittance windows are employed. If the medium is generally opaque to the passage of radiation but selectively transmits a particular small range of radiation, it is said to operate as a window in that range. A low thermal transmittance window reduces the heat radiation through the window. To achieve low thermal transmittance window therefore, surface coatings that transmits at $0.3 \leq \lambda \leq 3$ µm and reflects at $3 \leq \lambda \leq 100$ µm may be used. This means that maximum use is made of the solar energy in the short wavelengths range while the transmittance of thermal radiation is minimized.

9.4 Solar control and low thermal emittance materials

A thin homogeneous metal film is found capable of combining transmission in short wavelengths up to about 50% with high reflectance in long wavelengths (Okujagu, 1997; Wooten, 1972). The required thickness of such film, using copper, silver and gold is about 20mm. if the films are thinner, they will break up into discreet islands of strong absorptance of visible wavelengths.

Enhancement of luminous transmittance to more than 80% without significantly impairing the low thermal transmittance can be achieved by embedding the metal in anti-reflecting dielectric with high refractive index layers, such as T_i and O_2. In the alternative to the metal base coatings, we may use dope Oxide semiconductor. However a wide band gap is needed in the semiconductor to permit high transmittance in the luminous and solar spectral range. To make the material metallic, electrically conducting and infrared reflecting for wavelengths exceeding a certain plasma wavelength, it requires doping to a significantly high level. Semiconductors suitable for this are: oxides based on zinc, cadmium, tin, lead and thallium and their alloys.

10. Conclusion

Energy is necessary for the growth of any nation, and for improving the standard of living of the nation. Therefore energy has to be made available and cheap by the nation for rapid and quality growth of the economy.

Fossil fuels energy, the main source of energy for the world for now, is unable to meet the world's demands of energy and it is, at the same time, rapidly depleting, hence the fever of the world's search for alternative sources of energy. Each country therefore faces the challenges of developing her energy resources.

The renewable energy sources, some of which are: **wind, marine, geothermal, biomass, bio-dieses, hydro-power, land fill and solar energy** have become object of research, because they could be the alternative dependable and feasible sources of energy that the world is looking for to meet her energy demands. They are truly possible alternative sources of energy if their technologies are developed and mastered. Out of them all, solar energy seems to be the most capable of meeting world energy demands if properly harnessed and made cheap. The amount of it received per second during the daylight on the earth's surface is 10,000 times more than the total energy requirement of the world today. The varieties of solar energy applications and advantages are enormous, only a few are mentioned and discussed very briefly in this chapter.

11. Summary of the chapter

The inadequacy and inability and the inherent danger in the use of fossil fuels energy and other conventional sources of energy to meet the worlds demands for energy both now and in the nearest future is highlighted and emphasized. The world eventually turning to the renewable energy sources, solar energy in particular, is inevitable, expected and wise. The inevitable impediment such as the earth's atmosphere and its effect on the passage of solar radiation, to the realization of full utilization of solar energy are identified.

The major possible uses that solar energy are put to are mentioned. A specialized and new area of utilizing solar energy, the area of spectral selectivity property of thin films of materials, are highlighted and discussed. Devices such as heat mirror "cold mirror" solar control windows, in buildings, which basic principle is spectral selectivity, to mention a few, are discussed.

12. References

Angstrom, A. K (1924). Solar and atmospheric radiation, *J. Roy Meteorol.Soc.* 50, pp. 121 -126.
Babatunde, E. B (1995) Correlation of fraction of sun shine hours with clearness index and cloudiness index at a tropical station Ilorin, *Nigeria. Nig.J. of Solar Energy* Vol 13, pp 22 -27

Babatunde E.B (1999). Direct solar radiation model at tropical *Station Ilorin. Nig.J. Renewable Energy.* Vol 7, 1 & 2 pp46-49

Babatunde, E. B and T. O. Aro, (1990): Characteristic variations of global (total) solar radiation at Ilorin, Nigeria. *Nig. J. solar energy*, 9,157-173.

Babatunde E.B and Aro T. O (1996). A multiple regression model for global Solar radiation at Ilorin, *Nigeria. Nig. J. Pure and Applied Sciences.* Vol 11, pp 471- 474.

Babatunde E.B and Aro T. O (2000)Variation characteristics of diffuse solar radiation at a tropical station Ilorin, *Nigeria. Nig. J. of Physics.* Vol. 12, 20 – 24. (NIP Nigeria)

Chuah, Donald G.S. and Lee, S.L. (1981). Solar radiation estimate in Malaysia. *Solar energy* 36, pp 33-40

Glover, J. and McCullouch, J.S.G (1958). The empirical relation between solar radiation and hours of sunshine. *Quart. J R. Met. Soc.84* pp 172 -175.

Granguist, C.G. (1985). Spectrally selective coatings for energy efficient windows paper presented at workshop on the Physics of non-conventional energy sources and material science for energy, *I.C.T.P., Trieste, Italy.*

Greenpeace (2001). Solar generation for the European PV industries association.

Lampert, C.M (1985). Workshop on the Physics of non- conventional energy sources and material science for energy, I.C.T.P., Trieste, Italy (143)

Lovel, M.C., Avery, A.G. and Vernon, M.W. (1976). Physical properties of materials, Von Nostrand Company New York.

Meinel, M.P and Meinel, A.B (1974). 1st course on solar energy conversion.

Okeke ,P.N and Soon, W.H (2004) Introduction to Astronomy and Astrophysics. SAN PRESS Ltd, 187 Agbani Rd Enugu, Nigeria.

Okujagu, C.U (1997). Effect of materials on the transmitting thin films,Nig. J. Phys., 59

Rietveld , M.R (1978). A new method for estimating the regression coefficients in the formula relating solar radiation to sunshine. Agr . Meteorol. 19, pp 243 – 252

Schulze, R.E (1976). A physically based method of estimating solar radiation from suncards. Agric. Meteorol. 16, pp 85-101

Seraphin, B.O(1979). In solar energy conversion, solid state physics aspects, B.O Seraphin (ed) topics in Applied Physics (Springer- Verlang Berlin), 63-76

Shears, R.D., Flocchini, R.G and Hatfield, J.L (1981). Technical note on correlation of total, diffuse, and direct solar radiation with the percentage of possible sunshine for Davis, California. Solar energy 27 (4), pp 357-360

Wooten, Fedric (1972). Optical properties of solids. Academic press New York.

Section 2

Solar Radiation
Fundamentals, Measurement and Analysis

Solar Radiation Models and Information for Renewable Energy Applications

E. O. Falayi[1] and A. B. Rabiu[2]

[1]*Department of Physics, Tai Solarin University of Education, Ijebu-Ode,*
[2]*Department of Physics, Federal University of Technology, Akure,*
Nigeria

1. Introduction

The Sun is a sphere of intense hot gaseous matter with a diameter of 1.39×10^9m and is about 1.5×10^{11}m away from the Earth. A schematic representation of the structure of the Sun is shown in Figure 1.1. The Sun's core temperature is about 16 million K and has a density of about 160 times the density of water. The core is the innermost layer with 10 percent of the Sun's mass, and the energy is generated from nuclear fusion. Because of the enormous amount of gravity compression from all the layers above it, the core is very hot and dense.

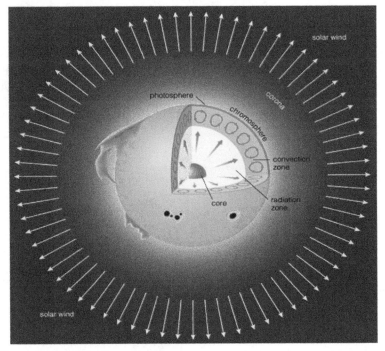

Fig. 1.1. The structure of the Sun

The layer next to it is the radiative zone, where the energy is transported from the sunspot interior to the cold outer layer by photons. Other features of the solar surface are small dark areas called pores, which are of the same order of magnitude as the convective cells and larger dark areas called sunspots, which vary in size. The outer layer of the convective cells is called the photosphere. The photosphere is the layer below which the Sun becomes opaque to visible light. Above the photosphere is the visible sunlight which is free to propagate into space, and its energy escapes the Sun entirely. The change in opacity is due to the decreasing amount of H− ions, which absorb visible light easily. The next layer referred to as the chromospheres, is a layer of several thousand kilometers in thickness, consisting of transparent glowing gas above the photosphere. Many of the phenomena occurring in the photosphere also manifest in the chromospheres. Because the density in the chromospheres continues to decrease with height and is much lower than in the photosphere, the magnetic field and waves can have a greater influence on the structure. Still further out is the corona which is of very low density and has a high temperature of about 1×10^{6}°K to 2×10^{6}°K.

The radiation from the sun is the primary natural energy source of the planet Earth. Other natural energy sources are the cosmic radiation, the natural terrestrial radioactivity and the geothermal heat flux from the interior to the surface of the Earth, but these sources are energetically negligible as compared to solar radiation. When we speak of solar radiation, we mean the electromagnetic radiation of the Sun. The energy distribution of electromagnetic radiation over different wavelength is called Spectrum. The electromagnetic spectrum is divided into different spectral ranges (Figure 1.2).

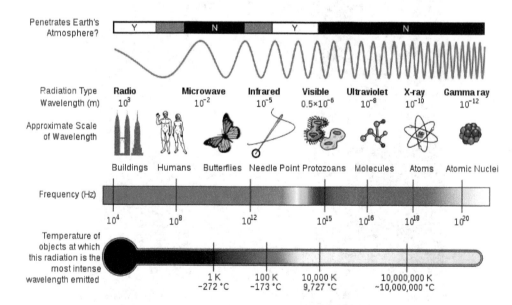

Fig. 1.2. Spectral ranges of electromagnetic radiation

Solar radiation as it passes through the atmosphere undergoes absorption and scattering by various constituents of the atmosphere. The amount of solar radiation finally reaching the surface of earth depends quite significantly on the concentration of airborne particulate matter gaseous pollutants and water (vapour, liquid or solid) in the sky, which can further attenuate the solar energy and change the diffuse and direct radiation ratio (Figure 1.3).

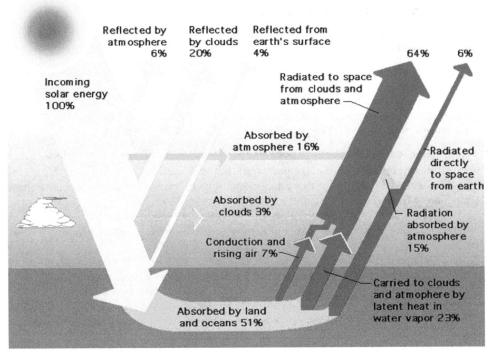

Fig. 1.3. Radiation balance of the atmosphere

The global solar radiation can be divided into two components: (1) diffuse solar radiation, which results from scattering caused by gases in the Earth's atmosphere, dispersed water droplets and particulates; and (2) direct solar radiation, which have not been scattered. Global solar radiation is the algebraic sum of the two components. Values of global and diffuse radiations are essential for research and engineering applications.

Global solar radiation is of economic importance as renewable energy alternatives. More recently global solar radiation has being studied due to its importance in providing energy for Earth's climatic system. The successful design and effective utilization of solar energy systems and devices for application in various facets of human needs, such as power and water supply for industrial, agricultural, domestic uses and photovoltaic cell largely depend on the availability of information on solar radiation characteristic of the location in which the system and devices are to be situated. This solar radiation information is also required in the forecast of the solar heat gain in building, weather forecast, agricultural potentials studies and forecast of evaporation from lakes and reservoir. However, the best solar radiation information is obtained from experimental measurement of the global and its components at the location. The use of solar energy has increased worldwide in recent years as direct and indirect replacements for fossil fuel, motivated to some degree by environmental concerns such were expressed in the Kyoto Protocol. As a result, a complete knowledge and detailed analysis about the potentiality of the site for solar radiation activity is of considerable interest.

1.2 Radiation fluxes at horizontal surface

The energy balance on a horizontal surface at the ground or on a solid body near the ground is given by

$$Q + K + H + L + W + P = 0 \tag{1.1}$$

Each term in this equation stands for an energy flux density or power density in Wm^{-2}. The vectorial terms in equation (1.1) are counted positive when they are directed towards the surface from above or below. The parameters have the following meaning.

Q =net total radiation=sum of all positive and negative radiation fluxes to the surface
K = Heat flux from the interior of the body (ground) to its surface
H = Sensible heat flux from the atmosphere due to molecular and convective heat conduction (diffusion and turbulence)
L = Latent heat flux due to condensation or evaporation at the surface.
W= Heat flux due to advection that is heat transported by horizontal air current. W is set zero if:
   a. the measuring surface is located at a horizontal and homogeneous plane of sufficient extension so that the so called Katabatic flow is negligible
   b. the measuring time is small compared to time of an air mass exchange.
P = Heat flux brought to the surface by falling precipitation. P is often not taken into consideration because the measurements are confined to times without precipitation (Kasten, 1983).

The net total radiation Q is at daytime, to be compensated by the heat fluxes K, H and L the net total radiation Q in equation (1.1) given

$$Q = (G - R) + (A - E) \tag{1.2}$$

Q is called the total radiation balance.
G= global radiation = sum of direct and diffuse solar radiation on the horizontal surface
R= reflected global radiation = fraction of G which is reflected by the body (ground)
A= atmospheric radiation = downward thermal radiation of the atmosphere (from atmosphere gases, mainly water vapour and from clouds)
E= terrestrial surface radiation = upward thermal radiation of the body (ground).
G and R are solar or shortwave radiation fluxes therefore

$$Q_s = G - R \tag{1.3}$$

Is called net solar or net global radiation, or short wave radiation balance. A and E are terrestrial or long wave radiation fluxes so that

$$Q_t = A - E \qquad (1.4)$$

Is called the long wave radiation balance and

$$-Q_t = E - A \qquad (1.5)$$

the (upward) net terrestrial surface radiation.

The short-wave radiation fluxes exhibit a pronounced variation during day light hours; the long-wave radiation fluxes vary but slightly because the temperature of atmosphere and ground vary during the day.

The ratio

$$Q_s = \frac{R}{G} \qquad (1.6)$$

is called short-wave radiation of the body

Terrestrial surface radiation E is composed of two terms:

1. The thermal radiation emitted by the body ground i.e.

$$E_1 = \alpha_t \sigma T^4 \qquad (1.7)$$

where α_t is called effective long-wave absorptance of the surface, slightly depending on temperature T. σ is called Stefan Boltzman constant = 5.6697 x 10[8] Wm-2K-4.

2. Reflected atmosphere radiation

$$E_2 = Q_t \cdot A \qquad (1.8)$$

where $Q_t = 1 - \alpha_t$ = effective long-wave reflectance of the surface. Thus E is strictly given by

$$E = E_1 + E_2 = \alpha_t \cdot \sigma T^4 + (1 - \alpha_t) \cdot A \qquad (1.9)$$

1.3 Solar declination angle

The angle that the Sun's makes with equatorial plane at solar noon is called the angle of declination. It varies from 23.45º on June 21 to 0º on September 21 to -23.45º on December 21, to 0º on March 21. It also defined as the angular distance from the zenith of the observer at the equator and the Sun at solar noon.

The axis of rotation is tilted at an angle of 23.45° with respect to the plane of the orbit around the Sun. The axis is orientated so that it always points towards the pole star and this accounts for the seasons and changes in the length of day throughout the year. The angle between the equatorial plane and a line joining the centres of the Sun and the Earth is called the *declination angle* (δ) Because the axis of the Earth's rotation is always pointing to the Pole Star the declination angle changes as the Earth orbits the Sun (Figure 1.4)

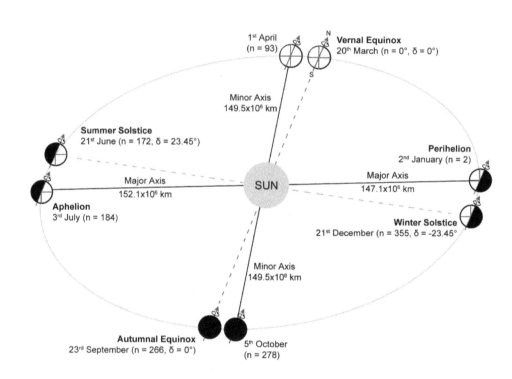

Fig. 1.4. Orbit of the Earth around the Sun

On the summer solstice (21st June) the Earth's axis is orientated directly towards the Sun, therefore the declination angle is 23.45° (Figure 1.4). All points below 66.55° south have 24 hours of darkness and all point above 66.55° north have 24 hours of daylight. The sun is directly over head at solar noon at all points on the Tropic of Cancer. On the winter solstice (21st December) the Earth's axis is orientated directly away from the Sun, therefore the declination angle is -23.45° (Figure 1.4). All points below 66.55° south have 24 hours of daylight and all point above 66.55° north have 24 hours of darkness. The sun is directly over head at solar noon at all points on the Tropic of Capricorn. At both the autumnal and vernal equinoxes (23rd September and 21st March respectively) the Earth's axis is at 90° to the line that joins the centres of the Earth and Sun, therefore the declination angle is 0° (Figure 1.4).

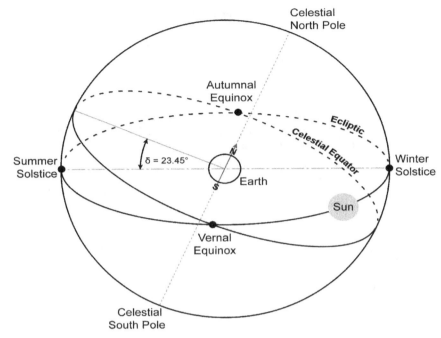

Fig. 1.5. The celestial sphere. Declination angle (δ) is the declination angle which is maximum at the solstices and zero at the equinoxes.

The equation used to calculate the declination angle in radians on any given day is

$$\delta = 23.45 \frac{\pi}{180} \sin\left[2\pi \left(\frac{284 + n}{365.25} \right) \right] \qquad (1.10)$$

where:
δ = declination angle (rads)
n = the day number, such that n = 1 on the 1st January and 365 on December 31st.

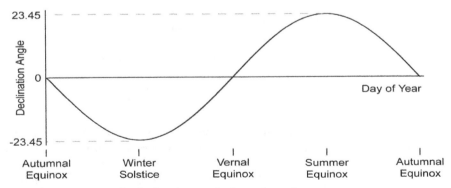

Fig. 1.6. The variation in the declination angle throughout the year.

The declination angle is the same for the whole globe on any given day. Figure 1.6 shows the change in the declination angle throughout a year. Because the period of the Earth's complete revolution around the Sun does not coincide exactly with the calendar year the declination varies slightly on the same day from year to year.

1.4 Solar hour angle

The hour angle is positive during the morning, reduces to zero at solar noon and increasingly negative when the afternoon progresses. The following equations can be used to obtain the hourly angle when various values of the angles are known.

$$\sin w = -\frac{\cos \alpha \sin A_z}{\cos \delta} \qquad (1.11)$$

$$\sin w = \frac{\sin \alpha - \sin \delta \sin \phi}{\cos \delta \cos \phi} \qquad (1.12)$$

Where
α = altitude angle
w = the hour angle
A_z = the solar azimuth angle
ϕ = observer angle
δ = declination angle

The hour angle is equals to zero at solar noon and since the hour angle changes at 15° per hour, the hour angle can be calculated at any time of day. The hour angles at sunrise (negative angle) and sunset (ws) is positive angle. They are important parameters and can be calculated from

$$\cos ws = -\tan \phi \tan \delta \qquad (1.13)$$

$$ws = \cos^{-1}\left(-\tan \phi \tan \delta\right) \qquad (1.14)$$

$$L = \frac{2}{15}\cos^{-1}\left(-\tan \phi \tan \delta\right) \qquad (1.15)$$

This L is known as the Length of the day also known as the maximum number of hour of insolation.

1.5 Solar constant

The solar constant is defined as the quantity of solar energy (W/m²) at normal incidence outside the atmosphere (extraterrestrial) at the mean sun-earth distance. Its mean value is 1367 W/m². The solar constant actually varies by +/- 3% because of the Earth's elliptical orbit around the Sun. The sun-earth distance is smaller when the Earth is at perihelion (first week in January) and larger when the Earth is at aphelion (first week in July). Some people, when talking about the solar constant, correct for this distance variation, and refer to the solar constant as the power per unit area received at the average Earth-solar distance of one "Astronomical Unit" or AU which is 1.49 x 10^8 million kilometres (IPS and Radio Services).

2. Empirical equations for predicting the availability of solar radiation

2.1 Angstrom-type model

Average daily global radiation at a specific location can be estimated by the knowledge of the average actual sunshine hours per day and the maximum possible sunshine hour per day at the location. This is done by a simple linear relation given by Angstrom (1924) and modified by (Prescott, 1924).

$$\frac{G}{G_O} = a + b\left(\frac{S}{S_{max}}\right) \tag{2.1}$$

In Nigeria, the hourly global solar radiation were obtained through Gun Bellani distillate, and were converted and standardized after Folayan (1988), using the conversion factor calculated from the following equations.

$$G = (1.35 \pm 0.176)H_{GB} KJ / m^2 \tag{2.2}$$

Where G is the monthly average of the daily global solar radiation on a horizontal surface at a location (KJ/m²-day), G_0 is the average extraterrestrial radiation (KJ/m²-day). S is the monthly average of the actual sunshine hours per day at the location. S_{max} monthly average of the maximum possible sunshine hours per day, n is mean day of each month.

$$G_o = \frac{24 \times 3600}{\pi} Gsc\left(1 + 0.033Cos\frac{360n}{365}\right)\left(Cos\phi Cos\delta SinW_S + \frac{2\pi W_S}{360} Sin\phi Sin\delta\right) \tag{2.3}$$

$$S_{max} = \frac{2}{15}\cos^{-1}\left(-\tan\phi\tan\delta\right) \tag{2.4}$$

Several researchers have determined the applicability of the Angstrom type regression model for predicting global solar irradiance (Akpabio et al., 2004; Ahmad and Ulfat, 2004; Sambo, 1985; Sayigh, 1993; Fagbenle, 1990; Akinbode, 1992; Udo, 2002; Okogbue and Adedokun, 2002; Halouani et al., 1993; Awachie and Okeke, 1990; El –Sebaii and Trabea; 2005, Falayi and Rabiu, 2005; Serm and Korntip, 2004; Gueymard and Myers, 2009; Skeiker, 2006; Falayi et al., 2011). Of recent (Akpabio and Etuk 2002; Falayi et al., 2008; Bocco et al., 2010; Falayi et al., 2011) have developed a multiple linear regression model with different variables to estimate the monthly average daily global. Also, prognostic and prediction models based on artificial intelligence techniques such as neural networks (NN) have been developed. These models can handle a large number of data, the contribution of these in the outcome can provide exact and adequate forecast (Krishnaiah, 2007; Adnan, 2004; Lopez, 2000; Mohandes et al., 2000).

2.2 Method of model evaluation

2.2.1 Correlation coefficient (r)

Correlation is the degrees of relationship between variables and to describe the linear or other mathematical model explain the relationship. The regression is a method of fitting the linear or nonlinear mathematical models between a dependent and a set of independent variables. The square root of the coefficient of determination is defined as the coefficient of correlation r. It is a measure of the relationship between variables based on a scale ± 1. Whether r is positive or negative depends on the inter-relationship between x and y, i.e. whether they are directly proportional (y increases and x increases) or vice versa (Muneer, 2004).

2.2.2 Correlation of determination (r^2)

The ratio of explained variation, $(G_{pred} - G_m)^2$, to the total variation, $(G_{obser} - G_m)^2$, is called the coefficient of determination. G_m is the mean of the observed G values. The ratio lies between zero and one. A high value of r^2 is desirable as this shows a lower unexplained variation.

2.2.3 Root mean square error, mean bias error and mean percentage error

The root mean square error (RMSE) gives the information on the short-term performance of the correlations by allowing a term-by-term comparison of the actual deviation between the estimated and measured values. The lower the RMSE, the more accurate is the estimate. A positive value of mean bias error (MBE) shows an over-estimate while a negative value an under-estimate by the model. MPE gives long term performance of the examined regression equations, a positive MPE values provides the averages amount of overestimation in the calculated values, while the negatives value gives underestimation. A low value of MPE is desirable (Igbai, 1983).

$$MBE = \frac{1}{n}\left[\sum\left(G_{pred} - G_{obs}\right)\right] \qquad (2.5)$$

$$RMSE = \left\{\left[\frac{1}{n}\sum\left(G_{pred} - G_{obs}\right)^2\right]\right\}^{\frac{1}{2}} \qquad (2.6)$$

$$MPE = \left[\sum\left(\frac{G_{obs} - G_{pred}}{G_{obs}} \times 100\right)\right] / n \qquad (2.7)$$

3. Monthly mean of horizontal global irradiation

Monthly mean global solar radiation data leads to more accurate modelling of solar energy processes. Several meteorological stations publish their data in terms of monthly-averaged values of daily global irradiation. Where such measurements are not available, it may be possible to obtain them from the long-term sunshine data via models presented in Chapter 4.

3.1 Monthly variation of extraterrestrial and terrestrial solar radiation

In order to obtain the pattern variation of monthly mean values of extraterrestrial (G_O) solar radiation, equation (2.3) is used in calculating it for various locations for which the measured global insolation is available. The calculated values are without any atmospheric effects. Based on the calculated values of extraterrestrial horizontal insolation for locations and the measured global insolation on a horizontal surface for the same locations. Also Terrestrial solar radiations (G) obtained from Eq. 2.2 are plotted with Latitudes (selected stations) and months of the year are plotted using the same axes (Figures 3.1 and 3.2).

Stations	Latitudes (oN)	Longitude (oE)	Altitudes (m)
Ikeja	6.39	3.23	39.35
Ilorin	6.50	4.58	307.30
Ibadan	7.22	3.58	234
Port Harcourt	4.43	7.05	19.55
Benin	5.25	5.30	77.70

Table 3.1. Geographical coordinates and altitudes of studied stations

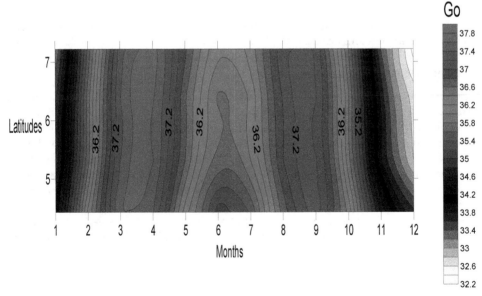

Fig. 3.1. Monthly variation of extraterrestrial solar radiation (G_O) for selected stations (Ikeja, Ilorin, Ibadan, Port Harcourt and Benin).

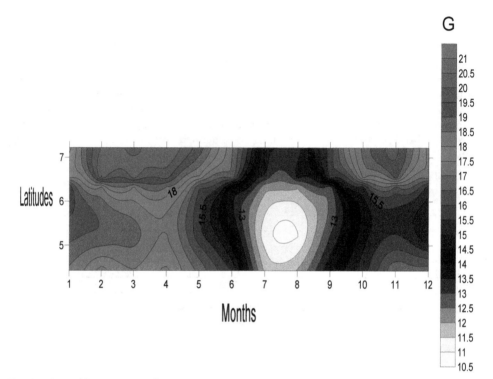

Fig. 3.2. Monthly variation of terrestrial solar radiation (G) for selected stations (Ikeja, Ilorin, Ibadan, Port Harcourt and Benin).

3.2 Monthly variation of Clearness Index

Clearness index (K_T) is defined as the ratio of the observation/measured horizontal terrestrial solar radiation (G), to the calculated/predicted horizontal extraterrestrial solar radiation (G_o). Clearness index is a measure of solar radiation extinction in the atmosphere, which includes effects due to clouds but also effects due to radiation interaction with other atmospheric constituents. To develop the model for the clearness index, the insolation on a horizontal surface for a few locations is measured over a period of time encompassing all seasons and climatic conditions. Different values of the clearness index at different stations may be as a result of different atmospheric contents of water vapour and aerosols. It can be seen from the above expressions that the extra-terrestrial horizontal insolation is a function of latitude and the day of year only. Hence, it can be calculated for any location for any given day. However, the calculated insolation does not take any atmospheric effects into account

$$K_T = \frac{G}{G_O} \tag{3.1}$$

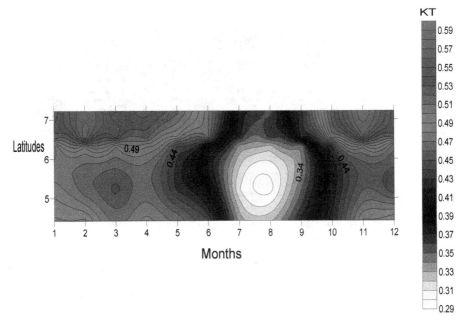

Fig. 3.3. Monthly variation of clearness index for selected stations (Ikeja, Ilorin, Ibadan, Port Harcourt and Benin).

3.3 Monthly variation of relative sunshine duration

The term sunshine is associated with the brightness of the solar disc exceeding the background of diffuse sky light, or, as is better observed by the human eye, with the appearance of shadows behind illuminated objects. According to WMO (2003), sunshine duration during a given period is defined as the sum of that sub-period for which the direct solar irradiance exceeds 120 Wm^{-2}. A new parameter describing the state of the sky, namely the sunshine number has been defined in Badescu (1999). The sunshine number is a Boolean quantity stating whether the sun is covered or not by clouds. Using the sunshine number, it strongly increases the models accuracy when computing solar radiation at Earth surface (Badescu, 1999). Relative sunshine duration is a key variable involved in the calculation procedures of several agricultural and environmental indices.

The relative sunshine duration is expressed as

$$R_s = \frac{S}{S_O} \tag{3.2}$$

Where S is the measured sunshine duration hours and S_0 the potential day length astronomical length. A high number of outliers in the data sets signify that the observation has high degree of variability or a large set of suspect data. Figure 3.3 shows that R_S is low between the months of June through October in Nigeria.

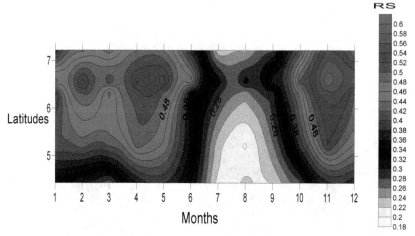

Fig. 3.4. Monthly variation of relative sunshine duration for selected stations (Ikeja, Ilorin, Ibadan, Port Harcourt and Benin).

3.4 Monthly variation of Clearness Index, relative humidity and temperature for Iseyin

There are other methods to estimate solar radiation. Satisfactory result for monthly solar radiation estimation was obtained by using atmospheric transmittance model, while other authors have used diffuse fraction and clearness index models. Parametric or atmospheric transmittance model requires details atmospheric characteristic information. Meteorological parameters frequently used as predictors of atmospheric parameters since acquiring detail atmospheric conditions require advance measurement. Meteorological parameters used in this section clearness index, sunshine duration, temperature and relative humidity data have been used to study monthly variation of atmospheric transmittance coefficient in parametric model. This kind of model is called meteorological model.

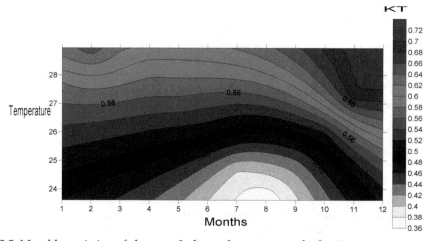

Fig. 3.5. Monthly variation of clearness Index and temperature for Iseyin

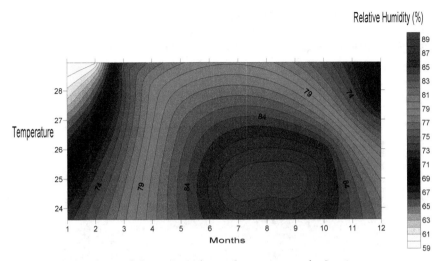

Fig. 3.6. Monthly variation of clearness Index and temperature for Iseyin

4. Variation of diffuse solar radiation

Several models for estimating the diffuse component based on the pioneer works of Angstrom (1924) and Liu and Jordan (1960) and developed by Klein (Klein, 1977). These models are usually expressed in either linear or polynomial fittings relating the diffuse fraction (H_d) with the clearness index and combining both clearness index (KT) and relative sunshine duration (Orgill and Hollands, 1977; Erbs et al., 1982; Trabea, 1992; Jacovides, 2006; Hamdy, 2007, Falayi et al., 2011) established hourly correlations between K_T and H_d under diverse climatic conditions. Ulgen and Hepbasli (2002) correlated the ratio of monthly average hourly diffuse solar radiation to monthly average hourly global solar radiation with the monthly average hourly clearness index in form of polynomial relationships for the city of Izmir, Turkey. Oliveira et al., (2002) used measurements of global and diffuse solar radiations in the City of Sao Paulo (Brazil) to derive empirical models to estimate hourly, daily and monthly diffuse solar radiation from values of the global solar radiation, based on the correlation between the diffuse fraction and clearness index

The diffuse solar radiation H_d can be estimated by an empirical formula which correlates the diffuse solar radiation component Hd to the daily total radiation H. The ratio, H_d/H, therefore, is an appropriate parameter to define a coefficient, that is, cloudiness or turbidity of the atmosphere. The correlation equation which is widely used is developed by Page (Page, 1964).

$$\frac{H_d}{H} = 1.00 - 1.13K_T \tag{4.1}$$

Another commonly used correlation is due to Liu and Jordan (1960) and developed by Klein (Klein, 1977) and is given by

$$\frac{H_d}{H} = 1.390 - 4.027K_T + 5.53\left(K_T\right)^2 - 3.108\left(KT\right)^3 \tag{4.2}$$

We engaged both Page (1964) and Klein (1977) models to study the variation of diffuse solar radiation for Ikeja, Ilorin, Ibadan, Port Harcourt and Benin. Large variations in the intensities of diffuse radiation due to cloudiness have been indicated as stated earlier.

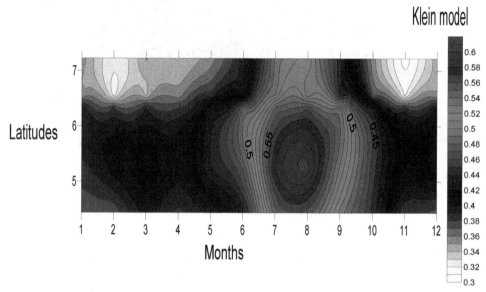

Fig. 4.1. Monthly variation of diffuse solar radiation using Klein model for selected stations (Ikeja, Ilorin, Ibadan, Port Harcourt and Benin)

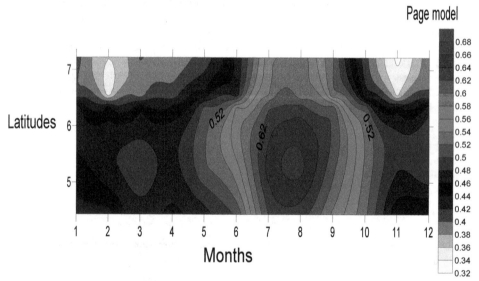

Fig. 4.2. Monthly variation of diffuse solar radiation using Page model for selected stations (Ikeja, Ilorin, Ibadan, Port Harcourt and Benin).

The results of the variation are plotted in Figures 4.1 and 4.2 exhibit the trend variation of diffuse solar radiation. The maxima of diffuse radiation for the month of July - September are quite appreciable. This means that there was a high proportion of cloudy days and relatively low solar energy resource in July –September across the locations, and there was high proportion of sunshine days and relatively abundant solar energy resource between the month of April and October across the stations. This wet season is expected due to poor sky conditions caused atmospheric controls as the atmosphere is partly cloudy and part of solar radiation are scattered by air molecules. The presence of low values of diffuse solar radiation in Figures 4.1 and 4.2 will be very useful for utilizing it for solar concentrators, solar cookers and solar furnaces etc.

5. Conclusion

The global solar radiation incident on a horizontal or inclined surface is estimated by establishing the sky conditions. Monthly variation of clearness index (KT), diffuse ratio (KD), Temperature and the relative sunshine duration (RS) were employed in this study. Klein and Page model were used in this study to examine the variation of diffuse solar radiation for Iseyin, as no station in Iseyin measures diffuse solar radiation.

6. Acknowledgement

The authors are grateful to the management of Nigeria Meteorological Agency, Oshodi, Lagos State for making the data of global solar radiation, sunshine duration, minimum and maximum temperature and relative humidity available.

7. References

Adnan Sozen, Erol Arcakliogulu and Mehmet Ozalp, 2004. Estimation of Solar Potential in Turkey by artificial neural networks using meteorological and geographical data. Energy Conversion and Management, 45: 3033-3052.

Ahmad F, Ulfat, I (2004). Empirical models for the correlation of monthly average daily solar radiation with hours of sunshine on a horizontal surface at Karachi, Pakistan. Turkish J. Physics, 28: 301-307.

Akinbode FO (1992). Solar radiation in Mina: A correlation with Meteorological data, Nig. J. Renewable Energy, 3: 9-17.

Akpabio L.E, Udo S.O, Etuk S.E (2004). Empirical correlation of global solar radiation with Meteorological data for Onne, Nigeria. Turkish J.Physics, 28: 222-227.

Akpabio LE, Etuk SE (2002). Relationship between solar radiation and sunshine duration for Onne Nigeria. Turkish J. Physics, 27: 161-167.

Almorox J, Benito M, Hontoria C (2005). Estimation of monthly Angstrom – Prescott equation coefficients from measured daily data in Toledo, Spain. Renewable Energy Journal, 30: 931-936.

Angstrom A.S (1924). Solar and terrestrial radiation meteorological society 50: 121-126.

Awachie I.R.N, Okeke C.E (1990). New empirical model and its use inpredicting global solar irradiation. Nig. J. Solar Energy, 9: 143-156.

Badescu, V., 1999. Correlations to estimate monthly mean daily solar global irradiation: application to Romania," Energy, vol. 24, no. 10, pp. 883–893.

Bocco Mónica, Enrique Willington, and Mónica Arias. Comparison of regression and neural networks models to estimate solar radiation. Chilean Journal of Agriculture Research: 70(3): 428-435

Burari, F.W, Sambo A.S (2001). Model for the prediction of global solar radiation for Bauchi using Meteorological Data. Nig. J. Renewable Energy, 91: 30-33.

Chandel SS, Aggarwal RK, Pandey AN (2005). New correlation to estimate global solar radiation on horizontal surface using sunshine duration and temperature data for Indian sites. J. solar engineering, 127(3): 417-420.

Che H.Z, Shi G.Y, Zhang X.Y, Zhao J.Q, Li Y (2007). Analysis of sky condition using 40 years records of solar radiation data in china. Theoretical and applied climatology. 89: 83-94.

Duffie J.A, Beekman W.A (1994). Solar Engineering of Thermal Processes, 2nd Edition. John Wiley, New York.

EL-Sebaii A.A and Trabea AA (2005). Estimation of global solar radiation on horizontal surfaces over Egypt, Egypt J. Solids, 28:163- 175.

Erbs DG, Klein S. A and Duffie J. A., 1982. Estimation of the diffuse radiation fraction for hourly, daily and monthly average global radiation. Sol Energ y,28:293–302.

Fagbenle R.O (1990). Estimation of total solar radiation in Nigeria using meteorological data Nig. J. Renewable Energy,14: 1-10.

Falayi E.O, Rabiu AB (2005). Modeling global solar radiation using sunshine duration data. Nig. J. Physics, 17S: 181-186.

Falayi, E.O., J.O. Adepitan and A.B. Rabiu, 2008. Empirical models for the correlation of global solar radiation with meteorological data for Iseyin. Nigeria Int. J. Phys. Sci., 3(9): 210-216.

Falayi, E.O., Rabiu, A.B and Teliat, R.O. 2011. Correlations to estimate monthly mean of daily diffuse solar radiation in some selected cities in Nigeria. Advances in Applied Science Research, 2 (4):480-490. Pelagia Research Library.

Folayan CO (1988). Estimation On Global Solar Radiation Bound for some Nigeria Cities. Nigeria J. Solar Energy, 3: 3-10.

Gueymard, C.A., Myers, D.R., 2009. Evaluation of conventional and high-performance routine solar radiation measurements for improved solar resource, climatological trends, and radiative modeling. Solar Energy 83, 171–185.

Halouani M, Nguyen CT, Vo Ngoc D (1993). Calculation of monthly average global solar radiation on horizontal for station five using daily hour of bright sunshine, Solar Energy,pp 50: 247-255.

Hamdy K. E., (2007).Experimental and theoretical investigation of diffuse solar radiation: Data and models quality tested for Egyptian sites. Energy 32, 73–82

http://www.ips.gov.au/Category/Educational/The%20Sun%20and%20Solar%20Activity/General%20Info/Solar_Constant.pdf

Igbal M (1993). An Introduction to solar radiation Academic press. New York pp. 59-67.

Jacovides, C.P., Tymvios, F.S., Assimakopoulos V.D and Kaltsounides, N.A., 2006. Comparative study of various correlations in estimating hourly diffuse fraction of global solar radiation. Renewable Energy 31, 2492–2504.

Klein S. A. Calculation of monthly average insulation on tilted surfaces. Sol Energy 1977; 43(3):153–68.

Krishnaiah, T., Srinivasa Rao, S., Madhumurthy, K., Reddy, K.S., Neural Network Approach for Modelling Global Solar Radiation. Journal of Applied Sciences Research, 3(10): 1105-1111, 2007

Liu B.Y.H and Jordan R.C. 1960. The interrelationship and characteristic distribution of direct, diffuse and total solar radiation. Sol Energy 1960; 4:1–19.

Lopez, G., M. Martinez, M.A. Rubio, J. Torvar, J. Barbero and F.J. Batlles, 2000. Estimation of hourly diffuse fraction using a neural network based model. In the Proceedings of the 2000 Asamblea Hispano-Portugusea de Geodisia y Geofisica, Lagos, Portugal, pp: 425-426.

Mohandes, M., A. Balghonaim, M. Kassas, S. Rehman and T.O. Halawani, 2000. Use of radial basis functions for estimating monthly mean daily solar radiation. Solar Energy, 68(2): 161-168.

Munner, T., Solar Radiation and Daylight Models. Second edition 2004, Grear Britain, 61-70.

Okogbue E.C, Adedokun J.A (2002). On the estimation of solar radiation at Ondo, Nigeria. NiG. J. Physics, 14: 97-99.

Oliveira A.P, Escobedo J.F, Machado A.J, Soares J. 2002. Correlation models of diffuse solar radiation applied to the city of Sao Paulo, Brazil. Appl Energy, 71: 59–73.

Orgill J.F and Holland K.G.T. 1977.Correlation equation for hourly diffuse radiation on a horizontal surface. Sol Energy,19:357–9.

Page, J.K., 1964. The estimation of monthly mean values of daily total short - wave radiation on vertical and inclined surfaces from sunshine records for latitude 40°N-40°S. Proceeding of the UN Conference on New Sources of Energy, Paper S/98.

Prescott JA (1940). Evaporation from a water surface in relation to solar radiation. Tran. R. Soc. S. Austr. 64: 114-118.

Sambo AS (1985). Solar radiation in Kano. A correlation with meteorological data. Nig. J. Solar Energy, 4: 59-64.

Sayigh AA (1993). Improved Statistical procedure for the Evaluation of solar radiation Estimating models. Solar Energy, 51: 289-291.

Serm J and Korntip T (2004). A model for the Estimation of global solar radiation from sunshine duration in Thailand. The joint international conference on Suitable energy and environment (SEE), pp. 11-14.

Skeiker K (2006). Correlation of global solar radiation with common geographical and meteorological parameters for Damascus province, Syria. Energy conversion and management, 47: 331-345.

Sodha M.S, Bansal N.K, Kumar K.P and Mali A.S (1986). Solar Passive building: Science and design persanon Press, pp. 1-16.

Trabea A. A., 1999. Multiple linear correlations for diffuse radiation from global solar radiation and sunshine data over Egypt. Renewable Energy 17: 411–20.

Udo SO (2002). Contribution to the relationship between solar radiation of sunshine duration to the tropics, A case study of experimental data at Ilorin Nigeria, Turkish J. Physics 26: 229-336.

Ulgen, K. and A. Hepbasli, 2002. Estimation of solar radiation parameters for Izmir, Turkey. Int. J. Energy Res., 26: 807-823.

WMO (2008) Guide to meteorological instruments and methods of observation, WMO-No.8. http://www.wmo.int/

Zhou J, Yezheng W, Gang Y (2004). Estimation of daily diffuse solar radiation in China. Renewable energy, 29: 1537-1548.

3

Impact of Solar Radiation Data and Its Absorption Schemes on Ocean Model Simulations

Goro Yamanaka, Hiroshi Ishizaki, Hiroyuki Tsujino,
Hideyuki Nakano and Mikitoshi Hirabara
Meteorological Research Institute, Japan Meteorological Agency
Japan

1. Introduction

Since absorption of solar radiation plays a major role in heating the upper ocean layers, it is essential for modeling physical, chemical and biological processes (e.g., ocean general circulation or marine carbon cycle). In order to simulate the upper ocean thermal structures as realistically as possible, an ocean general circulation model (OGCM) requires accurate solar radiation data, used as the surface boundary condition. In this sense, it is important to recognize the quality of the solar radiation data being expected or suitable for OGCMs beforehand. The appropriate choice of absorption schemes of solar radiation is also important for ocean modeling in the upper ocean. The absorption of solar radiation is greatly affected by many factors, such as the wavelength of sunlight, the zenith angle and ocean optical properties in the ocean interior. Many absorption schemes have attempted to mimic these processes, but the impact of those schemes on the upper ocean thermal structures is not yet fully understood.

The aim of this study is to determine the importance of solar radiation in ocean modeling. In particular, we examine the impact of both prescribed solar radiation data and its absorption schemes on OGCM simulations. The knowledge obtained here is expected to be useful for ocean modeling studies, as well as for understanding the upper ocean thermal structure.

This article is organized as follows. Section 2 examines the impact of solar radiation flux on ocean model simulation, focusing on discrepancies between simulated and observed sea surface temperature (SST) variations. Yamanaka (2008) discussed such discrepancies over the tropical Indian Ocean, whereas this study deals extensively with discrepancies over the tropical Indo-Pacific Ocean. Section 3 introduces three types of absorption schemes in solar radiation into an ocean model and examines the impact of those schemes on ocean model simulation. Section 4 summarizes this study.

2. Impact of incident solar radiation data on an ocean model simulation

2.1 Brief introduction

Indian Ocean SSTs have notably increased since the late 20th century (Lau & Weng, 1999). Figure 1 clearly shows that the positive SST anomaly has dominated especially after the mid-1980s.

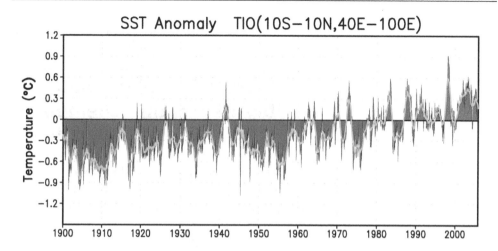

Fig. 1. Time series of the SST anomaly [°C] averaged in the tropical Indian Ocean (10°S-10°N, 40°E-100°E). The SST data set is based on COBE-SST (Ishii et al., 2005). The red (blue) shaded area denotes positive (negative) anomalies. The base period is from 1971 to 2000.

The warming of the tropical Indian Ocean is likely caused by climate variations, but may also, in turn, trigger some impacts on surrounding regions. Some studies using an atmospheric general circulation model (AGCM) with the prescribed SST suggest that the increasing trend in Indian Ocean SSTs can impact the climate. For example, Hoerling et al. (2004) indicated that local increases in precipitation associated with the warming of the Indian Ocean resulted in a remote response to the mid and high latitudes through the release of latent heat, and contributed to an increased trend of the North Atlantic Oscillation (NAO). Also, the warming of the Indian Ocean enhances the anti-cyclonic circulation anomaly at the lower level of the troposphere over the Philippines during the mature phase of El Nino (Watanabe & Jin, 2002), which has a major impact on the East Asian climate (Wang et al., 2000).

In order to understand the warming mechanism of the Indian Ocean, it is necessary to clarify the observation-based surface heat balance over the Indian Ocean. However, due to lack of long-term observation, many studies used OGCMs to diagnose the surface heat balance (e.g., Du et al., 2005; Murtugudde and Busalacchi, 1999).

The importance of the OGCM study is to know to what extent variations of the Indian Ocean are simulated by the model. Figures 2a and 2d show the anomaly correlation between observed and simulated SSTs, which is one of the means by which the model's performance may be determined. It is found that the anomaly correlation over the tropical Indian Ocean is below 0.6 and is less than that in other areas, such as the tropical Pacific and the mid and high latitudes. Poor simulation of the tropical Indian Ocean makes it difficult to analyze the surface heat balance over that area. However, the cause of this poor simulation is not yet fully understood.

This section aims to investigate the cause of the poor simulation, basically following Yamanaka (2008) and extensively looking at the tropical Indo-Pacific Ocean.

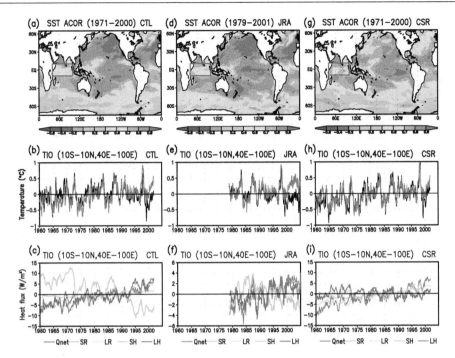

Fig. 2. (a) Annual mean anomaly correlation between the observed and the simulated SST anomalies for CTL. The statistical period is from 1971 to 2000. (b) Time series of the observed (red) and simulated (black) SST anomalies [°C] averaged in the tropical Indian Ocean (10°S-10°N, 40°E-100°E) for CTL. The base period is from 1971 to 2000. (c) Time series of sea surface heat flux anomalies [W m^{-2}] averaged in the tropical Indian Ocean (10°S-10°N, 40°E-100°E) for CTL. Lines denote net surface heat flux (red), solar radiation (green), long wave radiation (yellow), sensible heat flux (aqua), and latent heat flux (blue). The base period is from 1971 to 2000. (d)-(f) Same as (a)-(c) but for JRA. The base period is from 1979 to 2004. (g)-(i) Same as (a)-(c) but for CSR.

2.2 Model and methodology

We used a version of the Meteorological Research Institute community ocean model (MRI.COM) (Ishikawa et al., 2005), which is a z-coordinate primitive-equation model. The model domain is near global, from 75°S to 75°N. The horizontal resolution is 1° in longitude and 1° in latitude (0.3° near equator). The model has 50 vertical levels, with 24 levels in the top 200m.

Two sets of daily atmospheric reanalysis data are used as the surface boundary condition: ECMWF 40-year reanalysis data (hereafter ERA-40) (Uppala et al., 2005) from 1960 to 2001, and Japan Meteorological Agency 25-year reanalysis data (hereafter JRA-25) (Onogi et al., 2007) from 1979 to 2004. We used the bulk formula for the surface fluxes by Kara et al. (2000). After the model was integrated for 102 years as spin-up, three experiments were conducted with different interannual atmospheric forcing data. In CTL, the model was driven by atmospheric variables derived from ERA-40; in JRA, the model was driven by those derived from JRA-25. In CSR, the atmospheric forcing was the same as that in CTL, except that solar radiation data included only seasonal variations (no interannual or longer variations).

For comparison, we used the COBE-SST data set of in-situ measurements of SST (Ishii et al., 2005). The reanalyzed solar radiation data derived from ERA-40 and JRA-25 reanalysis were compared with satellite-based estimates of solar radiation: International Satellite Cloud Climatology Project (ISCCP) solar radiation data derived from the Common Ocean-ice Reference Experiment (hereafter CORE/ISCCP) (Large & Yeager, 2004). Also, the reanalyzed precipitation data derived from ERA-40 and JRA-25 were compared with two observation-based estimates of precipitation: Climate Prediction Center (CPC) Merged Analysis of Precipitation combined with NCEP/NCAR R1 reanalysis (hereafter CMAP) (Xie & Arkin, 1996), and Global Precipitation Climate Project Version 2 (hereafter GPCP) (Adler et al., 2003).

All data was converted to monthly means before further analysis. Monthly mean data for ERA-40 surface flux was produced using the daily mean data based on 36 hour forecast data at each 12UTC initials.

2.3 Results

2.3.1 The simulated Indian Ocean with the prescribed solar radiation

Figure 2b shows the time evolution of simulated SST anomalies over the tropical Indian Ocean ($10°$S-$10°$N, $40°$E-$100°$E) for CTL. The model was successful in simulating the interannual SST variation of 4 to 5 years associated with ENSO, but failed to capture the long-term warming trend found in the observed SSTs. For example, the simulated SST anomaly was slightly higher than the observed one in the 1960s, whereas it was substantially lower than observed after the late 1990s, indicating a cooling bias. A similar tendency was also found in JRA (Fig. 2e), where the simulated SST anomaly in the Indian Ocean has been gradually cooler than the observed one since the late 1980s, and the difference has increased since 2000. This result implies that the poor simulation of the Indian Ocean SSTs in both experiments is due to the cooling bias, especially in the late 1990s.

Cooling of the model Indian Ocean after 1990 was observed not only at the surface, but also below the surface. Figure 3a shows the mixed layer change between 10 years in CTL. The deepening of the mixed layer depth (MLD) was found to be wide in the tropical Indian Ocean. Figures 3b and 3c show vertical profiles of temperature and potential density at the equatorial Indian Ocean (EQ, $90°$E) between January 1990 and January 2000 in CTL. Temperature decreased up to about 60 m depth, and the MLD increased to 100 m depth during that ten-year period. This simulated cooling trend in the upper ocean differs from the trend observed by Levitus et al. (2005), in which significant warming near the surface accompanied cooling in the upper thermocline (Han et al., 2006). The deepening of the mixed layer may have altered the surface heat balance of the Indian Ocean in the model.

Next, we examine surface fluxes used as the surface forcing for the model. Figure 2c shows the time duration of each component (net flux, solar radiation, long wave radiation, sensible heat flux, and latent heat flux) of the model surface flux anomalies. It is noted that the solar radiation anomaly (green line) in CTL exhibits a significant decreasing trend. From the 1960s to early 1970s, the solar radiation anomaly was positive, corresponding to the simulated warmer SST anomaly. After the mid-1990s, the negative solar radiation anomaly became

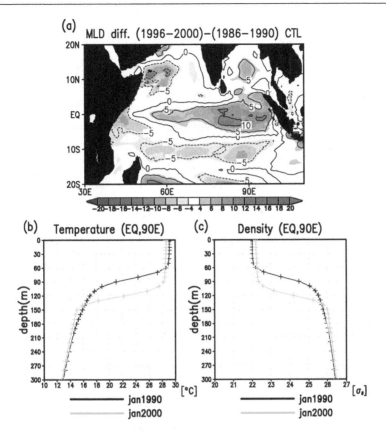

Fig. 3. (a) Mixed layer depth difference [m] between the 1996-2000 mean and the 1986-1990 mean. The shaded area denotes where the difference is positive. (b) Vertical profile of temperature [°C] in the equatorial Indian Ocean in CTL on January 1990 (black) and on January 2000 (green). (c) Same as (b) but for potential density [σ_θ].

dominant, corresponding to the cold bias of the model in this period. Sensible and latent heat fluxes partly weakened the decreasing trend caused by solar radiation because they were restored to the observed atmospheric variables based on the bulk formula. However, a decreasing trend remained in the simulated SSTs. A similar decreasing trend in the solar radiation anomaly was also found in JRA (Fig. 2f).

In order to clarify the role of the reanalyzed solar radiation data on the cooling bias in the simulated SSTs, an additional experiment (CSR) was carried out, where the atmospheric forcing was the same as at CTL except that the daily-mean climatological solar radiation data was used. Figure 2h shows that the simulated SSTs in CSR agree better with the observed SSTs (e.g., improvement in both the warming bias in the 1960s and the cooling bias in the late 1990s, compared to those in CTL). Also, the warming of the Indian Ocean in the 1990s was roughly captured in CSR even under climatological solar radiation forcing. According to the sea surface flux anomalies (Fig. 2i), the variability of the net heat flux was controlled by long wave radiation on a longer time scale, as well as by latent heat flux on an interannual time scale. These results suggested that increases in downward long wave radiation contributed to the simulated warming of the Indian Ocean in the 1990s in CSR.

Improvement of the bias in the simulated SSTs of the Indian Ocean was expected to result in better performance of the simulated SSTs. Figure 2g shows the annual mean anomaly correlation between the observed and the simulated SST anomalies in CSR. It was found that removal of the variations of the reanalyzed solar radiation on an interannual or longer timescale improved the simulated SST variability, especially in the tropical Indian Ocean. The SST skill increased by 0.1 to 0.3 in this region, compared to that of CTL. In the central to eastern equatorial Pacific, however, no significant change in SST variability was observed between CTL and CSR. It is suggested that the SST variability in the central to eastern equatorial Pacific is determined mainly by wind stress, rather than solar radiation. However, in the mid and high latitudes, the SST skill is significantly reduced in CSR, implying that SST variability in these regions is determined by variation in solar radiation.

These results strongly suggest that the cooling of simulated Indian Ocean SSTs is primarily caused by the atmospheric reanalysis data used as the surface boundary condition. Next, we examine why the atmospheric reanalysis products display decreasing trends in solar radiation.

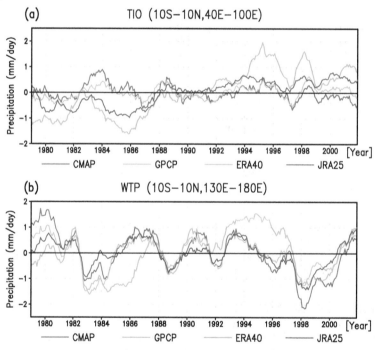

Fig. 4. Time series of 13-month running mean precipitation anomalies [mm/day] averaged over (a) the tropical Indian Ocean (10°S-10°N, 40°E-100°E) and (b) the western tropical Pacific (10°S-10°N, 130°E-180°) for CMAP (red), GPCP (acua), ERA-40 (green), and JRA-25 (blue).

2.3.2 Spurious trends included in atmospheric reanalysis data

Figure 4 shows a time series of precipitation anomalies averaged over the tropical Indian Ocean and the western tropical Pacific, based on the reanalysis data with the CMAP and GPCP data sets as reference. Over the tropical Indian Ocean, ERA-40 precipitation (green line) is generally greater than the observed precipitation, and an increasing trend is clearly seen since 1979. JRA-25 precipitation (blue line) also exhibits a similar increasing trend, though not as great as that of the ERA-40 precipitation. On the other hand, the observed precipitations in CMAP and GPCP exhibit no significant increasing trend, and also may be the observed cloud amount. In contrast, over the western tropical Pacific, the observed precipitation and the JRA-25 precipitations show a slight decreasing trend, though the ERA-40 precipitation indicates no significant trend.

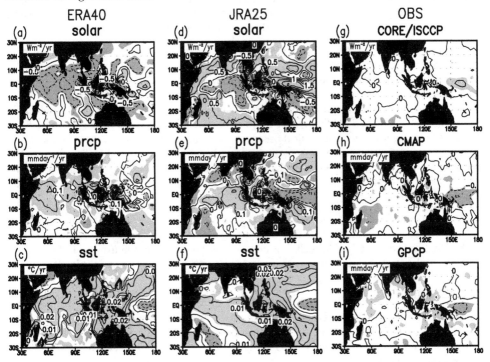

Fig. 5. Average trends from 1979 to 2001 in the reanalysis products (ERA-40 and JRA-25) and observed data (CORE/ISCCP, CMAP and GPCP) for solar radiation (upper), precipitation (middle), and the prescribed SST for the reanalyses (lower) over the Indian Ocean. The contour interval is 0.5 Wm^{-2}/year for solar radiation, 0.1 mm day^{-1}/year for precipitation, and 0.01 °C/year for SST. The red (blue) shaded areas denote where the rate of change is positive (negative) with statistical significance.

Figure 5 shows average trends during the period 1979 to 2001 in the atmospheric reanalysis products (ERA-40 and JRA-25) with the prescribed SSTs for the reanalyses, in addition to CORE/ISCCP solar radiation and CMAP precipitation. Over the tropical Indian Ocean, the reanalysis products exhibit decreasing trends in solar radiation, and their spatial patterns are almost the reverse of the precipitation patterns. This feature is generally found in all the reanalysis products, though it seems more pronounced in JRA-25 and ERA-40 than in NCEP/NCAR 40-year reanalysis (Kalnay et al., 1996) and NCEP-DOE AMIP-II reanalysis (Kanamitsu et al., 2002) as described in Yamanaka (2008). The increasing trend in precipitation roughly corresponds to the increasing trend in SSTs prescribed as the lower boundary condition for the reanalyses. Hence, the decrease in solar radiation may be associated with increase in precipitation directly over the region of the most rapidly warming SST. In contrast, no increasing trend in solar radiation over the Indian Ocean was observed in the CORE/ISCCP data. Also, the CMAP data showed no increasing trend in precipitation nor a decreasing trend over the southern Indian Ocean. Over the western tropical Pacific, the situation was almost the same; the area with a slightly decreasing trend in precipitation or a slightly increasing trend in solar radiation corresponds to the cooling SST region, suggesting a linkage between the trends in atmospheric reanalysis and the SST.

Several problems in the atmospheric reanalyses may have caused this spurious increasing trend in precipitation over the tropical Indian Ocean. One problem may arise from bias in the assimilation, for example, ERA-40 has rainfall problems over tropical oceans from the early 1990s, associated with the bias of satellite radiance corrupted by the Pinatubo eruption (Dee et al., 2008), and JRA-25 has major discontinuous changes associated with transition from TOVS to ATOVS in November 1998 (Tsutsui & Kadokura, 2008). Another problem may come from the bias in the model. Over the tropical oceans, where in-situ observations are infrequent and sparse, a reanalysis dataset would be equivalent to AGCM outputs where SST is given as the lower boundary condition (Arakawa & Kitoh, 2004). Hence, responding to the warming of the Indian Ocean, AGCM tends to enhance convective activities and thus to increase precipitation and cloud amounts.

As a result, the decrease in the solar radiation caused a cooling trend of the simulated SSTs in the Indian Ocean, which is inconsistent with the observed SSTs (Fig. 1). This is supported by the fact that the area with a relatively low skill of simulated SSTs in the tropical Indian Ocean approximately corresponds to that with a decreasing trend in solar radiation (Fig. 2a).

2.3.3 Discussion

We found that the poor simulation of the Indian Ocean SST was due to the atmospheric reanalysis data (ERA-40 and JRA-25) used as the surface boundary condition for OGCM, which included decreasing trends in solar radiation there. This decreasing trend in solar radiation was related to the increasing trend in precipitation over the Indian Ocean, which was partially as a response to the local warming of the SSTs.

The spurious trends in the atmospheric reanalysis products constitute a crucial problem for long-term ocean modeling studies, because surface flux data based on the atmospheric reanalysis products are widely used as the surface boundary conditions for OGCMs. Thus, caution is necessary when using atmospheric reanalysis data as the surface boundary conditions for OGCMs. One approach to avoid the unrealistic cooling of the model Indian Ocean is to use the CORE/ISCCP solar radiation. The CORE/ISCCP solar radiation data do not exhibit significant decreasing trend over the tropical Indian Ocean (Fig. 5), although it should be noted that the CORE/ISCCP solar radiation data included no interannual variations before mid-1983, because of the limited availability of satellite data. In fact, a recent study demonstrated that the ocean model driven by the CORE forcing (Large & Yeager, 2009) reasonably simulated long-term variations in the tropical Indian Ocean (Tsujino et al., 2011).

Several studies suggest that there may be no increase or even decrease in precipitation over the Indian Ocean. Copsey et al. (2006) reported a rise in sea surface pressure, as a proxy for precipitation, over the Indian Ocean between 1950 and 1996. Deser & Phillips (2006) concluded that there was no significant increase in precipitation over the Indian Ocean, based on analysis of the cloud amount and wind convergence over the ocean. Norris (2005) suggested a negative trend in upper level cloud cover in the equatorial Indian Ocean between 1952 and 1997. Further study based on observation is needed to clarify the long-term trend of precipitation in the Indian Ocean.

3. Impact of absorption schemes in solar radiation on an ocean model simulation

3.1 Brief introduction

The optical properties of seawater, which dominate the distribution of the penetration and absorption of the given outer radiation, are primarily determined by the phytoplankton biomass, measured by chlorophyll-a concentration in seawater, with their accompanying retinue of dissolved and particulate materials of biological origin (Case 1 Waters) (e.g., Morel, 1988; Morel & Prieur, 1977). Many studies have focused on the development of shortwave penetration schemes including the effects of chlorophyll-a concentration either in bulk or spectral formulae (e.g., Manizza et al., 2005; Morel, 1988; Morel & Antoine, 1994; Ohlmann, 2003; Ohlmann et al., 2000; Ohlmann & Siegel, 2000) and on the effects of these parameterizations on the ocean dynamics and thermodynamics through forced ocean model experiments (e.g., Anderson et al., 2007; Manizza et al., 2005; Murtugudde et al., 2002; Nakamoto et al., 2001). While the direct effect of including the chlorophyll-a concentration increased absorption in shallower layers, one of the most pronounced changes was the indirect effect: an increased cooling in SST in the eastern equatorial Pacific. This increased cooling resulted from increased upwelling through changes in the equatorial current system (Gnanadesikan & Anderson, 2009; Manizza et al., 2005; Murtugudde et al., 2002; Nakamoto et al., 2001; Sweeney et al., 2005).

The solar zenith angle or the solar altitude affects penetrating radiation and the vertical distribution of heating by absorbing the radiation in a water column under clear sky condition. Some shortwave penetration schemes explicitly examined the effects of solar altitude on this (e.g., Morel & Antoine, 1994; Ohlmann, 2003). Ishizaki & Yamanaka (2010) (hereafter referred to as IY10) examined the impact of sun altitude on ocean radiant heating, assuming that all sunlight is direct solar rays. They introduced sun altitude into the simple radiation formulation of Paulson & Simpson (1977) (hereafter referred to as PS77) with diurnal changing incident angle, and studied the sensitivity of an ocean model to the formulation. Introduction of the solar angle caused the effective attenuation depth for the diurnal-mean penetrating radiation shallower than that of the downward vertical radiation, and caused the locus of radiation absorption to shift upward. This was qualitatively the same as including chlorophyll-a concentration, resulting in the same indirect effect of cooling in SST in the eastern equatorial Pacific.

Here we examined the impact of solar radiation absorption schemes on ocean model simulation. We considered three absorption schemes. The first is a conventional scheme based on PS77, in which sunlight has diurnal constant intensity and is vertically downward.

The second is the above-mentioned IY10 scheme, in which sunlight has a diurnal changing incident angle, leading to vertical change in the diurnal-mean attenuation rate of the sun light. The third scheme introduces the effect of chlorophyll-a concentration by Morel & Antoine (1994)'s (hereafter referred to as MA94) formulation, in addition to the second scheme. We confirmed the impact of these three schemes on the mean ocean state, especially focusing on the effective euphotic layer depth, temperature and current fields.

3.2 Formulations of three absorption schemes

3.2.1 Basic assumption

We made the following assumptions for formulating the changing solar altitude angle of the sun: (a) All sunlight consists solely the direct rays without any scattered light. (b) The sun is a point source of light, i.e., the visual angle of the sun is zero. (c) The ratio of the actual to the mean earth-sun separation is assumed to be unity, that is, the earth's revolution orbit is perfectly circular. (d) The declination angle, δ, of the sun (i.e., its latitude on the celestial sphere) is constant on a diurnal time scale. (e) The effective radiation intensity I_{org} of sunlight on a plane perpendicular to the ray is diurnal constant, regardless of the sun altitude, and is calculated from the diurnal-mean irradiance I_{DM} given as a boundary condition (IY10). (f) The refractive index of seawater γ is a constant, i.e., $\gamma = 1.34$. (g) The optical characteristics of seawater are homogeneous with depth, in horizontal direction, and over time, and are assumed to be of Jerlov (1968) Water Type I (PS77) for the first (PS77) and the second (IY10) absorption scheme. This water has an e-folding depth (attenuation depth) of 23 m for the shorter wavelength part (PS77). (h) Sea surface albedo α is set to be a constant, 0.066, independent of sun altitude.

3.2.2 Formulation

According to PS77, incoming solar radiation is divided into two parts: the longer-wavelength (infrared (IR)), which is absorbed immediately at the sea surface and the shorter-wavelength (visible plus ultra-violet (visible-UV)), which penetrates a relatively long distance, which is expressed as

$$I/I_0 = R exp(-z/\zeta_1) + (1 - R) exp(-z/\zeta_2) \tag{1}$$

where I_0 and I are the irradiances just under the sea surface and at depth z, respectively; R is the ratio of the IR part to the total at the surface; and ζ_1 and ζ_2 are the e-folding depths (attenuation depths) of the IR and visible-UV parts, respectively. We take this formulation as our first absorption scheme with R = 0.58, $\zeta_1 = 0.35$ m, and $\zeta_2 = 23$ m (for Water Type I) (PS77) and call it the "PS77-scheme".

For the second absorption scheme, the incoming radiation with incident angle A at the sea surface penetrates the sea with refracted angle A'. The sun altitude A is given by the observer's latitude θ, declination of the sun δ ($-23.5° < \delta < 23.5°$), and the local time t as

$$sinA = sin\delta sin\theta - cos\delta cos\theta cos\omega t \tag{2}$$

where ω is the diurnal angular velocity of the sun for the observer, i.e., $\omega = 2\pi/24$ hours. The relationship between A and A' is given by Snell's law:

$$cosA/cosA' = \gamma(= 1.34) \tag{3}$$

so that

$$sin A' = ((\gamma^2 - 1) + sin^2 A)^{1/2}/\gamma \qquad (4)$$

The minimum of A' is 41.7° for $A = 0°$. The path length is expressed as $z/sin A'$ where z is the depth, so that

$$I/I_0 = R exp(-z/(\zeta_1 sin A')) + (1 - R) exp(-z/(\zeta_2 sin A')) \qquad (5)$$

where the values of R, ζ_1, and ζ_2 are the same as those of the first scheme. We call this the "IY10-scheme". The practical calculation procedure of solar radiation from a given diurnal-mean irradiance I_{DM} is given in IY10.

For the third absorption scheme, MA94's formulation is used with a climatological chlorophyll-a data at 1 m depth. Chlorophyll-a data is derived from a Sea-viewing Wide Field-of-view Sensor (SeaWiFS; http://oceancolor.gsfc.nasa.gov/SeaWiFS). Their formulation is expressed as

$$I/I_0 = F_{IR} exp(-z/(Z_{IR} sin A')) + F_{VIS}(V_1 exp(-z/Z_1) + V_2 exp(-z/Z_2)) \qquad (6)$$

where F_{IR} is the fraction of the infrared (IR) radiation (wavelengths > 0.75μm) to the total, F_{VIS} is that for the visible- UV radiation ($F_{IR} + F_{VIS} = 1$), and $Z_{IR} = 0.267$ m, is the attenuation length of the IR radiation. The visible-UV part consists of two exponentials with the partitioning factors and attenuation depths, V_1, V_2 ($V_1 + V_2 = 1$), Z_1, and Z_2, depending on the chlorophyll-a concentration C. Here, the term including Z_1 in (6) represents the longer wavelength range of the visible-UV part, with Z_1 being of a few meters over the whole range of the chlorophyll-a concentration. They are expressed by polynomials of $log_{10}C$, and values of their coefficients and the functional forms of the four parameters are given for the range of C between 0.02 and 20 mg m^{-3} in MA94. Two sets of coefficients of the polynomials are given, one for uniform pigment profiles and the other for nonuniform ones. Here we use (6) with the variable solar incident angle even for the visible-UV part:

$$I/I_0 = F_{IR} exp(-z/(Z_{IR} sin A')) + F_{VIS}(V_1 exp(-z/(Z_1 sin A')) + V_2 exp(-z/(Z_2 sin A'))) \qquad (7)$$

The value of F_{VIS} depends on the atmospheric condition and the solar zenith angle; MA94 gives the values 0.54 - 0.57 for clear skies and 0.60 for overcast skies. Here, however, we use $F_{VIS} = 1 - R$ ($F_{IR} = R = 0.58$) for the sake of consistency with the first and second absorption scheme. The polynomials for non-uniform pigment profiles are used because we used the chlorophyll-a data obtained by satellites (Morel & Berthon, 1989). We call this a modified MA94-scheme, that is, the "mMA94-scheme".

3.3 General features of optical property of sea surface layer

Before describing the implementation of the above schemes in an ocean model, we theoretically discuss the annual-mean of the diurnal-mean euphotic layer depth, attenuation depth (e-folding depth), and absorption of the penetrating radiation (only for the visible-UV part). To calculate the diurnal-mean radiation, the time step was taken as 1 min with the incident angle ($sin A$) calculated by (2) at every time step. Here, the radiation intensity of sunlight is assumed to be the solar constant 1.37 kW m^{-2} only in this subsection and the sea surface irradiance I_0 is the intensity multiplied by $sin A$.

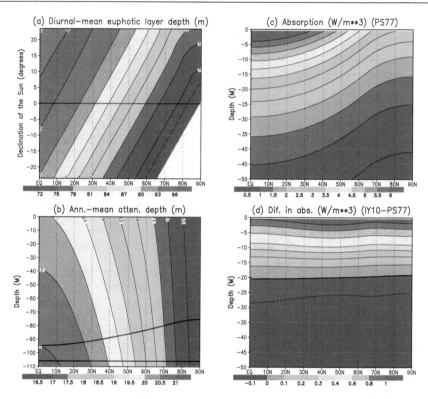

Fig. 6. (a) Diurnal-mean euphotic layer depth d_e [m] for IY10, for the sun declination δ and the northern latitude θ. (b) Meridional section of the effective attenuation depth ζ [m] defined in each 1 m-layer for annually averaged diurnal-mean penetrating irradiance of IY10. (c) Absorption of annually averaged irradiance of PS77 [W m^{-3}]. (d) Difference in absorption between IY10 and PS77 [W m^{-3}] (IY10 - PS77). The thick lines in (b) indicate the annual mean euphotic layer depth for IY10 and PS77 (constantly 105.9 m).

3.3.1 Difference between IY10 and PS77 schemes

The diurnal-mean euphotic layer depth d_e for the IY10-scheme is a function of θ and δ (Fig. 6a), which was numerically obtained as the depth where diurnal-mean irradiance of the visible-UV part of the incident radiation became 1 % of its surface value. Depth d_e ranged from less than 71 m in winter at high latitudes to more than 96 m at equatorial equinox through the Tropic of Cancer at summer solstice. For the PS77-scheme, d_e theoretically had a constant value of 105.9 m; thus, the difference ranged from 10 m to 35 m (9 - 33 % of 105.9 m).

Figure 6b shows the vertical structure of the effective vertical attenuation for the annual-mean radiation, defined in each 1m-layer and expressed by the e-folding depth $\zeta(\zeta(z) = 1/ln(I(z - 0.5)/I(z + 0.5)))$. The maximum vertical difference in ζ of about 1 m is seen in the upper 100 m layer at the equator, while ζ is almost vertically homogeneous at the high latitudes. The vertical structure originates from the diurnal and seasonal variations of the solar altitude. For PS77, ζ is a constant 23 m. Also shown in Fig. 6b are the annual-mean euphotic layer depths for the PS77 (constant, 105.9 m) and IY10 (sinusoidal) schemes. The range of the latter (75 - 94 m) is somewhat narrower than that in Fig. 6a because of the annual averaging process.

Panels c and d of Fig. 6 show absorption of the annually averaged irradiance of the PS77 scheme expressed by its vertical convergence, and the difference in absorption between IY10 and PS77 (IY10-PS77), respectively. The absorption pattern naturally indicates latitudinal variation for both PS77 and IY10 (not shown for IY10), but the difference between the two hardly has any latitudinal variation (Fig. 6d), with the zero line staying at about 20 m at all latitudes. Introducing the solar altitude variation results in more warming at levels shallower than 20 m and more cooling below that level.

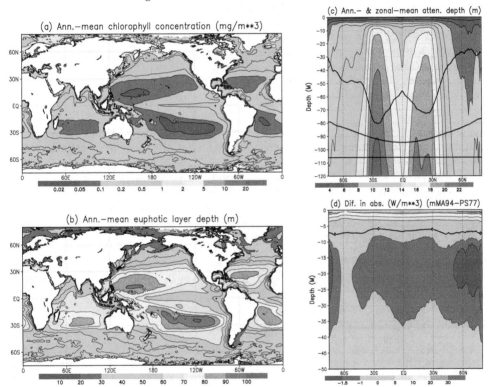

Fig. 7. (a) Annual-mean chlorophyll concentration (SeaWifs) [mg m^{-3}]. (b) Annual-mean euphotic layer depth [m] base on MA94 with diurnal variation of sun altitude (mMA94). (c) Effective attenuation depth ζ (m) for zonally averaged annual-mean penetrative irradiance of mMA94. (d) Difference in absorption between mMA94 and PS77 [W m^{-3}] (mMA94 - PS77). The thick lines in (c) indicate the zonally averaged annual-mean euphotic layer depth for mMA94 (uppermost), IY10 (middle) and PS77 (lowest, constantly 105.9 m).

3.3.2 Difference between mMA94 and PS77 schemes

Annual mean chlorophyll-a concentration (SeaWiFS) is shown in Fig. 7a, which was obtained by averaging its monthly mean values. The values are very low in the subtropical circulations in both hemispheres. In contrast, high values are seen near the continental coasts. The corresponding annual mean euphotic layer depth based on the mMA94 scheme (7) (Fig. 7b) exhibits a pattern similar to that of chlorophyll-a. It has large values exceeding 80 m in the center of every subtropical circulation area, and exceeds 100 m in the South Pacific, which is greater than the largest value in IY10 (94 m, Fig. 6b). Very low values of less than 20 m are often observed along the continental coast, corresponding to high chlorophyll-a concentrations. Along the equator chlorophyll-a is relatively high and the euphotic layer depth is relatively shallow (50 - 60 m).

The effective attenuation depth calculated based on zonally averaged annual-mean penetrating irradiance is presented in Fig. 7c, with the zonal-mean euphotic layer depth for PS77, IY10, and mMA94. Figure 7d shows the difference in absorption between mMA94 and PS77 (mMA94 - PS77). In contrast with IY10, the effective attenuation depth (Fig. 7c) is very small (less than 8 m) in the top several-meter layer over the whole range of latitude. This is due to the fact that the attenuation depth Z_1 in the mMA94 scheme (7) is a few meters over the whole range of chlorophyll-a concentration. Corresponding to the low effective attenuation depth, the absorption is very large in the top layer in mMA94 compared to that in PS77 (Fig. 7d).

In the deeper layers below, 10 m, where the last term in (7) seems to be dominant, two peaks of dome-shaped distribution of high attenuation depth (low attenuation) correspond to the horizontal pattern of low chlorophyll-a and large euphotic layer depth (Figs. 7a and b). At the equator the two domes are divided, in contrast with IY10, where a single dome is centered at the equator (Fig. 6b). The high concentration of chlorophyll-a along the equator characterizes these optical properties and structures in the deeper part of the surface layer. The zonal-mean euphotic layer depth of mMA10 also has two peaks in magnitude, but much less than that of IY10 as a whole (Fig.7c). The zero line for the absorption difference (Fig. 7d) is at about 7 m over the whole range of latitude. Below that level, it has a vertical structure with a minimum in magnitude of 20 m, but its latitudinal variation seems to be very weak. The magnitude in the absorption difference for (mMA94 - PS77) is one order greater than that for (IY10 - PS77) (Fig. 6d).

3.4 Model and experiments

The OGCM used in this section is MRI.COM3 (Tsujino et al., 2010) that is, a free-surface, depth-coordinate ocean ice model. The model has a global domain with a tripolar grid (Murray, 1996) that consists of a spherical, latitude-longitude grid south of 64°N and a bipolar grid with generalized orthogonal coordinates with polar singularities in Siberia (64°N, 80°E) and Canada (64°N, 100°W).

The horizontal resolution is 1° in longitude and 0.5° in latitude south of 64°N. There are 50 vertical levels with a bottom boundary layer (Nakano & Suginohara, 2002). The surface layer thickness is 4 m, and the model has 30 levels in the upper 1000 m.

The mixed layer scheme is based on Noh & Kim (1999). The generalized Arakawa scheme (Ishizaki & Motoi, 1999) was used to calculate the momentum advection terms. A numerical advections scheme based on conservation of second-order moments (SOM) (Prather, 1986) was used for advection of tracers. Isopycnal diffusion (Redi, 1982) and eddy-induced transport parameterized as isopycnals layer thickness diffusion (Gent & McWilliams, 1990) are used as sub-grid-scale mixing.

The surface boundary conditions are based on the surface atmospheric condition by Large & Yeager (2009) and provided as the Coordinate Ocean-ice Reference Experiment (CORE) forcing dataset (CORE.v2). A detrended 59-year interannual forcing dataset from 1948 to 2006 was used for spin-up. More details about the model settings may be found in Tsujino et al. (2011).

The model was integrated for about 1,350 years (23 cycles) from an initial state by using the detrended CORE data, and reached a quasi-steady state. In the spin-up period, we used the PS77 scheme as the absorption scheme of solar radiation.

Three experiments were then carried out to examine the impact of the three absorption schemes described in the previous subsection. The first experiment used the conventional PS77 scheme and was called "CTL". The second experiment used the IY10 scheme and was called "SLR". The third experiment used the chlorophyll-a dependent scheme based on mMA94 scheme and was called "CHL". In the third experiment, chlorophyll-a data was derived from the monthly mean satellite-based observation (SeaWiFS: Fig. 7a). Each experiment started from a quasi-steady state and was integrated for five additional cycles (295 years) using the detrended CORE data. The yearly mean data over the fifth cycle were used for analysis.

3.5 Results

This subsection describes the impact of the three absorption schemes on ocean simulations. In particular, we focus on the oceanic structures of the tropical Pacific, where the impact of those schemes is most clearly found.

Figure 8a shows the SST and surface current differences between SLR and CTL. When the IY10 scheme was introduced, the SST increased slightly to about $0.1°C$ in the western tropical Pacific, the Indian Ocean and the subtropics. In contrast, the SST decreased to about $0.3°C$ in the central and eastern equatorial Pacific east of $175°E$ within the north-south 5 degrees band, together with the coastal areas. This SST contrast in the tropical Pacific Ocean has already been reported by IY10. Introducing the solar angle shifted the locus of radiation absorption upward, resulting in warming in the SST in all regions, except the eastern equatorial Pacific, where the indirect effect led to the cooling in the SST. When the mMY94 scheme was introduced (Fig. 8c), the pattern of the SST contrast in the tropical Pacific was almost the same, but the magnitude increased. The SST decreased in the eastern equatorial Pacific reaching even about $1°C$ around $120°W$. The impact of the chlorophyll-a concentration (CHL) on the SST in the equatorial Pacific was about three times greater than that of the solar angle (SLR).

The IY10 scheme caused the westward current surface anomalies in the equatorial Pacific (Fig. 8a). The direction of the surface current anomalies turned pole ward apart from the equatorial region, corresponding to a divergent flow. This result is consistent with the impact of the IY10 scheme as described by IY10. These surface current anomalies are associated with increased upwelling and changes in the equatorial current system. Introduction of the chlorophyll-a concentration (CHL) produced effects similar to those of SLR, but with greater amplitude of the current anomaly (Fig. 8b). The amplitude is almost twice that of SLR, and the direction of the surface current is more pole ward. Thus, the divergent flow in the eastern equatorial Pacific is more enhanced in the CHL run than in the SLR run.

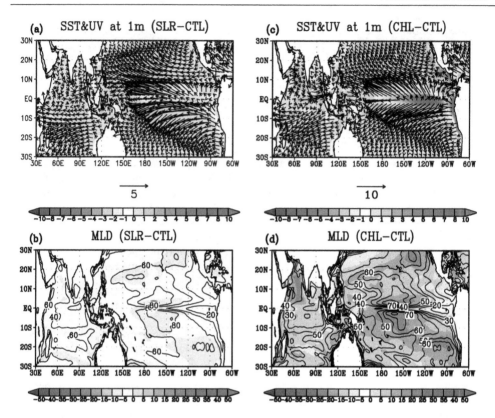

Fig. 8. (a) SST [0.1°C] (color) and surface current [cm/s] (vector) differences between SLR and CTL. (b) Mixed layer depth [m] difference between SLR and CTL (shaded) and mixed layer depth [m] for SLR (contour) . (c)-(d) Same as (a)-(b) but for CHL.

Introduction of the solar angle changed the vertical profile heating, due to solar radiation. The surface layer received more heating, while the subsurface layer received less (Fig. 6d, Fig. 7d). This vertical contrast of heating made the MLD shallower, over most regions (Fig. 8c). In the tropical Pacific, the MLD decreased to about 5 m in the central equatorial Pacific where the mean MLD was large. This situation was strengthened when the mMA94 scheme was used (Fig. 8d). Introducing the chlorophyll-a concentration made the MLD more than 20 m shallower. High chlorophyll-a concentration along the equator characterizes optical properties and structures in the deeper part of the surface layer, as mentioned in 3.3.2. The decrease in the MLD was also significant in the Arabian Sea, where chlorophyll-a concentration was relatively high (Fig. 7a). These changes in the MLD reflected the absorption differences among the mMA94, IY10, and PS77 schemes (Fig. 7d).

Changes in SSTs and surface currents of the equatorial Pacific due to the ocean radiant schemes were associated with a change in the shallow meridional circulation of the tropical Pacific, called the subtropical cell (STC). The STC played an important role in connecting subduction regions of the subtropical gyre with upwelling regions in the tropics. When the solar angle was introduced (Fig. 9a), the pole ward surface current in the upper 30 m, and

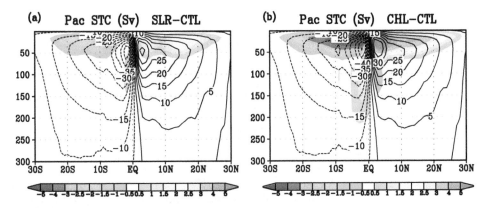

Fig. 9. (a) Meridional mass transport [Sv] zonally averaged in the Pacific Ocean for SLR (contour). The shaded area denotes the difference between SLR and CTL. (b) Same as (a) but for CHL.

the equator ward surface current at depths of 30 to 60 m were enhanced. This results in enhanced meridional circulation up to about 2.5 Sv ($1Sv=10^6$ m^3/s) in the North Pacific. Since the maximum transport of the STC is about 35 Sv in the mean state, the STC was strengthened by about 7 %. When the effect of chlorophyll-a concentration was introduced (Fig. 9b), similar current changes occured in the upper 70 m; however, the STC was further enhanced by about 7 Sv, corresponding to more than 20 % of the mean state. Thus, a more advanced ocean radiant scheme leads to more enhanced STC.

These results are understood from the following. In the tropical Pacific, the upper meridional transport (M_y) is expressed as (Sweeney et al., 2005):

$$M_y = \int_{D_{ML}}^{\eta} -\frac{1}{\rho_0 f}\frac{\partial p}{\partial x}dz + \frac{\tau_x}{\rho_0 f} \tag{8}$$

where f is a Coriolis parameter, τ_x is a zonal wind stress, $\partial p/\partial x$ is a zonal pressure gradient, ρ_0 is sea water density, η is a surface elevation, and D_{ML} is a MLD. This equation means that the upper meridional transport is given by the difference between the pole ward Ekman transport and the equatorward geostrophic transport. The MLD is reduced by the introduction of the solar angle, or the effect of the chlorophyll-a distribution. The Ekman transport is the same in the three runs, because the employed surface wind stress is the same. Also, the difference between zonal pressure gradients in each run is small (Sweeney et al., 2005). Hence, the decreased MLD leads to the reduced meridional geostrophic transport. As a result, the pole ward transport increases, leading to an enhanced STC. The enhanced STC produces a divergent flow at the surface and strengthens the equatorial upwelling, and the resulting cold water from the deep layer cools the SST in the eastern equatorial Pacific.

To summarize, the impact of the changes in absorption schemes of the solar radiation on the tropical Pacific occurs not only due to the local heating of the solar radiation itself as a direct effect, but also by the dynamical response as an indirect effect. Introducing the chlorophyll-a distribution and varying the solar angle enhances the shallow meridional circulation (STC), which leads to nontrivial changes in the tropical oceanic structure.

4. Summary

This article examined the impact of solar radiation data and its absorption schemes on ocean model simulation. Both are essential for modeling the upper ocean thermal structure.

Section 2 investigated the discrepancy between observed and OGCM-simulated anomalies in recent SSTs of the tropical Indian Ocean. Observed SSTs indicate warming beginning in the late 1990s, whereas simulated SSTs exhibit cooling over the same period. Examination of surface heat fluxes in the OGCM showed that the simulated SST cooling was caused primarily by a decreasing trend in the reanalyzed solar radiation used as the surface boundary condition. In the atmospheric reanalysis, the decrease in solar radiation was attributed to an increase in cloud cover, deduced from precipitation data, and in part, responsible for the observed local warming of the Indian Ocean SSTs prescribed as the lower boundary condition. Observation-based estimates of precipitation, however, showed no significant increasing trend; thus, no increase in cloud cover was indicated. Caution is necessary when atmospheric reanalysis data are used for surface boundary conditions for OGCMs.

Section 3 examined three absorption schemes for ocean model simulations: (1) a conventional scheme (Paulson & Simpson, 1977), (2) an introduction of varying solar angle (Ishizaki & Yamanaka, 2010), and (3) an introduction of the effect of local heating by chlorophyll-a concentration (Morel & Antoine, 1994) together with the second scheme. Introducing the new scheme resulted in a significant change especially in the equatorial Pacific, where the MLD decreased by about 10 m, and the surface current field showed a divergent flow. Associated with the surface current field, the equatorial upwelling was enhanced and the STC transport intensifies by more than 20 % in the Pacific. These changes in SLR run are explained by a dynamical response of the equatorial Pacific to the change in MLD (Sweeney et al., 2005).

These results indicate that both the solar radiation data and the employed absorption scheme of solar radiation were important, especially in the tropical ocean. Careful attention must be paid to the treatment of solar radiation data and the absorption scheme of radiation for ocean modeling.

Further observation-based studies are needed to clarify the long-term trend of precipitation in the Indian Ocean. In addition, from the standpoint of ocean modeling, further progress on reanalysis products is desired to improve sea surface fluxes, for example by including air-sea interaction processes (e.g., Fujii et al., 2009), which are lacking in the current atmospheric reanalyses.

Although this study set the ratio of the longer-wavelength (IR) to the total at the surface to a constant (R in (1)), atmospheric models generally treat direct rays and scattered light separately, and provide spectral intensities of radiation. Thus, the absorption of radiation in the sea can be accurately calculated by using coupled models. This is the next step of this study.

5. Acknowledgements

We thank Dr. T. Toyoda for providing us the SeaWifs data. Comments from the editor were helpful in improving the manuscript. This work was funded by Meteorological Research Institute, and was partly supported by the Grant-in-Aid for Science Research 22540455 from the Ministry of Education, Culture, Sports, Science and Technology, Japan.

6. References

Adler, R. F.; Huffman G. J.; Chang, A.; Ferraro, R.; Xie, P.; Janowiak, J.; Rudolf, B.; Schneider, U.; Curtis, S.; Bolvin, D.; Gruber, A.; Susskind, J. & Arkin, P. (2003). The Version 2 Global Precipitation Climatology Project (GPCP) Monthly Precipitation Analysis (1979-Present). *J. Hydrometeor.*, Vol. 4, 1147-1167.

Anderson, W. G.; Gnanadesikan, A.; Hallberg, R.; Dunne, J. & Samuels, B. L. (2007). Impact of ocean color on the maintenance of the Pacific Cold Tongue. *Geophys. Res. Lett.*, Vol. 34, L11609, doi:1029/2007GL030100.

Arakawa, O. & Kitoh, A. (2004). Comparison of local precipitation-SST relationship between the observation and a reanalysis dataset. *Geophys. Res. Lett.*, Vol. 31, L12206, doi:10.1029/2004GL020283.

Copsey, D.; Sutton, R. & Knight, J. (2006). Recent trends in sea level pressure in the Indian Ocean region. *Geophys. Res. Lett.*, Vol. 31, L12206, doi:10.1029/2004GL020283.

Dee, D.; Uppala, S.; Kobayashi, S.; Lindskog, M. & Simmons, A. (2008). Developments in bias correction for reanalysis. *Proceedings of the 3rd WCRP International Conference on Reanalysis*, p58.

Deser, C. & Phillips, A. (2006). Simulation of the 1976/77 climate transition over the North Pacific: Sensitivity to tropical forcing. *J. Clim.*, Vol. 19, 6170-6180.

Du, Y.; Qu, T.; Meyer, G.; Masumoto, Y. & Sasaki, H. (2005). Seasonal heat budget in the mixed layer of the southeastern tropical Indian Ocean in a high-resolution ocean general circulation model. *J. Clim.*, Vol. 110, C04012, doi:10.1029/2004JC002845.

Fujii, Y.; Nakaegawa, T.; Matsumoto, S.; Yasuda, T.; Yamanaka, G. & Kamachi, M. (2009). Coupled climate simulation by constraining ocean fields in a coupled model with ocean data. *J. Clim.*, Vol. 22, No. 20, 5541-5557.

Gent, P. R. & McWilliams, J. C. (1990). Isopycnal mixing in ocean circulation model. *J. Phys. Oceanogr.*, Vol. 20, 150-155.

Gnanadesikan, A. & Anderson, W. G. (2009). Ocean water clarity and the ocean general circulation in a coupled climate model. *J. Phys. Oceanogr.*, Vol. 39, doi:10.1175/2008JPO3935.1, 314-332.

Han, W.; Meehl, G. & Hu, A. (2006). Interpretation of tropical thermocline cooling in the Indian and Pacific oceans during recent decades. *Geophys. Res. Lett.*, Vol. 33, L23615, doi:10.1029/2006GL027982.

Hoerling, M. P.; Hurrell, J. W.; Xu, T.; Bates, G. T. & Phillips, A. S. (2004). Twentieth century North Atlantic climate change: Part II. Understanding the effect of Indian Ocean warming. *Clim. Dyn.*, Vol. 23, No. 3-4, 391-405.

Ishii, M.; Shoji, A.; Sugimoto, S. & Matsumoto, T. (2005). Objective Analyses of SST and marine meteorological variables for the 20th Century using ICOADS and the Kobe Collection. *Int. J. Climatol.*, Vol. 25, 865-879.

Ishikawa, I.; Tsujino, H.; Hirabara, M.; Nakano, H.; Yasuda T. & Ishizaki, H. (2005). Meteorological Research Institute Community Ocean Model (MRI.COM) Manual. (in Japanese). *Technical reports of the Meteorological Research Institute*, Vol. 47, 1-189.

Ishizaki, H. & Motoi, T. (1999). Reevaluation of the Takano-Onishi scheme for momentum advection on bottom relief in ocean models. *J. Atmos. Ocean Technol.*, Vol. 16, 1994-2010.

Ishizaki, H. & Yamanaka, G. (2010). Impact of explicit sun altitude in solar radiation on an ocean model simulation. *Ocean Modelling*, Vol. 33, 52-69.

Jerlov, N. G. (1968). *Optical Oceanography*, Elsevier, pp. 194.

Kalnay, E.; Kanamitsu, M.; Kistler, R.; Collins, W.; Deaven, D.; Gandin, L.; Iredell, M.; Saha, S.; White, G.; Woollen, J.; Zhu, Y.; Chelliah, M.; Ebisuzaki, W.; Higgins, W.; Janowiak, J.; Mo, K. C.; Ropelewski, C.; Wang, J.; Leetmaa, A.; Reynolds, R.; Jenne, R. & Joseph, D. (1996). The NCEP/NCAR 40-years reanalysis project. *Bull. Am. Meterol. Soc.*, Vol. 77, 437-471.

Kanamitsu, M.; Ebisuzaki, W.; Woolen, J.; Yang, S-K.; Hnilo, J. J.; Fiorino, M. & Potter, G. L. (2002). NCEP-DOE AMIP-II reanalysis (R-2). *Bull. Am. Meterol. Soc.*, Vol. 83, 1631-1643.

Kara, A. B.; Rochford, P. A.; & Hulburt, H. E. (2000). Efficient and accurate bulk parameterizations of air-sea fluxes for use in general circulation models. *J. Atmos. Ocean Technol.*, Vol. 17, 1421-1438.

Large, W. G. & Yeager, S. G. (2004). Diurnal to decadal global forcing for ocean and sea-ice models: the data sets and flux climatologies. *NCAR Technical Note*, NCAR/TN-460+STR.

Large, W. G., & Yeager, S. G., (2009). The global climatology of an interannualy varying air sea flux data set. *Clim. Dyn.*, Vol. 33, 341-363. doi:10.1007/s00382-008-0441-3.

Lau, K. M. & Weng, H. Y. (1999). Interannual, decadal-interdecadal, and global warming signals in sea surface temperature during 1955-97. *J. Clim.*, Vol. 12, No. 5, 1257-1267.

Levitus, S.; Antonov, J. & Boyer, T. (2005). Warming of the world ocean, 1955-2003. *Geophys. Res. Lett.*, Vol. 32, L02604, doi:10.1029/2004GL021592.

Manizza, M.; Quere, C. L.; Watson, A. J. & Buitenhuis, E. T. (2005). Bio-optical feedbacks among phytoplankton, upper ocean physics and sea-ice in a global model. *Geophys. Res. Lett.*, Vol. 32, L05603.

Morel, A. (1988). Optical modeling of the upper ocean in relation to its biogenous matter content (Case I waters). *J. Geophys. Res.*, Vol. 93, 10749-10768.

Morel, A. H. & Antoine, D. (1994). Heating rate within the upper ocean in relations to its biooptical state. *J. Phys. Oceanogr.*, Vol. 24, 1652-1665.

Morel, A. H. & Berthon, J. F. (1989). Surface pigments, algal biomass profiles, and potential production of the euphotic layer: Relationships reinvestigated in view of remote sensing applications. *Limnol. Oceanogr.*, Vol. 34, No. 8, 1545-1562.

Morel, A. H. & Prieur, L. (1977). Analysis of variations in ocean color. *Limnol. Oceanogr.*, Vol. 22, 709-722.

Murray, R. J. (1996). Explicit generation of orthogonal grids for ocean models. *J. Comput. Phys.*, Vol. 126, 251-273.

Murtugudde, R. & Busalacchi, A. J. (1999). Interannual variability of the dynamics and thermodynamics of the tropical Indian Ocean. *J. Clim.*, Vol. 12, 2300-2326.

Murtugudde, R.; Beauchamp, J.; McClain, C. R.; Lewis, M. & Busalacchi, A. J. (2002). Effects of penetrative radiation on the upper tropical ocean circulation. *J. Clim.*, Vol. 15, 470-486.

Nakamoto, S.; Kumar, S.; Oberhuber, J.; Ishizaka, J.; Muneyama, K. & Frouin, R. (2001). Response of the equatorial Pacific to chlorophyll pigment in a mixed layer isopycnal ocean general circulation model. *Geophys. Res. Lett.*, Vol. 28, 2021-2024.

Nakano, H. & Suginohara, N. (2002). Effects of bottom boundary layer parameterization on reproducing deep and bottom waters in a world ocean model. *J. Phys. Oceanogr.*, Vol. 32, 1209-1227.

Noh, Y. & Kim, H-J. (1999). Simulations of temperature and turbulence structure of the oceanic boundary layer with the improved near-surface process. *J. Geophys. Res.*, Vol. 104, 15621-15634.

Norris, J. R. (2005). Trends in upper-level cloud cover and surface divergence over the tropical Indo-Pacific Ocean between 1952 and 1997. *J. Geophys. Res.*, Vol. 110, D21110, doi:10.1029/2005JD006183.

Ohlmann, J. C. (2003). Ocean radiant heating in climate models. *J. Clim.*, Vol. 16, 1337-1351.

Ohlmann, J. C.; Siegel, D. A. & Mobley, C. D. (2000). Ocean radiant heating. Part I: Optical influences. *J. Phys. Oceanogr.*, Vol. 30, 1833-1848.

Ohlmann, J. C. & Siegel, D. A. (2000). Ocean radiant heating. Part II: Parameterizing solar radiation transmission through the upper ocean. *J. Phys. Oceanogr.*, Vol. 30, 1833-1848.

Onogi, K.; Tsutsui, J.; Koide, H.; Sakamoto, M.; Kobayashi, S.; Hatsushika, H.; Matsumoto, T.; Yamazaki, N.; Kamahori, H.; Takahashi, K.; Kadokura, S.; Wada, K.; Kato, K.; Oyama, R.; Ose, T.; Mannoji, N. & Taira, R. (2007). The JRA-25 reanalysis. *J. Meteor. Soc. Japan*, Vol. 85, 369-432.

Paulson, C. & Simpson, J. (1977). Irradiance measurements in the upper ocean. *J. Phys. Oceanogr.*, Vol. 7, 952-956.

Prather, M. J. (1986), Numerical advection by conservation of second-order moments. *J. Geophys. Res.*, Vol. 91, 6671-6681.

Redi, M. H. (1982), Oceanic isopycnals mixing by coordinate rotation. *J. Phys. Oceanogr.*, Vol. 12, 1154-1158.

Sweeney, C.; Gnanadesikan, A.; Griffies, S. M.; Harrison, M. J.; Rosati, A. J. & Samels, B. L. (2005). Impacts of shortwave penetration depth on large-scale ocean circulation and heat transport. *J. Phys. Oceanogr.*, Vol. 35, 1103-1119.

Tsujino, H.; Motoi, T.; Ishikawa, I.; Hirabara, M.; Nakano, H.; Yamanaka, G.; Yasuda, T. & Ishizaki, H. (2010). Reference manual for the Meteorological Research Institute Community Ocean Model (MRI.COM) version 3. *Technical reports of the Meteorological Research Institute*, Vol. 59, 1-241.

Tsujino, H.; Hirabara, M.; Nakano, H.; Yasuda, T.; Motoi, T. & Yamanaka, G. (2011). Simulating present climate of the global ocean-ice system using the Meteorological Research Institute Community Ocean Model (MRI.COM): simulation characteristics and variability in the Pacific sector. *J. Oceanogr.*, Vol. 67, 449-479. doi:10.1007/s10872-011-0050-3.

Tsutsui, J. & Kadokura, S. (2008). Multiple regression analysis of the JRA-25 monthly temperature. *Proceedings of the 3rd WCRP International Conference on Reanalysis*, pp. 28.

Uppala, S. M.; Kållberg, P. W.; Simmons, A. J.; Andrae, U.; da Costa Bechtold, V.; Fiorino, M.; Gibson, J. K.; Haseler, J.; Hernandez, A.; Kelly, G. A.; Li, X.; Onogi, K.; Saarinen, S.; Sokka, N.; Allan, R. P.; Andersson, E.; Arpe, K.; Balmaseda, M. A.; Beljaars, A. C. M.; van de Berg, L.; Bidlot, J.; Bormann, N.; Caires, S.; Chevallier, F.; Dethof, A.; Dragosavac, M.; Fisher, M.; Fuentes, M.; Hagemann, S.; Hólm, E.; Hoskins, B. J.; Isaksen, L.; Janssen, P. A. E. M.; Jenne, R.; McNally, A. P.; Mahfouf, J-F.; Morcrette, J-J.; Rayner, N. A.; Saunders, R. W.; Simon, P.; Sterl, A.; Trenberth, K. E.; Untch, A.; Vasiljevic, D.; Viterbo, P. & Woollen, J. (2005). The ERA-40 re-analysis. *Quart. J. Roy. Meteor. Soc.*, Vol. 131, 2961-3012.

Wang, B.; Wu R. & Fu, X. (2000). Pacific-East Asian teleconnection: How does ENSO affect East Asian climate ? *J. Clim.*, Vol. 13, 1517-1536.

Watanabe, M. & Jin, F.-F. (2002). Role of Indian Ocean warming in the development of Philippine Sea anticyclone during ENSO. *Geophys. Res. Lett.*, Vol. 29, 1478, doi:10.1029/2001GL014318.

Xie, P. & Arkin, P. (1996). Analysis of global monthly precipitation using gauge observations, satellite estimates, and numerical model predictions. *J. Clim.*, Vol. 9, 840-858.

Yamanaka, G. (2008). Discrepancies between observed and ocean general circulation model-simulated anomalies in recent SSTs of the tropical Indian Ocean caused by apparent trends in atmospheric reanalysis data. *Geophys. Res. Lett.*, Vol. 35, doi:10.1029/2008GL034737.

4

The Relationship Between Incoming Solar Radiation and Land Surface Energy Fluxes

Edgar G. Pavia

Centro de Investigación Científica y de Educación Superior de Ensenada
Mexico

1. Introduction

Incoming solar radiation (R) is the driver of the land surface energy fluxes: latent heat (E) or soil evaporation (i.e. the natural transfer of water from the topsoil to the atmosphere, although it might include also condensation), sensible heat (H) so-called because it can be "felt" (i.e. it is related to temperature differences between the surface and the atmosphere), and the ground heat flux (G) so-called because it is restricted to the interior of the ground (i.e. it is related to temperature differences between ground layers). All this seems rather obvious during the daytime, when R provides the energy input and apparently the output of E, H and G balance it. However the situation is less clear at night when R is nil but E, H and G may not vanish, while the energy balance must be kept. To understand this simple idea lets consider the following example: under certain conditions, like wet soil in low- and mid-latitudes, E may be considered almost as proportional to R; that is $E \sim (a_1 \times R) + OT$, where a_1 is a proportionality factor (not necessarily a constant) and OT are other terms (in this case: H and G, which are usually smaller than E and ($a_1 \times R$), but there could be other terms). If we are able to estimate E, R, H and G, or these terms are somehow known, we can tentatively solve for a_1, which could characterize the relationship between R and these fluxes, and the term ($a_1 \times R$) is called the net radiation (R_n); i.e. the part of R which is actually balanced by E, H and G. The problem, however, is not trivial because even if we restrict ourselves to this simplified case, and we could measure R and E, the smaller terms H and G would have to be assessed as well. Nevertheless we believe that this difficulty may be partially overcome by empirically modeling E. Recall that calculating a_1 with observations of total and net solar radiation: $a_1 = (R_n)_{obs}/R_{obs}$ may not be appropriate for our purposes, because it would not consider E, H and G, and a_1 is but an element of a vector **a** yet-unknown. Therefore our goal is to develop a full energy flux model to show that indeed the relationship between R and surface fluxes may be achieved in this empirical way. That is, in this work we will attempt to approximate these surface energy fluxes by simultaneously modeling them based on a simplified energy balance. Although, due to their importance in many environmental issues (from crop-field irrigation (Brisson & Perrier, 1991), (Allen et al., 1998), to the study of the global water cycle (Huntington, 2006)) one usually first looks for R_n in order to evaluate surface fluxes; here we will attempt to model E, H and G in order to estimate $R_n = (a_1 \times R)$. However even if our model is successful, that is even if in general it correlates well with observations, we will examine the situations in which the model fails, the energy balance does not hold, and a simple relationship between R and surface fluxes cannot be established. Thus the limitations of this study should serve as a motivation for future work.

1.1 Modeling approach

The simplest way to determine if a model is appropriate is to compare it with observations. Even though assessing E in general is difficult because it depends not only on the ambient conditions, but also on soil composition and moisture content, here we use observations of soil evaporation (E_{obs}) obtained through micro-lysimetry (Figure 1). That is, we will try to model E from E_{obs} and calculate their correlation coefficient $r(E_{mod}, E_{obs})$ to evaluate the appropriateness of our model: $E_{mod} = (R_n - H - G)_{mod}$. Although diverse efforts have been devoted to model E for different applications (Penman, 1948), (Priestley & Taylor, 1972), (Twine et al., 2000), (Brutsaert, 2006), (Agam et al., 2010), all these efforts possess different limitations and degrees of difficulty. In other words, there is no general way to model E which is practicable for all situations (Crago & Brutsaert, 1992), and thus we must develop an *ad hoc* E-model to estimate a_1 for our particular case. In this sense we will focus on a relatively simple case, the diurnal variation of bare soil evaporation when water is not a limiting factor (for example wet sand with substantially more than 5% of water (Pavia & Velázquez, 2010)). That is, when the main diurnal surface energy balance is between R_n and E: $R_n \sim E$. Previous works in cases similar to the present one have confirmed that daytime E is highly correlated with R (Pavia, 2008); therefore we should expect our model to reflect daytime better than nighttime conditions. We will perform an experiment with an evaporating tray containing a small amount of wet sand (~35 Kg maximum), so that E_{obs} should be easier to measure throughout the day than R_n. Our hypothesis is that we can obtain E_{mod} from a small number of standard meteorological observations and experimentally-obtained variables, which are chosen by their assumed relationship to energy terms; namely R, air temperature (T_a), surface temperature (T_o), soil temperature (T_s) and observed soil evaporation (E_{obs}). Therefore we will try to fit E_{obs} to a linear combination of terms derived from the above variables. Specifically $E_{mod} = E_{obs} \sim L(R, \Delta T_a, \Delta T_s)$, where, as it will be explained in the next section, the model E (E_{mod}) is achieved from R, $\Delta T_a = T_o - T_a$, $\Delta T_s = T_o - T_s$ and E_{obs}, through a multiple regression procedure yielding a vector **a** which includes a_1 among other parameters. This approach is physically-motivated by the primary land surface energy balance:

$$R_n = E + H + G,\tag{1}$$

Where R_n would be approximated by the R term (Gay, 1971) and the sensible heat flux H and the ground heat flux G would be similarly approximated by the ΔT_a and ΔT_s terms, respectively. Therefore it is anticipated that the multiple-regression parameter-vector **a** resulting from our model may give a preliminary assessments of the relative importance of R on each of these surface energy flux densities.

2. Methods

In this section we describe the original technique to find the relationship between R and the surface energy fluxes. This includes the experimental evaluation of E, the approximation made of $H \sim \Delta T_a$ and $G \sim \Delta T_s$ from the observed temperatures, and the multiple regression method to optimize these approximations.

2.1 The experiment

A 27-d experiment was performed from 12 February to 11 March 2011, in Ensenada, Mexico (31° 52′ 09″ N, 116° 39′ 52″ W) at 66 m above mean sea level. It consisted of a bird-guarded wet-sand evaporating tray (equipped with temperature sensors at depths $z_o = 0.02$ m, for T_o, and $z_1 = 0.07$ m for T_s) set on an electronic scale to register the varying weight (w) next to a meteorological station recording R and T_a among other variables (see Figure 1). All variables are registered at $\Delta t = 300$ s intervals, and the total number of samples is N = 7776. See (Pavia & Velázquez, 2010) for more details on similar experiments.

Fig. 1. The experimental setup: the meteorological station, the evaporative tray and the weighing scale used in the study.

2.2 The empirical approach

We begin by calculating a time series of weight-change time-rates $\Delta w_i = (w_{i-1/2} - w_{i+1/2}) / \Delta t$ [Kg s^{-1}], where $w_{i-1/2}$ and $w_{i+1/2}$ represent smooth averaged weight values (e.g. precipitation has been filtered out), which is used to obtain a time series of observed evaporation, $(E_{obs})_i = \lambda \times \Delta w_i / A$ [W m^{-2}], where $\lambda = 2.45 \times 10^6$ [J Kg^{-1}] is the latent heat of water vaporization, and $A = 0.23$ m^2 is the evaporating surface area. Then we fit $(E_{obs})_i$ to the corresponding series of R_i, $(\Delta T_a)_i$, and $(\Delta T_s)_i$, that is:

$$(E_{mod})_i = a_1 R_i + a_2 (\Delta T_a)_i + a_3 (\Delta T_s)_i; \quad i = 1,2,\ldots,7776. \tag{2}$$

And the problem is now reduced to finding the values of a_1, a_2, and a_3.

2.3 The statistical method

A simple technique to try to solve the above problem is a least-square multiple regression procedure, which in this case is formulated as follows. First we construct the vector:

$$\mathbf{y} = \left[(E_{obs})_1\ (E_{obs})_2\ \cdots\ (E_{obs})_{7776} \right], \tag{3}$$

and the matrix:

$$\mathbf{X} = \begin{bmatrix} R_1 & R_2 & \cdots & R_{7776} \\ (\Delta T_a)_1 & (\Delta T_a)_2 & \cdots & (\Delta T_a)_{7776} \\ (\Delta T_s)_1 & (\Delta T_s)_2 & \cdots & (\Delta T_s)_{7776} \end{bmatrix}. \tag{4}$$

Then we posit that $\mathbf{y}_{mod} = \mathbf{aX}$, where $\mathbf{a} = [a_1\ a_2\ a_3]$ is the coefficients-vector to be found. Using (3) and (4) this is done by minimizing $Z \equiv (\mathbf{y} - \mathbf{aX})(\mathbf{y} - \mathbf{aX})^T$; that is $\partial Z / \partial \mathbf{a} = 0$, which finally yields $\mathbf{a} = \mathbf{yX}^T (\mathbf{XX}^T)^{-1}$ and consequently $\mathbf{y}_{mod} = [(E_{mod})_1\ (E_{mod})_2\ \cdots\ (E_{mod})_{7776}]$.

3. Results and discussion

The above procedure gave $a_1 = 0.48$, $a_2 = -3.77$ [W m^{-2} K^{-1}], $a_3 = -14.25$ [W m^{-2} K^{-1}], which are used in (2) to evaluate E_{mod} [W m^{-2}]. The comparison of the evolution of E_{mod} and E_{obs} is presented in Figure 2. These two series have a correlation coefficient $r(E_{mod}, E_{obs}) = 0.90$, which indicates that our method has been rather successful to model E. In addition we will try to relate each term of E_{mod} to surface energy fluxes using (1); that is $a_1 R = E_{obs} - a_2 \Delta T_a - a_3 \Delta T_s$, or $0.48 R = E_{obs} + 3.77 \Delta T_a + 14.25 \Delta T_s$. The most important term of the model is $0.48 \times R$, because most of E occurs during the daytime. This means that here the net radiation is principally proportional to the absorbed radiation: $R_n \sim b_1 (1 - \alpha) R + b_2 T_a + b_3 T_o$, where $b_1 = a_1 / (1 - \alpha)$, $b_2 = 0$, $b_3 = 0$, and α is the wet-sand albedo. Moreover if $0.10 < \alpha < 0.25$ we obtain reasonable values for b_1: $0.53 < b_1 < 0.64$ (Gay, 1971), (Stathers et al., 1988). Obviously this assumption is not valid during the nighttime, when $E \sim 0$ but not nil. Likewise we may consider the sensible heat flux to be approximated by $H = -a_2 \times \Delta T_a$.

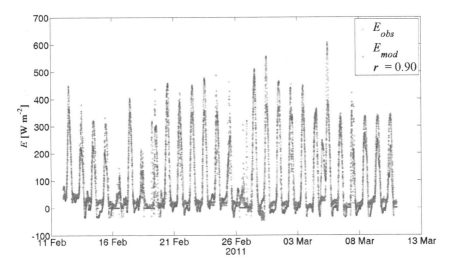

Fig. 2. Comparison between modeled soil evaporation E_{mod} and observed soil evaporation E_{obs}.

Comparing this approximation with its theoretical expression (Stathers et al., 1988) $H = (\rho\, c_p\, /r_{aH}) \times \Delta T_a$, where $\rho \sim 1.0$ [Kg m^{-3}] is the density of air, $c_p \sim 1000.0$ [J Kg^{-1} K^{-1}] is the specific heat capacity of air at constant pressure, and r_{aH} is the aerodynamic resistance to heat transfer between the surface and $z \sim 2.0$ m (the height at which T_a is measured), we straightforwardly get $r_{aH} = -(\rho\, c_p\, /\, a_2) \sim 265$ [m^{-1} s]. This value is a very good approximation to the one obtained from its simplest theoretical form, i. e. for stable atmospheric conditions (Webb, 1970):

$$r_{aH} = \frac{[\ln\ (z/z_T) + 4.7 \times (z\,/\,L)] \times [\ln\ (z/z_M) + 4.7 \times (z\,/\,L)]}{k^2 u} \sim 293\ [\mathrm{m}^{-1}\mathrm{s}], \qquad (5)$$

where we have chosen in (5) the following values, $z_T = 0.0002$ m for the surface roughness length for sensible heat transfer, $L = 10$ m for the Monin-Obukhov length, $z_M = 0.0005$ m for the surface roughness length for momentum, $u = 2.0$ [m s^{-1}] for the mean wind speed at $z = 2$ m height, and $k = 0.4$ is the von Kármán constant (Stathers et al., 1988). Similarly if we approximate the soil heat flux obtained by integrating the heat conduction equation (Peters-Lidard et al., 1998): $G = (\kappa\ \partial T/\partial z)_0 \sim \kappa\ (\Delta T_s\ /\Delta z)$, where κ is the soil thermal conductivity and $\Delta z = (z_1 - z_0) = 0.05$ m, and compare it with our estimate of $G = -a_3\ \Delta T_s$ we straightforwardly obtain $\kappa = (14.25 \times 0.05) = 0.7125$ [W m^{-1} K^{-1}], which is a reasonable value, although somewhat low since for water $\kappa = 0.6$ [W m^{-1} K^{-1}] and for soil minerals $\kappa = 2.9$ [W m^{-1} K^{-1}] (see Table I of Peters-Lidard et al. (1998)). Therefore we may consider that as $E_{obs} \sim E$, $a_1\, R \sim R_n$, $a_2\, \Delta T_a \sim H$, and $a_3\, \Delta T_s \sim G$, the surface energy balance is approximately satisfied ($R_n \sim E + H + G$). And if we calculate the mean diurnal variations during our 27-d observation period (defined positive toward the surface):

$$< F >_i = \sum_{j=1}^{27} (F_i)_j, \quad i = 1, 2, \ldots, 288, \tag{6}$$

where F is any of the approximated energy fluxes considered here, we observe that, as expected, most of the time the magnitudes of $< H >$ and $< G >$ are smaller than those of $< E >$ and $< R_n >$ (up to about one order of magnitude during the daytime). However their progresses during the day are more telling (see Figure 3); that is the $< R_n >$ maximum around noon, the $< G >$ minimum at mid-morning and the $< H >$ minimum in the afternoon; all suggest that our empirical approach is appropriate. Perhaps it can be improved with better observations, but these results are definitely encouraging.

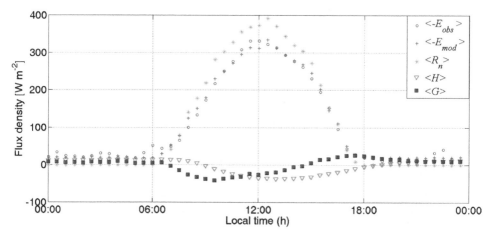

Fig. 3. The diurnal variation of the different approximate energy fluxes (6) defined positive towards the surface. Note that evaporations are plotted with a negative sign. For clarity fluxes are plotted at 5Δt intervals.

Nevertheless we must acknowledge the limitations of our empirical model. Since in this case E so strongly depends on R and only marginally on ΔT_a and ΔT_s, we cannot include more terms, for example terms related to wind speed and relative humidity (which are also related to R), because that could render the model unstable as these terms do not significantly contribute to explain variance. For example if we focus on 5 March 2011 (see Figure 4), a particularly windy and dry day apparently resulting from a brief Santa Ana event (see Raphael (2003) for a description of this kind of events), we observe that E_{mod} underestimates E_{obs}, especially during the nighttime early hours. In this situation E_{mod} can only be appropriately modeled if we could include in our algorithm wind and humidity observations, which as mentioned before is not possible. Yet the model clearly indicated that in this case other evaporative causes, besides the ones related to energy fluxes, were also related to R and playing a role in E. And, on the other hand, when we tested our model with independent data (Figure 5); that is using the current values of the vector $\mathbf{a}$ with new observations R_i, $(\Delta T_a)_i$, and $(\Delta T_s)_i$ for the period 17-29 May 2011, we found that now the model overestimates the observations: $(\Sigma\ E_{obs}) / (\Sigma\ E_{mod}) \sim 0.7$.

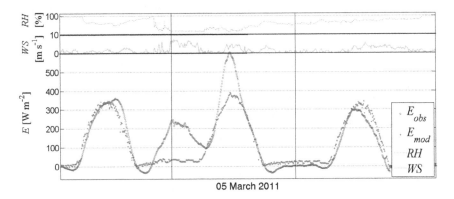

Fig. 4. Close up centered on 05 March 2011. Wind speed *WS* and relative humidity *RH* are also shown schematically.

Although correlation were still high, $r(E_{mod}, E_{obs})= 0.83$, in this situation E_{obs} were limited by the lack of moisture in the wet sand. Here again the model clearly indicated that in this case the sand was drier than when we first calculated **a**, as the average weight of the evaporating tray in this case was 21.0 Kg, compared to 25.5 Kg in the original experiment.

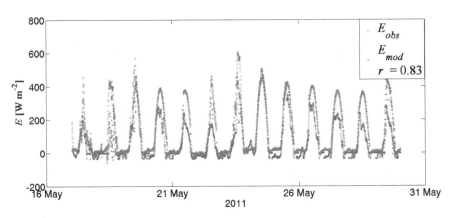

Fig. 5. Same as Figure 2, but for the test period with independent data. Sand was drier after 26 May.

Finally we compared our method (Figure 6) with a previous technique developed for modeling 7-h (08:30 to 15:30 h, local time) total soil evaporation (Pavia, 2008). In this case evaporation, in mm, is given by:

$$E^{(1)}_{mod} = 0.8 \times \left[0.1525 \times (\overline{T}_a - 18) + 0.0053 \times (\overline{R} - 404) \right] + 2.2 \text{ [mm]};\qquad(7)$$

where the overbar indicates dimensionless mean values during the 7-h observing period, and the corresponding values for our model are computed by:

$$E^{(2)}{}_{mod} = \sum_{k=1}^{84}(E_{mod})_k \times \Delta t / \lambda \ [\text{mm}], \tag{8}$$

where $k = 1$ corresponds to 08:30 h local time. The higher correlation given by the second model: $r(E^{(2)}{}_{mod}, E_{obs}) = 0.9$ versus $r(E^{(1)}{}_{mod}, E_{obs}) = 0.7$ of the first model, furthermore suggests that the new model improves the predictions.

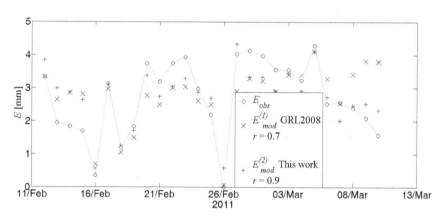

Fig. 6. Comparison of the 7-h total E obtained with present model $E^{(2)}{}_{mod}$ and that obtained with the model GRL2008 of Pavia (2008) $E^{(1)}{}_{mod}$.

4. Conclusions

The main objective of this work, which is the optimal estimation of **a** by the empirical modeling of soil evaporation, has been achieved (see Figures 2 and 6). This vector represents the relationship between solar radiation and surface energy fluxes. Nevertheless it has a drawback, since a_1 is proportional to R it is pointless when the incoming solar radiation is nil. However this empirical approach, physically motivated by the surface energy balance, yields promising results by still suggesting an energy balance at night; i.e. when $R = 0$. For example, we conclude that in this case the net radiation $R_n = a_1 R \sim b_1 (1 - a) R$ is largely a function of the absorbed solar radiation, because here we are dealing with substantially wet sand and most of the evaporation occurs during the day (see Figure 7); but we also conclude that the sensible heat flux $H = a_2 \times \Delta T_a \sim (\rho\ c_p\ /r_{aH}) \times \Delta T_a$, since the value obtained here for the aerodynamic resistance to heat transfer $r_{aH} = 265\ \text{m}^{-1}\ \text{s}$ is very close to its theoretical estimation $r_{aH} = 293\ \text{m}^{-1}\ \text{s}$ obtained with (5) (see Figure 8). And, similarly, we conclude that the ground heat flux $G = a_3 \Delta T_s \sim \kappa\ (\Delta T_s\ /\Delta z)$, since the value obtained here for the thermal conductivity $\kappa = 0.7125\ [\text{W m}^{-1}\ \text{K}^{-1}]$ is within the expected range (Peters-Lidard et al., 1998) of values: 0.6 to 2.9 $[\text{W m}^{-1}\ \text{K}^{-1}]$ (see Figure 9).

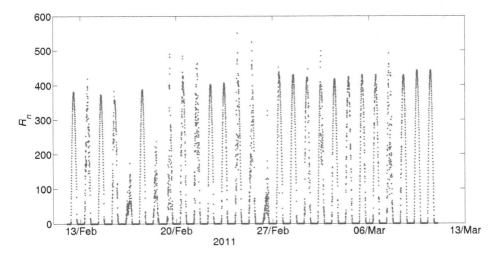

Fig. 7. Time series of the modeled net radiation.

The shapes of the progresses (see Figure 3) of their mean diurnal values (6) furthermore support these conclusions.

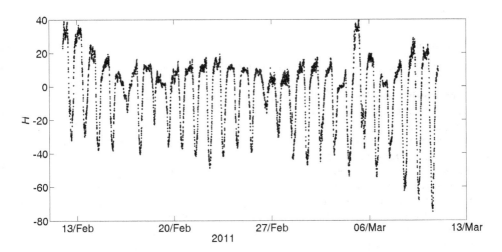

Fig. 8. Time series of the modeled sensible heat.

However our empirical model is limited because statistically it is not possible to have more than a few terms. Considering wind speed and relative humidity terms in our algorithm may result in better predictions during Santa Ana events. Considering single temperature terms may improve the net radiation term $R_n = b_1 (1 - \alpha) R + b_2 T_a + b_3 T_o$, as b_2 and b_3 become non-zero. This in turn may improve the estimations of the H and G terms, which may result in better predictions when the wet sand becomes drier, for example. Efforts to overcome these limitations are in progress, i.e. trying to model the difference between evaporation and net radiation $(E_{mod} - R_n) = L(T_a, \Delta T_a, \Delta T_s)$ or $L(T_s, \Delta T_a, \Delta T_s)$, since T_a and T_s are correlated. Nevertheless the present empirical approach provides an interesting alternative to more sophisticated methods.

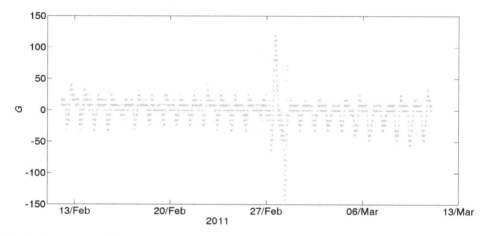

Fig. 9. Time series of the modeled ground heat flux.

5. Summary

The relationship between incoming solar radiation and the surface energy fluxes E, H and G has been investigated by empirically modeling E through a multiple regression method. We propose this new empirical model of wet sand evaporation, which gives excellent results when moisture is not a limiting factor and wind and air humidity are not extreme (see Figure 10), as a means to establish this relationship (represented here by **a**). The algorithm was physically motivated by the surface energy balance $R_n = E + H + G$; i.e. we do not consider other terms (i.e. relative humidity or wind speed). In this sense we measured R, T_a, T_o, T_s, and E_{obs}, in order to model E from R, $\Delta T_a = T_o - T_a$, and $\Delta T_s = T_o - T_s$. Namely $E_{mod} = a_1 R + a_2 \Delta T_a + a_3 \Delta T_s$; where E_{mod} is the model E, and the coefficients a_1, a_2, and a_3 are determined through multiple regression. Therefore the model provides also a preliminary assessment of the relative importance of energy fluxes. That is, making $E = E_{obs}$, $R_n = a_1 R$, $H = a_2 \Delta T_a$, and $G = a_3 \Delta T_s$, we get $a_1 R = E_{obs} - a_2 \Delta T_a - a_3 \Delta T_s$. Comparison of model results with observations may serve to identify the active role of other variables (wind speed or air humidity) on evaporation, when the model underestimates observations; or the departure from saturation of the evaporating media, when the model overestimates observations. These two cases represent extreme situations when the relationship between solar radiation and surface energy fluxes can not be established by this simple model.

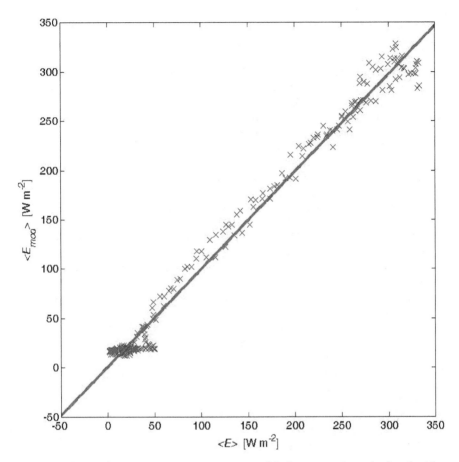

Fig. 10. Mean observed evaporation versus mean modeled evaporation calculated with equation (6). The slope of the linear fit is ~1.0 (red line).

6. Acknowledgment

I thank S. Higareda and I. Velázquez for innumerable helps. This research was funded by the Mexican CONACYT system. This work is to recognize Prof. Ignacio Galindo's 50 years dedicated to scientific research, mainly on solar radiation.

7. References

Agam, N.; Kustas, W. P.; Anderson, M. C.; Norman, J. M.; Colaizzi, P. D.; Howell, T. A.; Prueger, J.H.; Meyers, T. P. & Wilson, T.B. (2010) Application of the Priestley–Taylor approach in a two-source surface energy balance model. *Journal of Hydrometeorology* Vol. 11, pp. 185–198, doi: 10.1175/2009JHM1124.1

Allen, R. G.; Pereira, L. S.; Raes, D. & Smith, M. (1998) *Crop evapotranspiration. Guidelines for computing crop water requirements* (*Irrigation and drainage paper 56*), FAO, Rome, Italy.

Brisson, N. & Perrier, A. (1991) A semiempirical model of bare soil evaporation for crop simulation models. *Water Resources Research* Vol. 27, pp. 719-727.

Brutsaert, W. (2006) Indications of increasing land surface evaporation during the second half of the 20th century. *Geophysical Research Letters* Vol. 33, L20403, doi:10.1029/2006GL027532.

Crago, R. D. & Brutsaert, W. (1992) A comparison of several evaporation equations. *Water Resources Research* Vol. 28, pp. 951-954.

Gay, L. W. (1971) The regression of net radiation upon solar radiation. *Archiv für Meteorologie, Geophysik und Bioklimatologie. Serie B* Vol. 19, pp. 1-14.

Huntington, T. G. (2006) Evidence for intensification of the global water cycle: Review and synthesis. *Journal of Hydrology* Vol. 319, pp. 83-95.

Pavia, E. G. (2008) Evaporation from a thin layer of wet sand. *Geophysical Research Letters* Vol. 35, L08401, doi:10.1029/2008GL033465.

Pavia, E. G. & Velázquez, I. (2010) Does wet sand evaporate more than water in the Guadalupe Valley? In *The Ocean, The Wine, and The Valley: The Lives of Antoine Badan*, Pavia, E. G.; Sheinbaum, J. & Candela, J. (eds), CICESE, Ensenada, Mexico, ISBN 978-0-557-94026-4, pp. 331.

Penman, H. L. (1948) Natural evaporation from open water, bare soil and grass. *Proceedings of the Royal Society* Vol. 193, pp. 120-145.

Peters-Lidard, C. D.; Blackburn, E.; Liang, X. & Wood, E. F. (1998) The effect of soil thermal conductivity parameterization on surface energy fluxes and temperatures. *Journal of Atmospheric Sciences* Vol. 55, pp. 1209-1224.

Priestley, C. H. B. & Taylor, R. J. (1972) On the assessment of surface heat flux and evaporation using large-scale parameters. *Monthly Weather Review* Vol. 100, pp. 81-92.

Raphael, M. N. (2003) The Santa Ana winds of California. *Earth Interactions* Vol. 7, pp. 1-13.

Stathers, R. J.; Black, T. A.; Novak, M. D. & Bailey, W. G. (1988) Modelling surface energy fluxes and temperatures in dry and wet bare soils. *Atmosphere-Ocean* Vol. 26, pp. 59-73.

Twine, T. E.; Kustas, W. P.; Norman, J. M.; Cook, D. R.; Houser, P.R.; Meyers, T. P.; Prueger, J. H.; Starks, P.J. & Wesely, M. L. (2000) Correcting eddy-covariance flux underestimates over a grassland. *Agricultural and Forest Meteorology* Vol. 103, pp. 279-300. doi:10.1016/S0168-1923(00)00123-4

Webb, E. K. (1970) Profile relationships: The log-linear range and extension to strong stability. *Quarterly Journal of the Royal Meteorological Society* Vol. 96, pp. 67-90.

5

Interannual and Intraseasonal Variations of the Available Solar Radiation

Kalju Eerme
Tartu Observatory
Estonia

1. Introduction

The availability of solar radiation is important climate and welfare related environmental factor like temperature or moisture. Living organisms are adapted to the local annual cycles as well as interannual and intraseasonal variations of environmental factors. The importance of appropriate models for global solar radiation has increased in recent decades due to wider use of solar energy applications, including photovoltaic power generation (Šuri et al., 2007, Tiwari and Sodha, 2006). Better understanding of the influence of spectral composition of ground-level solar radiation on terrestrial and aquatic ecosystems has increased the interest in the availability of solar direct irradiance (Lohmann et al., 2006). The annual total does not contain enough information for applications of solar radiation data. Seasonal or even monthly time resolution is often necessary. The annual cycle of availability of solar energy is determined by the annual cycle of noon solar elevation angle, with significant contribution from cloudiness (Tooming, 2002) and atmospheric transparency (Russak, 1990, 2009). Cloud cover at moderate latitudes tends to be thicker and more frequent in late autumn and early winter and less frequent in spring and summer. In the dark half-year not only absolute but also relative availability of solar radiation is smaller.

The attempts of measuring and recording ground-level solar radiation have started about 100 years ago, but usually these activities remained episodic. Most available solar radiation data sets cover significantly shorter time intervals than those of temperature or precipitation. Due to relatively short time series the reasons for variation and regular changes of ground-level solar radiation are still not completely understood.

The present chapter in considering local seasonal and monthly relative availability of solar radiation is based on Estonian solar radiation data. The longest and most complete data set on solar radiation in Estonia has been collected at a typical Estonian rural site at the Tartu-Tõravere Meteorological Station (58o.16′N, 26o.28′E, 70 m a.s.l.). The attempts of recording sunshine duration were made since 1906 (Kallis et. al., 2005). First regular measurements of solar irradiance were performed in late 1930s and continued after 1950 (Ohvril et. al., 2009). Before 1965 the station was based closer to town than its present site. Simultaneous measurements at both sites during one year did not reveal systematic differences. The landscapes at both sites are similar. The Tartu-Tõravere site as well as that before 1965 can be considered typical for Northern Europe. At other geographical regions the contrast between summer and winter as well as the seasonal impacts of cloud cover and aerosols may be significantly different. Here, the variations of solar ground-level integral global and direct irradiance on seasonal and monthly scales are examined. The continuous record of pyranometer-measured daily global radiation extends back to 1953 and that of pyrheliometer-measured direct irradiance back to 1955. The study is based on this long-term data set for years 1955-2010 when both quantities are available. The data set is supported by the conventional meteorological data and visual cloud inspection data.

Much of information in meteorological and climatological studies is obtained from measurement data applying statistical methods. The aim of exploratory data analysis (EDA) is to get an insight into the possible processes behind the variations in the collected data. Often the seasonal or monthly data are analyzed for their trends in time. In EDA, mainly the numerical summary measures of collected data sets, characterizing central tendency, spread and symmetry of data samples during their time evolution are used (Wilks, 2006). Quite often the conventional mean is used as a central tendency measure without checking how adequate it is. To get realistic insights into the processes the chosen characteristics must be robust. Robustness means insensitivity to deviations from the assumptions made. Suitability of different central tendency and spread characteristics of the recorded daily sums is compared in the case of skewed probability density distributions and the appropriate characteristics of seasonal and monthly relative solar radiation are found. Major features of variation and trends in the availability of solar radiation in 1955-2010 are studied.

2. Daily relative global irradiance G/G_{clear} and relative direct irradiance I'/I'_{clear}

The conventionally measured daily energy amounts of global G and direct solar irradiance I' falling on a horizontal surface are presented in physical units of MJ/m^2. For the direct irradiance perpendicular to solar rays, irradiance I is measured and its values transformed to the horizontal surface $I,'$ are also made available for each day. Until 1996 the Yanishevski AT-50 actinometers and Savinov-Yanishevski M-115 pyranometers were used but were since replaced by the Eppley Labor. Inc. pyrheliometers and Kipp & Zonen pyranometers. The absolute accuracy of the ventilated Kipp & Zonen pyranometers is about ±2% and that of the pyrheliometers ±1%. In the case of older instruments these uncertainties usually were doubled. In the past intercalibration of sensors was regularly performed in Voeikov Main Geophysical Observatory (St. Petersburg, Russia), whereas now it is done in World Radiation Center (Davos, Switzerland). Between the comparison campaigns, the absolute radiometer PMO-6 No R850405 is used as a secondary standard for regular assurance of the calibration. The previous standard, Ångström pyrheliometer M-59-8 No. J-1981, has been in use during more than 20 years and the scales of old and new standards have been in agreement with the World Radiation Reference within ±0.1% (Russak and Kallis, 2003).

The results of statistical treatment of values measured in physical units are illustrative in the case of interannual variations over longer time intervals. Due to the annual cycle even in clear conditions, the summer daily maxima of global solar irradiance at the study site, in absolute scale, are about 17.5 times higher than these of winter minima. The intraseasonal differences of the relative availability of solar radiation are emphasized if the daily values are presented relative to their climatological clear-sky background as the ratios G/G_{clear} and I'/I'_{clear}. The daily climatological clear-sky values G_{clear} and I'_{clear} are the assumed clear-day values for each calendar day corresponding to the typical conditions of atmospheric characteristics (Eerme et al., 2006; Eerme et al., 2010). Also they could be defined as those for the assumed dry atmosphere (Eerme et al., 2010). The clear-sky climatological daily sums could be calculated, using radiative transfer codes inserting realistic aerosol optical depth (AOD) data as well as realistic vertical profiles of temperature, water vapor and aerosol content. The clear-sky daily sums could be also interpolated from the observed data corresponding to average typical conditions by selecting the cloudless days proceeding from the recorded daily AOD values. We have used the latter version to avoid systematic differences in scales.The precipitable water vapour variations influence the clear-sky values of G_{clear} and I'_{clear} as well but the range of variations of AOD influence is about ten times larger than that of water vapour (Russak et al., 2005).

3. Composing of annual cloudless-sky cycles

The interpolated annual cycles of clear-sky daily G_{clear} and I'_{clear} for the study site (Eerme et al, 2006) are presented in Fig. 1.

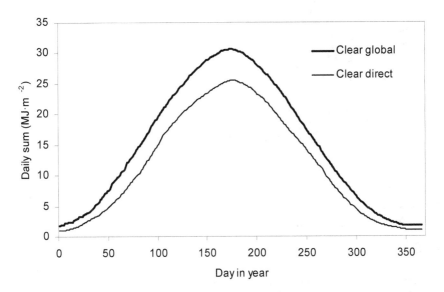

Fig. 1. Graph of the daily global and direct broad-band radiation in MJ·m-2 on cloudless day vs day in the year for Tartu-Tõravere Meteorological Station site

Selection of cloudless days was based on the daily sums of direct irradiance and sunshine duration with inclusion of the hourly values, if necessary, and on the cloud visual inspection data. The used data enabled us to confirm that the solar disk was not obscured by clouds but did not exclude a possibility of appearance non-obscuring clouds during the day. Such small cloud amounts have minor or practically no influence on the recorded daily values of solar radiation. The effect of variation of the distance between the Earth and the Sun is considered in the data. The smoothed annual cycles have been composed, using a moving average of 5 to 10 days with balanced positive and negative deviations of the AOD from its seasonal climatological value. For the period before 2002 the AOD values for broad-band solar radiation prepared by Russak (Russak, 2006; Russak et al., 2005, 2007) have been used. The data for years 1983-1985 and 1992-1995 were excluded for reason of containing significantly higher values from El Chichon and Pinatubo major eruptions than the usual contribution of large values. Major volcanic eruptions can be considered as stochastic natural fluctuations of atmospheric conditions in time scales of the performed study. The variations of global volcanic activity appear at much lower frequencies than the studied variations of available solar radiation.

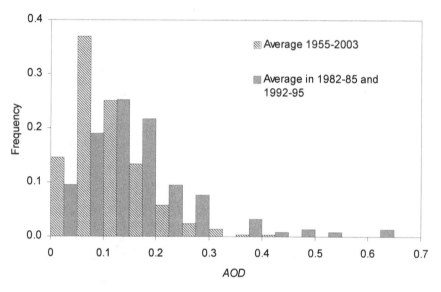

Fig. 2. Probability density distributions of broadband aerosol optical depth in 1955-2003 in normal conditions and in volcanically disturbed atmosphere in 1983-1985 and 1992-1995

Fig 2. illustrates the distribution of broad-band AOD in summer half-year in normal and volcanically disturbed conditions. Since February 2002 the AERONET Cimel-318a sun-photometer (http://aeronet.gsfc.nasa.gov) operates at Tartu-Tõravere Meteorological Station. The cloud corrected AOD data at level 2.0 are used. Reliable relationship was established between the broadband AOD and AERONET AOD at 500 nm (Teral et al., 2004). The summer half-year distribution of AOD at 500 nm is presented in Fig. 3. The major part of AOD data are recorded in April to September. In February and November as well as often in October and March the amount of data is too small for statistical conclusions. In December and January almost no data have been recorded due to very low noon solar elevation. Thus, in November to February the cycles of G_{clear} and I'_{clear} are less reliable than in March to October. It should be mentioned that at the study site about 80 % of global solar radiation and 87 % of direct irradiance are received during the bright period of the year from spring equinox to autumnal equinox. The average contribution, from the period November to February, to the annual amount of gobal irradiance is around 5.9 % and that of direct irradiance around 3.3 %.

The cloudless days exhibiting large deviations of G_{clear} and I'_{clear} from the current normal value were excluded. The seasonal probability density distributions (see Fig. 2 and Fig. 3) of the AOD are skewed with a sharp maximum at small values and long tail of large values (Eerme et al., 2006). In the case of such distribution the conventional mean is not an appropriate measure of central tendency because it is shifted toward larger values than the major body of distribution. Median coincides with the peak of distribution much better and its value is about 20-30 % less than the conventional mean. However, median does not consider the differences in both wings of distribution. In the cases of equal median the distribution of values in wings may be substantially different.

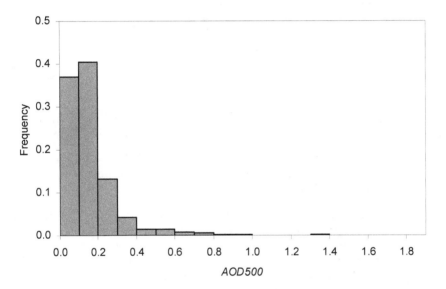

Fig. 3. Probability density distribution of AERONET measured AOD in summer half-year at 500 nm in 2002-2010

Proposed by British statistican Bowley and popularised in the classic book by Tukey (Tukey, 1977), trimean takes into consideration the distribution in wings through inclusion of the 0.25 and 0.75 quartilles. In calculating trimean these values are considered with single and median with double weight. For AOD as well as for later G/G_{clear}, three central tendency measures – conventional mean, median and trimean, have been calculated and compared. We have a reason to consider trimean as the most appropriate measure for central tendency of skewed distributions.

4. Probability density distributions of daily G/G_clear and measures of their central tendency and spread

The daily ratios of G/G_{clear} were studied statistically within four conventional seasons of the year. The winter season extends from Dec 22 to March 20, the spring season from March 21 to June 20, the summer season from June 21 to Sept 22 and the autumnal season from Sept

23 to Dec 21. The summer, half-year which includes both spring and summer seasons was also studied separately. Similar study was performed on monthly level. The maximum, mean and minimum values of calculated central tendency measures for relative global irradiance in all seasons are presented in Table 1. Also the spread characteristics, StDevTri and StDevMed, were calculated relative to the trimean and median like the conventional standard deviation (StDev) is calculated relative to the mean (Wilks, 2006). For almost all seasons and months the StDevTri is the smallest and StDevMed tends to be the largest.

Quantity	Winter	Spring	Summer	Autumn	Summer half-year
G/G_{clear}					
Mean	0.568	0.658	0.652	0.467	0.650
Median	0.530	0.653	0.646	0.412	0.675
Trimean	0.552	0.656	0.648	0.429	0.668
Min	0.406	0.563	0.567	0.371	0.586
Max	0.767	0.771	0.768	0.605	0.748
I'/I'_{clear}					
Mean	0.306	0.425	0.400	0.230	0.412
Min	0.163	0.249	0.283	0.090	0.305
Max	0.521	0.601	0.557	0.378	0.542

Table 1. Seasonal central tendency measures of the relative global irradiance G/G_{clear} and relative direct irradiance I'/I'_{clear} in 1955–2010

The seasonal and monthly probability density distributions of daily G/G_{clear} for every year were studied for their symmetry and deviation from the Gaussian distribution. The seasonal as well as monthly probability density distributions in most cases were flatter than the Gaussian distribution and skewed. Sharper distributions appeared less frequently. Both the negatively and positively skewed ones were found. In dark half-year the monthly and seasonal probability density distributions tend to be positively and in bright half-year negatively skewed. Median and trimean of distributions in separate years tend to be larger than conventional mean in the prevailing sunny conditions. In cloudy conditions, on the contrary, conventional mean tends to be larger than median and trimean. The difference between median and mean is usually larger than that between trimean and mean. Year-to-year variation of the ratios, Mean/Trimean and Mean/Median within seasons of the dark half-year was significantly larger than in the summer half-year.

In cloudy dark half-year, when small values of G/G_{clear} dominate, a few sunny days appearing in a month may cause positive skewness of distribution and enlarge the value of the conventional mean. The biases between mean and trimean, and between mean and median may exceed 20 %, while the majority of G/G_{clear} are very small values. In the months of bright half-year, a few very small values due to heavily cloudy days shift the conventional mean toward lower value. The annual monthly values of mean, median and trimean of G/G_{clear} and their year-to-year variation were studied. The seasonal and half-yearly differences between trimean, median and conventional mean are small because during the long periods different situations are encountered, and the distribution becomes more symmetric than it is for shorter intervals. Using the conventional mean as a reference leads to underestimation of the contrast between the winter and summer months availability of solar irradiance.

The ratio of Mean/Trimean of G/G_{clear} is positively correlated to the skewness of distribution during a whole year. The monthly coefficients of linear correlation vary between 0.55 and 0.88. In September to February the ratio, Mean/Trimean is positively correlated with the kurtosis of distribution, and in March to August the correlation was negative. In March to August the main tendency and the spread measures were negatively correlated. The monthly coefficients varied between –0.20 and –0.65. It means that instead of large values of G/G_{clear} dominating, the contribution of large deviations is small. In October to December positive correlation, with coefficients 0.60 and 0.65 respectively were obtained. It means that the probability of large deviations increases with increasing contribution of relatively small values.

5. Year-to-year variations of half-yearly, seasonal and monthly availability of global irradiance

The annual availability of solar radiation at the latitude of study was determined by the contribution of summer half-year since the contribution of winter half-year was only about 20 %. At the same time the interannual variations tend to be larger in winter months. The variation in every summer half-yearly total, during the 56 years considered remains within ±10 % about the average, except for the two extremely sunny years 1963 and 2002, when the totals exceeded the average by 15 % and reached 75 % of the climatological cloudless-sky value.

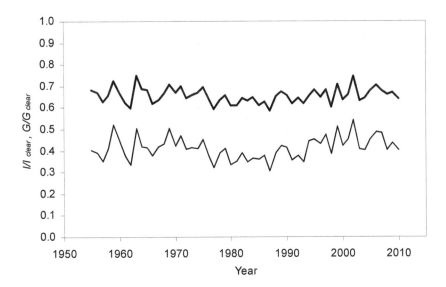

Fig. 4. Time series of summer half-year average G/G_{clear} (above) and I'/I'_{clear} (below) in 1955-2010

In most years, the deviations from the mean are significantly less than 10 %. For the summer half-yearly totals, the difference between conventional mean and trimean is about 0.3 % and that between conventional mean and median about 0.6 %. The conventional mean could be considered here as acceptable as the measure of the general trend. The year-to-year variations are most strongly correlated with cloudiness. The coefficients of linear correlation of the average G/G_{clear} with the average total cloud amount, low cloud amount and the number of overcast days in all seasons have been around −0.80. The linear correlation between the summer half-yearly sums of relative global and relative direct irradiance is equal to 0.90, which is much higher than that for the winter half-year, when it is only 0.60, indicating larger contribution from the direct radiation to the variation of global irradiance at higher solar elevations and snow-free conditions. At the seasonal level the correlation was the highest, 0.96, in summer, and somewhat lower, 0.92, in spring when the ground albedo was not stable.

In the time series of yearly averages of G/G_{clear} and I'/I'_{clear} in the summer half-year (Fig.4), a remarkable feature is observed: an interval of reduced values in the years 1976–1993, when 15 out of 18 values were lower than the 1955–2010 average. The period of low values was more contrasting in terms of direct than global irradiance. Searchers of linear trends could easily find a dimming trend between the 1960s and the mid-1980s and a brightening trend from the middle of the 1980s. Similar behaviour of annual totals from late 1950s and early 1960s has been obtained also in other sites of Northern Europe and European Russia (Wild et al., 2005; Chubarova, 2007). The finding of approximately 35-year periodicity of alternation of cloudy and bright conditions in Western Europe was attributed to sir Francis Bacon who declared it in first decade of 17th century. Later it was forgotten and found again in several cases. The dimming and following brightening presented in Fig. 4 is rather a manifestation of this periodicity. A little more than one full period is presented.

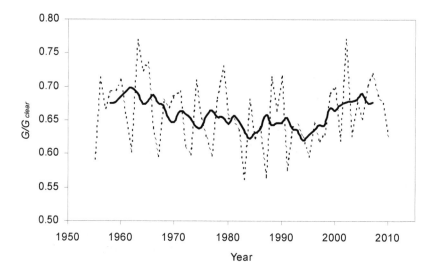

Fig. 5. Long-term variation of G/G_{clear} with year in spring, 1955–2010: Annual values and 7-year moving average

Considering the spring and summer seasons separately, some significant differences in the long-term behaviour of the available solar irradiance become evident. The smooth line representing 7-year moving average of interannual variation of G/G_{clear} in the spring season exhibits maxima around 1965 and 2000 and minima between 1980 and 1985 as well as around 1995 (Fig. 5).

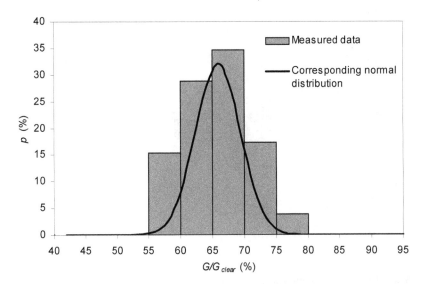

Fig. 6. Probability density histogram of G/G_{clear} in spring season

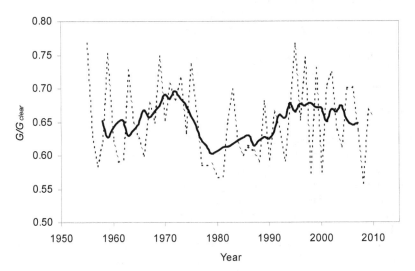

Fig. 7. Long-term variation of G/G_{clear} with year in summer, 1955–2010: Annual values and 7-year moving average

A 7-year moving average turned out to be effective for smoothing the short-term variations and emphasizing the expected trends. This has also been approved in the case of other Estonian meteorological and hydrological data (Järvet and Jaagus, 1996).

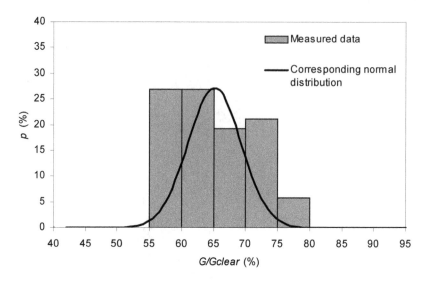

Fig. 8. Probability density histogram of G/G_{clear} in summer season

The probability density distribution of G/G_{clear} for the spring of years 1955–2010 (Fig. 6) is symmetric and the values were close to the average with the highest frequency. The major part , 65 %, of the annual values occurred between 0.60 and 0.70; 20% were above 0.70 and 15% were below 0.60. The observations by several authors of the long-term dimming trend up to the middle of the 1980s and those of the following brightening trend (Che et al., 2005; Wild et al., 2005; Sanchez-Lorenzo et al., 2007) were generally in agreement with the G/G_{clear} observed in the spring.

In the summer, the distribution was less symmetric and the amounts of very large and small values were about 27% for both. In summer periods of large and small values of G/G_{clear} (Fig. 7) were revealed. Large values dominated in 1966–1975 and more so from 1994. No obvious long-term trend of dimming or brightening was found. Up to the late 1960s each fourth summer was sunny. During the mentioned bright periods, approximately each second summer was sunny. The probability density distribution of summer G/G_{clear} shown in Fig. 8 does not fit the normal distribution and exhibits rather bimodal nature. This suggests that the existence of two different regimes associated with mean wet or dry summers (D'Andrea et al., 2006), being partly driven by soil moisture, may be the reason of such distribution observed in Estonia, and presumably in other parts of Northern Europe. Such behaviour is expected to result from multiple equilibria in the continental water balance containing the contributions of large scale weather patterns as well as of the local soil water contents. The spring and summer mean G/G_{clear} are poorly correlated with each other.

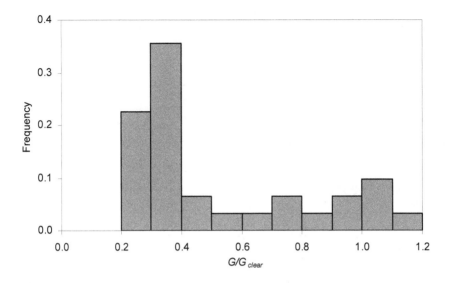

Fig. 9. Typical histogram of monthly distribution of relative daily sum of global irradiance G/G_{clear} in dark half-year

At the site, the daily mean of global irradiance in the summer half-year is approximately 0.65 of its climatological cloudless-sky value. It is almost equal to that of spring and summer. In two extremely fine weather years of 1963 and 2002, its value reached 0.75 of the clear-sky value. In three most cloudy summer half-years the value remained slightly below 0.60 of the climatological clear- sky value.

The monthly trimean values of G/G_{clear} presented in Table 2, varied over wide range. The ratio of maximal to minimal value varied from about 3.4 in December to 1.6 in June. The lowest values in November and December were about 0.20, while the highest was recorded in March and was about to 0.99.

Typical histogram of monthly distribution in dark half-year is presented in Fig 9. Sharp high value was recorded during the prevailing thick cloudiness. In November, about 50 % and in December and January about 25 % of the distributions were sharper than the Gaussian. In bright half-year the distribution was flatter than the Gaussian. In summer months sharp distributions may occur in the extremely fine weather conditions. Typical histogram of monthly distribution in bright half-year is shown in Fig 10.

In Table 3, the monthly average together with maximum and minimum values of the ratio Mean/Trimean are presented as well as the coefficients of linear correlation of this ratio with the kurtosis and skewness of distribution.

Variations in spread characteristics have been largest in StDevMed, but only moderately larger than those in two other spread characteristics. For all the months except November, the StDevTri in most cases happened to be the smallest. The ratio StDev/StDevTri exhibits sharper and more symmetric distributions than the ratio of StDevMed/StDevTri. Its values remained in the range between 0.98 and 1.02 with a very few exceptions lower than 0.98 and only one exception above 1.02 out of 648. The distribution of the ratio StDevMed/StDevTri exhibits only one case of value below 0.98, but the tail often reaches the range between 1.06 and 1.10.

Month	G_{clear} MJ · m^{-2}	G/G_{clear} trimean			I'_{clear} MJ · m^{-2}	I'/I'_{clear}		
		Min	Mean	Max		Min	Mean	Max
Jan	87.0	0.251	0.484	0.726	50.6	0.038	0.184	0.389
Feb	184.9	0.304	0.554	0.773	117.8	0.108	0.262	0.522
March	416.5	0.350	0.640	0.989	291.1	0.095	0.365	0.756
Apr	626.1	0.390	0.634	0.851	491.7	0.117	0.373	0.625
May	830.4	0.465	0.700	0.899	675.0	0.244	0.451	0.765
June	904.9	0.528	0.693	0.852	748.6	0.222	0.435	0.630
July	878.7	0.484	0.687	0.889	732.9	0.199	0.424	0.696
Aug	698.8	0.428	0.662	0.878	566.7	0.153	0.405	0.739
Sept	458.4	0.384	0.593	0.774	366.7	0.114	0.323	0.597
Oct	263.0	0.271	0.447	0.760	188.7	0.070	0.248	0.519
Nov	113.4	0.200	0.345	0.646	73.2	0.020	0.130	0.355
Dec	58.5	0.208	0.442	0.705	36.2	0.028	0.144	0.307

Table 2. The monthly values of assumed normal G_{clear} in physical units and recorded minimal, mean and maximal values of the trimean of monthly relative global G/G_{clear}. Monthly assumed normal I'_{clear} and recorded minimal, mean and maximal values of I'/I'_{clear} in 1955-2010

6. Year-to-year variations of seasonal and monthly direct irradiance

The major contribution to the values of G/G_{clear} came from the variations of the relative direct irradiance I'/I'_{clear}. In all seasons and months, overcast days occurred when no direct irradiance was recorded. In late autumn and early winter majority of days were overcast. In December of one of the years under study, sunshine was available only in one day out of 31. In such situations it is impossible to study the probability density distributions of I'/I'_{clear} like it was done for the global irradiance G/G_{clear}. The monthly and seasonal values of I'/I'_{clear} presented in Table 1 and Table 2 were obtained by dividing the recorded sum of I' by the integrated I'_{clear} for the same period.

The sunniest season during the study was spring, when 0.43 of the direct irradiance relative to the assumed clear sky conditions was available on average. Similarly during autumn, darkest period, less than 0.25 was available. For the whole summer half-year, the I'/I'_{clear} was 0.41 on average. The range of interannual variations has been the largest in autumnal period when the largest value exceeds about four times the smallest one. In the winter period this ratio was close to three, and in spring and summer it was close to two. The monthly I'/I'_{clear} varied from its overall maximum 0.765, encountered in one May to overall minimum 0.02 encountered in one November. The typical monthly values have been between 0.40 and 0.45 in the brightest months, May to August, and between 0.135 and 0.185 in the darkest ones, November to January. The range of interannual variations of monthly I'/I'_{clear} was the smallest in June as it was for G/G_{clear}.

Month	Ratio Mean/Tri			Correl of Mean/Tri		Correl. of Kurt Skew	% of Posit Skew
	Max	Mean	Min	Kurtosis	Skewness		
Jan	1.28	1.085	0.97	0.68	0.81	0.83	90
Feb	1.23	1.045	0.93	0.57	0.88	0.64	69
March	1.25	0.995	0.88	-0.15	0.62	-0.25	52.4
Apr	1.09	0.985	0.90	-0.30	0.82	-0.56	26
May	1.10	0.97	0.90	-0.24	0.75	-0.62	12
June	1.03	0.97	0.92	-0.28	0.75	-0.68	9.5
July	1.03	0.975	0.90	-0.03	0.59	-0.73	4.8
Aug	1.03	0.975	0.93	-0.22	0.64	-0.77	11.9
Sept	1.11	1.005	0.92	0.45	0.84	0.49	47.6
Oct	1.41	1.07	0.91	0.30	0.67	0.72	81
Nov	1.34	1.14	1.00	0.44	0.69	0.91	100
Dec	1.32	1.095	0.97	0.28	0.55	0.91	100

Table 3. Monthly ratios Mean/Tri of G/G_{clear} (first 3 columns) and their linear correlation with kurtosis and skewness of the distribution of daily values G/G_{clear} (columns 4 and 5); linear correlation of kurtosis and skewness (column 6) and percent of positively skewed distributions (column 7)

Sunshine duration was recorded in more sites than direct irradiance, and the records often covered longer time intervals. Due to this advantage, sunshine duration has been used as a proxy for reconstruction of global irradiance (Ångström, 1924; Zekai, 1998; 2001; Cancillo et al., 2005) and also direct irradiance (Power, 2001). Direct irradiance is a more appropriate measure of potential sunshine effects than sunshine duration because sunshine episodes prefer certain solar elevation ranges. In May to September frequent convective clouds reduce available sunshine during noon hours when the solar elevation is high and the monthly relative direct irradiance is usually about 75-80 % of relative sunshine duration at the study site.

7. Year to year variation of seasonal and monthly totals and trends

At the study site the summer half-year of 1955-2010 contributed 80 % of the annual global irradiance, about 87 % of direct irradiance and about 75 % of sunshine duration on average.

In the dark winter period the amplitude of variations was larger before year 1977, including the extreme lowest value recorded in 1961 and the extreme highest one in 1963.

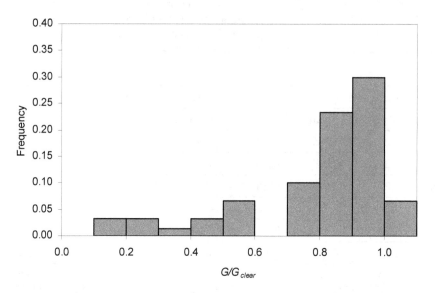

Fig. 10. Typical histogram of monthly distribution of relative daily sum of global irradiance G/G$_{clear}$ in bright half-year

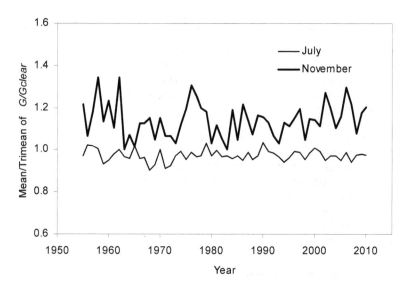

Fig. 11. Time evolution of the ratio Mean/Trimean of G/G$_{clear}$ in July and in November 1955-2010

Often linear trends are used for characterizing the tendencies of variation in time. We have no reason to expect some overall trend of increase or decrease in G/G_{clear} during the whole observed time interval of 1955-2010 in any month. However, there were shorter intervals manifesting quite linear tendency of brightening or dimming. Atmosphere is essentially a non-linear system exhibiting quasiperiodic and nonsymmertic variability in different timescales. Ascending and descending branches of those cycles could be successfully approximated by linear trends.

Month	Mean	Trimean	Median	Time interval
Jan	-0.00416	-0.00454	-0.00418	1963-2008
Feb	0.00299	0.00348	0.00405	1955-1986
March	-0.00599	-0.00691	-0.00755	1955-1995
Apr	-0.00250	-0.00220	-0.00205	1955-1984
	0.00320	0.00454	0.00563	1985-2010
May	0.00090	0.00113	0.00139	1955-2010
June	-0.00109	-0.00117	-0.00123	1955-2010
July	0.00028	0.00138	0.00228	1955-1978
	0.00257	0.00288	0.00329	1979-2010
Aug	-0.01095	-0.01092	-0.01046	1995-2010
Sept	-0.00295	-0.00362	-0.00368	1955-1994
Oct	0.00213	0.00309	0.00353	1955-1989
	-0.00433	-0.00523	-0.00598	1985-2010

Table 4. Slopes of linear trends of G/G_{clear} by the conventional mean, trimean and median during time intervals allowing linear approximation

Usually the slopes of calculated linear trend in G/G_{clear}, both positive and negative, happened to be larger for the trimean and median than for the conventional mean. A sample of estimated linear trends in all the three general trends measures for time intervals, manifesting the trends most evidently is presented in Table 4. The time evolutions of the ratio of Mean/Trimean in summer and winter months are different. In Fig. 11 the time evolutions of the ratio for relatively sunny month, July and cloudy month, November are presented. In November to March the dimming trends prevail at the site in all months and negative slopes for trimean and median are larger than those for the conventional mean. In January there appeared an overall positive tendency in the ratio Mean/Trimean in 1967-2008. Relatively large positive peaks of the ratio have been encountered in separate years with an increasing frequency after 1988. As was noticed above, those large values of ratio appear when G/G_{clear} values were small, presenting sharp and strong positive skewed distribution. In February they were about two times smaller than those in January, manifesting also overall negative trend. During the last two decades, on the average, the ratios Mean/Trimean were larger, and the values of all three central trend measures of G/G_{clear} were smaller than before.

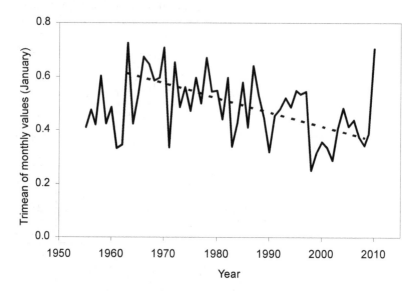

Fig. 12. Trend of dimming for January in 1963-2008

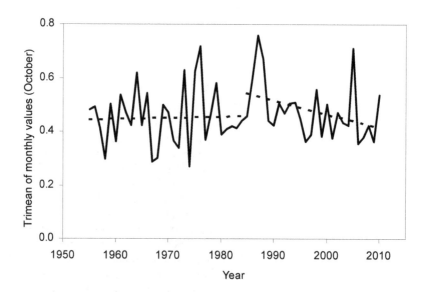

Fig. 13. Switching from brightening to dimming trend in October

In March, a dimming trend was obvious in 1955-1995 and in April of 1955 – 1977 the trend also appeared, but then has changed to the brightening in 1977-1995. In May to August small positive trends in G/G_{clear} were observed. In June the positive trend in 1955-1973 has changed to negative in 1973-2003. July and August exhibit small overall positive trends due to the more frequent fine weather conditions since 1994. In September the trend of dimming in 1955-1992, has changed to the brightening later. In October quite significant tendency of brightening has changed to more strong dimming during the last two decades (see Fig. 13). In November, the dimming tendency was observed during the whole observation period. The negative slope of the trend has increased significantly after 1983. In December, similar increase of the linear dimming trend happened around 1977.

8. Conclusion

The relative availability of solar radiation at moderate and subpolar latitudes manifests no significant correlation between seasons. The weather regimes often change significantly during each of the conventional four seasons. The annual available amount of solar energy at these latitudes depends on the contributions of spring and summer seasons. For solar radiation applications study of the interannual variations and trends on the seasonal and even on the monthly level is necessary. Detailed data on interannual variation of the available solar energy within seasonal and often even monthly limits are useful for estimation of the biospheric responses, local potential for recreation services as well as for food and clean energy production.

The results obtained at one Estonian site could be considered as an example manifesting that the variations and trends of the relative availability of solar radiation may be substantially different in different seasons and months. Statistical analysis of trends and interannual variations needs at first finding the most appropriate general trend measures based on the probability density distributions of daily amounts of relative global irradiance within selected time intervals. In the case of symmetric distributions the conventional mean is a good measure but in the case of skewed distributions using of it leads to distorted interseasonal proportions and trends.

Using conventional mean instead of more robust trimean, leads to overestimation of the monthly general trend of G/G_{clear} in October to February by 4.5 % to 14% on average, depending on the month. In cases of thick clouds domination in separate years, the values of monthly mean may be by 30-40 % larger than trimean and median. In some cases the values of conventional mean were found to be up to 10 % lower than the trimean and median. In bright half-year, April to August, the conventional mean was consistently about 2.5 % lower than the trimean, and using it leads to some underestimation of central tendency in G/G_{clear} . Using conventional mean as a measure of monthly central tendencies is related to apparent reduction of the contrast between relative solar energy supply in summer and winter.

Monthly relative availabilities of broadband solar irradiance at the study site and presumably also in its wider neighbourhood varied in wide ranges. The ranges of variations were smaller in months of summer half-year and larger in autumnal and winter seasons. The average availability of global irradiance in summer half-year, and separately in spring and summer seasons, was 0.65 and that of direct irradiance 0.41 of the assumed normal cloudless weather amount. In spring the relative availability of direct irradiance has been to some extent larger than in summer and its interannual variation smaller.

Slopes of the linear trends were found to be smallest in the case of using conventional mean as a central tendency measure. It leads to underestimation of the dimming and brightening trends in the cases when linear approximation is appropriate for their description. In summer months, in most cases, small brightening trends were found. In June it changed to rather a dimming trend around 1973. In winter months there appears an overall dimming tendency of G/G_{clear} in recent two decades. Using the conventional mean leads to underestimation of that trend by up to 20-30 % as compared to the version calculated on the basis of trimean.

9. References

Ångström A. 1924. Solar and terrestrial radiation. *Q. J. Roy. Meteor. Soc.* 50, 121–126

Cancillo M. L., Serrano A., Ruiz A., Garcia J. A., Anton M., and Vaquero J. M. 2005. Solar global radiation and sunshine duration in Extremadura (Spain). *Physica Scripta*, T118, 24-28

Che H. Z., Shi G.Y., Zhang X.Y., Zhao J.Q., Li Y. 2007. Analysis of sky conditions using 40 year records of solar radiation data in China. *Theor Appl Climatol 89*, 83–94

Chubarova N. 2007. UV variability in Moscow according to long-term UV measurements and reconstruction model. In: *Proceedings of the conference "One cetury of UV radiation research"*,18-20 September, Davos, Switzerland, 17-18

D'Andrea F., Provenzale A., Vautard R., De Noblet-Decoudre N. 2006. Hot and cool summers: Multiple equilibria of the continental water cycle. *Geophys Res Lett 33*, L24807

Eerme K., Veismann U., Lätt S. 2006. Proxy-based reconstruction of erythemal UV doses over Estonia for 1955-2004. *Ann. Geophys.*, 24,1767-1782

Eerme K., Kallis A., Veismann U., Ansko I. 2010. Interannual variations of available solar radiation on seasonal level in 1955-2006 at Tartu Tõravere Meteorological station. *Theor. Appl. Climatol.*, 101, 371-379

Järvet A., Jaagus J. 1996. The impact of climate change on hydrological regime and water resources in Estonia. In: *Punning J-M (ed) Estonia in the system of global climate change. Publ Inst Ecol*, Tallinn, 4, 84–103

Kallis A., Russak V., Ohvril H. 2005. 100 Years of Solar Radiation Measurements in Estonia. In:World Climate Research Programme. Report of the Eighth Session of the Baseline Surface Radiation Network (BSRN), Workshop and Scientific Review (Exeter, UK, 26.30 July 2004), *WCRP Informal Report No. 4/2005*, C1.C4

Lohmann S., Schillings C., Mayer B., Meyer R. 2006. Long-term variability of solar direct and global radiation derived from ISCCP data and comparison with reanalysis data. *Solar Energy 80(11)*, 1390-1411

Ohvril H., Teral H., Neiman L., Kannel M., Uustare M., Tee M., Russak V., Okulov O., Jõeveer A., Kallis A., Ohvril H., Terez E. I., Terez G. A., Guschin G. K., Abakumova G. M., Gorbarenko E. V., Tsvetkov A. V., Laulainen N. 2009. Global Dimming and

Brightening Versus Atmospheric Column Transparency, Europe, 1906-2007. *J. Geophys. Res.* *114*, D00D12, Doi:10.1029/2008JD010644

Power H. C. 2001. Estimating clear-sky beam irradiation from sunshine duration. *Solar Energy*, *71*, 4, 217-224

Russak V. 1990. Trends in solar radiation, cloudiness and atmospheric transparency during recent decades in Estonia. *Tellus*, *42B*, 206-210

Russak V., Kallis A. (compilers), Tooming H. (ed). 2003. *Handbook of Estonian solar radiation climate*. EMHI, Tallinn, 384 pp (in Estonian)

Russak V., Ohvril H., Teral H., Jõeveer A., Kallis A., Okulov O. 2005. Multi-annual changes in spectral aerosol optical thickness in Estonia. In: *Abstracts of the European Aerosol Conference 2005* (28 August-2 September 2005, Ghent, Belgium), pp 399

Russak V. 2006. Changes in solar radiation in Estonia during the last half-century. In: *CD-ROM:EMS Annual Meeting. Sixth European Conference on Applied Climatology (ECAC2006)*. *Abstracts*, Vol. 3, ISSN 1812-7053

Russak V., Kallis A., Jõeveer A., Ohvril H., Teral H. 2007. Changes in spectral aerosol optical thickness in Estonia (1951–2004). *Proc. Estonian Acad. Sci. Biol. Ecol.*, *56*, 1, 69–76

Russak V. 2009. Changes in Solar Radiation and Their Influence on Temperature Trend in Estonia (1955-2007). *J. Geophys. Res.* *114*, 1-6, D00D01, Doi: 10.1029/2008JD010613

Sanchez-Lorenzo A., Brunetti M., Calbó J., Martin-Vide J. 2007. Recent spatial and temporal variability and trends of sunshine duration over the Iberian Peninsula from a homogenized data set. *J Geophys Res 112*, D20115

Šuri M., Huld T..A., Duniop E..D., Ossenbrink H..A. 2007. Potential of solar electricity generation in the European Union member states and candidate countries. *Solar Energy 81(10):* 1295–1305

Teral H., Ohvril H., Okulov O., Russak V., Reinart A., Laulainen N.. 2004. Spectral aerosol optical thickness from solar broadband direct irradiance - summer 2002, Tõravere, Estonia. In: Abstracts of the European Aerosol Conference 2004 (6-10 September 2004, Budapest, Hungary), *J. Aerosol Science*, pp 547-548

Tiwari A. and Sodha M. S. 2006. Performance evaluation of solar PV7T systems: An experimental validation. *Solar Energy 80 (7)*, 751-759

Tooming H. 2002. Dependence of global radiation on cloudiness and surface albedo in Tartu, Estonia, *Theor. Appl. Climatol.*, *72*, 3-4, 165-172

Tukey J. W. 1977. Exploratory Data Analysis, Reading, Massachusetts, Addison Wesley, 688 pp

Wild M., Gilgen H., Roesch A., Ohmura A., Long C. N., Dutton E. G., Forgan B., Kallis A., Russak V., Tsvetkov A. 2005. From Dimming to Brightening: Decadal Changes in Solar Radiation at Earth's Surface. *Science 308*, 847-850

Wilks, D. S., 2006. Statistical methods in the atmospheric sciences. Second edition.Elsevier Inc. 627 pp

Zekai en. 1998. Fuzzy algorithm for estimation of solar irradiation from sunshine duration. *Solar Energy 63*, 1, 39-49

Zekai en. 2001. Angström equation parameter estimation by unrestricted method. *Solar Energy 71,* 2, 95-107

6

Variation Characteristics Analysis of Ultraviolet Radiation Measured from 2005 to 2010 in Beijing China

Hu Bo

State Key Laboratory of Atmospheric Boundary Layer Physics and Atmospheric Chemistry (LAPC), Institute of Atmospheric Physics, Chinese Academy of Sciences, Beijing, China

1. Introduction

Global ultraviolet radiation (UV) is a small fraction of the total extraterrestrial solar radiation outside the atmosphere. The amount of UV radiation reaching the Earth's surface comprises only a small fraction of global radiation, about 6-7% of global radiation is in the UV-A (320-400 nm) range and less than 1% is in the UV-B (280-320 nm) range. However, UV radiation plays an important role in many biological and photochemical reactions. UV of certain doses could lead to a variety of adverse health and environmental effects (National Radiological Protection Board, 2002; United Nations Environment Programme (UNEP), 1998; Slaper and Koskela, 1997; Outer, 2005). UV radiation also has impacts on the photodegradation of plastics, colorants, paints, and artificial and natural fibers as well as the formation and decomposition of photosensitive urban and industrial contaminants (Cañada et al., 2000). Understanding the amount of UV received by human, plant and animal organisms on the earth's surface is important to a wide range of field such as cancer research, forestry, tropospheric chemistry, agriculture and oceanography (Grants and Heisler, 1997; McKenzie et al., 1991). An understanding of changes in long-term, ground-level UV radiation is required to support assessments of UV radiation-induced health and environmental risks (Slaper and Koskela, 1997). UV radiation-measuring networks are extremely scarce, particularly in the arid and semi-arid, where the effects of UV radiation may be of great importance due to many hours of sunshine throughout the year. Despite its anthropogenic importance and impacts, concern about the amount of UV radiation reaching the Earth's surface has only recently developed, primarily as a result of the thinning ozone layer linked to the depletion of stratospheric ozone in the 1980s (Su et al., 2005).

Numerous factors can influence UV radiation incidence, including cloud characteristics, solar zenith angles, total ozone, aerosol pollution, and surface albedo. Altitude has an important effect on UV radiation (Piazena, 1996; Seckmeyer et al., 1997; McKenzie et al., 2001).Clouds are known to affect the attenuation of UV radiation differently, based on their location, percentage cover, optical thickness, water content, and droplet size distribution. In order to obtain more UV data for other study, lots of studies have focused on quantitative variations in UV radiation and the ratio of UV to global solar radiation (R_s). Additional studies have addressed long-term trends in the variations of UV through reconstructions of past UV radiation based on ground-based and satellite data (Kaurola et al., 2000; Fioletov et al., 2001; Lindfors et al., 2007; Feister et al., 2008;Hu et al., 2010a). In the last few decades, there has been a progressive increase and great concern in the amount of UV reaching the Earth's surface as a consequence of the thinning of the stratospheric ozone. Despite its anthropogenic importance and impacts, concern about the amount of UV radiation reaching the Earth's surface has only recently been developed, primarily as a result of the thinning of ozone layer linked to the depletion of stratospheric ozone in the 1980s (Su et al., 2005). UV radiation-measuring networks are extremely scarce, particularly in China.

The objective of this chapter, apart from showing seasonal variations of UV and UV/R_s values in Beijing, based on the reconstruction method, is to develop a long-term data set of UV radiation, and also study variation characteristics of UV in Beijing.

2. Methods

2.1 Site

Beijing, the capital of the People's Republic of China, is located at 39°56' N latitude and 116°20' E longitude. East, North and West of Beijing are surrounded by mountains. The climate of Beijing is an East Asia monsoon type, with cold and dry winters, and hot and humid summers. During winter, the Siberian air masses that move southward across the Mongolian Plateau are accompanied by cold and dry air. In summer, the air mass is hot owing to warm and humid monsoon winds from the southeast, bringing most of the annual precipitation in Beijing area.

The East Asian Monsoon season is the dominating climate of Beijing. In spring (March, April, May), the content of the water vapor in the atmosphere is low and there is little rain in this region. The rainy season begins in June, ends in August and then comes in the dry season. The main rainfall is in July. In autumn (September, October, November), the sky condition is always clear; for Beijing, it is often controlled by the anticyclone. In winter (December, January, February), Siberian anticyclones frequently take place in the Beijing area. Rainless and cloudless conditions are the dominant sky conditions in the region. In this paper, spring and winter are called the dry season because there are few rainfall events in these months. . Summer and autumn are called the humid season, for most of the rainfall occur in these two seasons.

The monsoon starts in July, and ends in October when the dry season begins. There are many active synoptic systems (depressions) in the humid period,. while stabilization system control prevails in the dry season and most days are clear in this period.

In the humid season, the southern wind from the ocean prevails bringing abundant vapor in to the atmosphere, thus the vapor content of the atmosphere is high in this season. Water vapor markedly affects the long wavelength radiation by absorbing them, leaving the UV spectral portion and the short wavelength spectral radiation for possible scattering and reflection. . Consequently, the general decrease in the global radiation could cause the ratio of UV to global radiation to increase as the water vapor increases.

In the dry season, the northern wind from the continent is the prevailing wind bringing dry and cold air. So the vapor content is low in the dry season and the ratio of UV to global radiation decreases as the water vapor decreases.

The seasonal variation pattern of the ratio is that the highest and lowest values appear in summer and winter respectively, and the ratio in autumn is smaller than that in spring. This variation trend is controlled by the water vapor content.

The experimental site is located in down town Beijing area, between the Fourth and the Third Ring Roads. This area is part of downtown Beijing. There are domestic dwellings to the north and south of this site ، and a freeway near to it's east. The measurements in this area can represent the average conditions for Beijing. The sampling instruments are installed on a flat platform at the top of the chemical laboratories' building, of the Institute of Atmospheric Physics, Chinese Academy of Sciences.

2.2 Instruments and measurments

A solar radiation observation system was set up on the top of the two-floor building (10 m). R_s, PAR, direct radiation, diffuse radiation, concentration and meteorology parameters(temperature, relative humidity, air pressure) measurements are being carried out in this observatory.. R_s is measured by using a Kipp&Zonen radiometer CM-21 (Delft, The Netherlands). R_s measurements have an estimated experimental error of 2-3%. UV radiation (290–400 nm) is measured using CUV3 radiometers (Kipp & Zonen, Delft, Netherlands) with an accuracy of 5%. All radiation values were recorded at 1-min intervals, and an hourly average value was obtained by integrating the 1-min values. Temperature and relative humidity were measured with HMP45D(Vaisala, Finland),the accuracy of temperature and relative humidity are 0.1 and 3% respectively. The solar radiation parameters observation system is completed with a data acquisition system (Vaisala M520, Finland).

All pyranometers were calibrated by using the 'alternate method' (*Bruce*, 1996). During the process of calibration, we were required to take on-site measurements of global, diffuse, and direct (pyrheliometer) sensor voltages in clear and sunny conditions. The pyrheliometer was calibrated against a reference pyranometer, which had been calibrated against a standard pyrheliometer (PMO6), i.e. absolute irradiance radiometer (Switzerland). This absolute irradiance radiometer is periodically calibrated every five years at the World Radiation Center in Davos, Switzerland.

Manufacturers usually calibrate UV sensors using standard lamps with a known spectral irradiance. The calibration of the UV3 sensor was conducted with a standard light source in standard spectral irradiance that can be traceable to the National Bureau of Standards lamp. A spectroradiometer measures a standard lamp spectral irradiance, and then retrieves the spectral sensitivity under standard lamp conditions. By using the same method, we could deduce the spectral sensitivity under sunshine conditions (equation 2).

$$K_f^D = \left. V_{D,\Delta\lambda} \middle/ E_{D,\Delta\lambda} \right. = \tau_{\Delta\lambda} \cdot S_{\Delta\lambda} \tag{1}$$

where K_f^p is the spectral sensitivity of the spectroradiometer under standard lamp conditions, $V_{D,\Delta\lambda}$ is the respond voltage in response to the standard lamp in $\Delta\lambda$, and $E_{D,\Delta\lambda}$ is the standard irradiance of the standard lamp.

$$K_f^S = {V_{S,\Delta\lambda}}\Big/{E_{S,\Delta\lambda}} = \tau_{\Delta\lambda} \cdot S_{\Delta\lambda} \qquad (2)$$

where K_f^S is the spectral sensitivity of the spectroradiometer under sunshine conditions, $V_{S,\Delta\lambda}$ is the respond voltage in response to the standard lamp in $\Delta\lambda$, and $E_{S,\Delta\lambda}$ is the standard irradiance of solar radiation. $\tau_{S,\Delta\lambda}$ is the transmittance of the fiber optic extension cord.

In narrow wavebands, the spectral sensitivity K_f^p is equal to K_f^S, and thus the spectroradiometer and standard lamp can be used to calibrate the quantum sensor. The spectral sensitivity K_f^p for each narrow waveband can be derived from the lamp spectral irradiance. Then, this spectroradiometer can be used to measure the sun irradiance and the integral sun spectral irradiance between 290-400nm to calculate UV. At the same time, the UV sensor to be calibrated measures the respond voltage. The difference between measured results of the quantum sensor calibrated by the spectroradiometer and the new quantum sensor is not more than 1%.

Quality control of the UV radiation measurements was based on two main principles: that observed UV radiation should be less than the extraterrestrial UV radiation at the same geographical location, and that the range of the ratio of UV radiation to R_s should be limited to values between 0.02 and 0.08. Elhadidy et al. (1990) noted that the UV radiation/R_s ratio can be as low as 0.02 on days experiencing high dust levels, while the UV radiation/R_s ratio at the top of atmosphere is 0.08 as derived from integral sun spectral irradiance (Geuymard, 2004). Therefore, the UV radiation/R_s ratio should be between 0.02 and 0.08. Values outside this range were flagged as questionable and excluded from the data set. About 1% of the measurements were eliminated based on these quality control processes. Quality control for R_s was similar to that for UV radiation, and the smallest acceptable value of the ratio of R_s to extraterrestrial global solar radiation was 0.03 (Geiger, et al., 2002).

The extraterrestrial UV_0 can be derived from equation 1 expressed as follow:

$$UV_0 = I_{scuv}(\frac{12}{\pi\rho^2})\int_{\omega_1}^{\omega_2} \sin\alpha \quad d\omega, \qquad (3)$$

where α is the solar declination, ω is the hour angle and ρ is correction factor of the Earth's orbit.

$$I_{scuv} = 78 \text{ W m}^{-2}, \qquad (4)$$

I_{scuv} has been obtained from the spectral values given by Frohilich and Wherli (Lenobe,1993).

K_s is the ratio of R_s to extraterrestrial broadband solar radiation (H_0) given as follow:

$$K_s = {R_s}\Big/{H_0}, \qquad (5)$$

All of the calibration works were done at the beginning and the ending of the data collection. At the start and end of data collection, all measurement sensors were cross - evaluated in Beijing. The maximum deviation of CM-11 sensors averaged 1.5% (the average deviation was 1.2%), and the maximum deviation of CUV3 sensors averaged 2.42% (the average deviation was 1.8%).

3. Six-year variation characteristics of UV radiation in Beijing

3.1 Time series of UV radiation and UV/R_s derived from measurement data

The measurement site is located between the northern Third and Fourth Ring Roads in the mega city of Beijing. This site belongs to the Institute of Atmospheric Physics (IAP), Chinese Academy of Sciences. There are domestic dwellings to the north and south of the site and a freeway close to its east side. Measurement instruments were installed on a flat platform on top of the roof (about 10 m) of a building. The Measurement's representation of this site was validated by comparing it to observations results from the China Meteorological Administration (CMA) site (54511) in southern region of Beijing. Figure 1 show the variation characteristics of measurement R_s in IAP and CMA station from January 2005 to December 2008 used for the comparison. From this figure, we find that R_s values at the IAP site were almost same as those at the CMA (54511) station; the largest deviation in average R_s between the two sites was 4.3% and the average deviation was 1.5% (Fig. 1). These results indicated that the measurements at the IAP station are a good representation of the Beijing urban area.

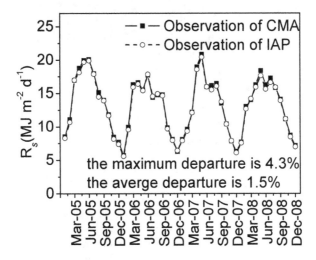

Fig. 1. Comparison of R_s measurements at CMA with that IAP from January 2005 to December 2008 in Beijing.(Hu et al., 2010 a)

The seasonal variation characteristics of UV radiation averaged over 4 years are presented in Fig. 2. UV shows the same seasonal features as that of R_s. The high values of UV and R_s both appear in summer and the low values appear in winter; values in spring and autumn are intermediate. For the diurnal variations, the maximums of Q_p and R_s both appeared around noon each day. There was a good correlation between these two solar-radiation components, by which one of them can be estimated from the other. For the annual variations, one year's variation mode is same as another year's. The minimum monthly average of daily values of UV radiation indicates, in the cool-dry season (i.e., winter period) an increasing trend in the spring with peaks in May. UV radiation levels decrease gradually after the May peak, falling to minimum level in the winter. The annual mean daily values of UV radiation for 6-year period were 0.39 ± 0.16 MJ m^{-2} d^{-1} with the lowest and highest UV radiation values being 1.06 MJ m^{-2} d^{-1} and 0.01 MJ m^{-2} d^{-1}, respectively.

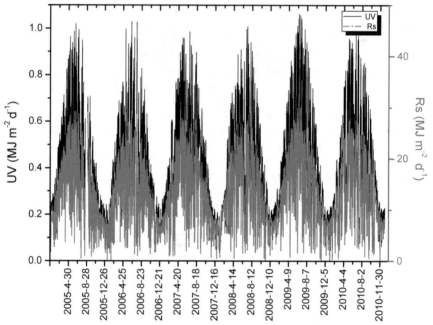

Fig. 2. Variation of UV in Beijing from 2005 to 2010

Figure 3 displays box plots of the monthly statistics of the UV/R_s level in Beijing. The different symbols represent the monthly mean, maximum, and minimum values. There is significant monthly variation in UV/R_s levels. The monthly mean UV/R_s level gradually increases from about 2.7% in November to about 3.7% in August after which it gradually decreases. This seasonal variation of UV/R_s is influenced by seasonal variations in the water vapor content and cloud amount. From spring, the content of column-integrated water vapor generally reaches its maximum in summer, with abundant rainfall, and decreases from August. This variation characteristic is similar to that of UV/R_s. The high values of K_t (the ratio of R_s to extraterrestrial irradiance) occurred in autumn and the low values occurred in summer; the values in spring and winter are intermediate. This seasonal variation pattern of K_t is similar to of column-integrated water vapor content. Under humid and cloudy conditions, the absorption of solar radiation in the infrared region is enhanced because of the increased water vapour content, whereas absorption in the UV region does not vary significantly.

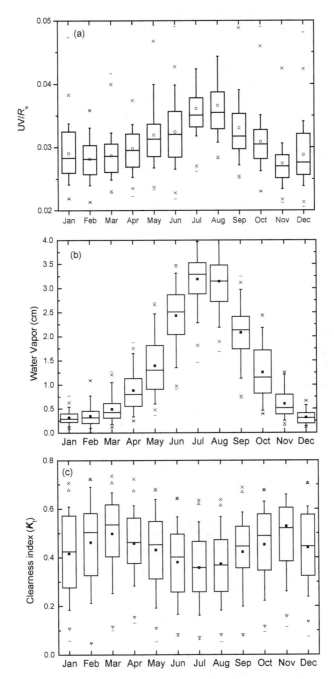

Fig. 3. The graphs of of (a) the monthly ratio of Ultraviolet radiation to solar radiation (b) water vapor content and (c) clearness index. (Hu et al., 2010 b)

In Figure 3, the central bar is the median and the lower and upper limits are the maximum and minimum, respectively.

Most studies use measured values to determine the relationship between UV and R_s and based on this, UV-estimating models are developed. The range of UV/R_s levels must be recalibrated to account for local climatic and geographical differences as well as atmospheric conditions (Udo, 2000). Our study uses direct-measurement UV data to investigate UV radiation properties under various atmospheric conditions in Beijing.

The clearness index, K_t, is defined as the ratio of the total irradiance to extraterrestrial solar irradiance, both defined on a horizontal surface. The K_t ratio is a general indicator of scattering and absorption processes due to aerosols, gases, and clouds that may interrupt the transmission of irradiance through the atmosphere (Liu and Jordan, 1960; Elhadidy et al., 1990).

Fig. 4 depicts hourly UV radiation as a function of the cosine of the solar zenith angle (μ).

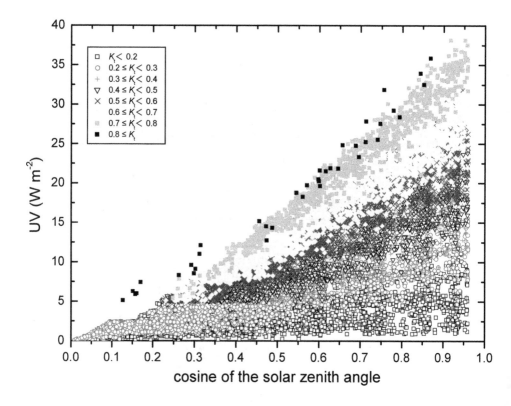

Fig. 4. UV as a function of the cosine of the solar zenith angleat different K_t. Different clearness index (Kt) values are represented by different colours. The values of UV radiation within a narrow range of clearness index values increase almost exponentially with the cosine of the solar zenith angle.

Different colour represents data with different K_t values. The results show that UV radiation increases almost exponentially with μ for a specified K_t. Long and Ackerman (2000) used a power law equation to describe the dependence of R_s on clear sky conditions. For a narrow range of K_t values, the relationship between UV radiation and μ can be very well described using the following power law equation:

$$UV = UV_{0m} \times \mu^b \tag{6}$$

where UV_{0m} is the maximum value of UV radiation per unit of μ.

Base on this, hourly UV radiation was estimated using the following equation. The method of establishing an empirical model for UV computation is explained in detail in Hu et al (2010 b).

$$UV = (0.95 + 74K_t - 71K_t^2 + 55\ K_t^3) \times \mu^{1.031}. \tag{7}$$

The daily values of R_s at the Beijing site were measured by the National Meteorological Center, China Meteorological Administration (CMA). The hourly values of R_s, however, were not possible to obtain, therefore, in order to obtain long-period trends in the variation of UV radiation in Beijing, equation 7 was modified as in equation 8 in order to use the daily values of R_s. However, the daily values of R_s could be used to compute daily values of UV using equation 4:

$$UV_{daily} = (8.4 + 3206.8\overline{K_t} - 2210.7\ \overline{K_t}^2 + 2074.8\ \overline{K_t}^3) \times (\overline{\mu})^{1.031} \times t_d, \tag{8}$$

where UV_{daily} is the daily amount of UV radiation, $\overline{K_t}$ is the ratio of daily R_s to daily extraterrestrial solar irradiance, $\overline{\mu}$ is the average of the cosine of the solar zenith angle from sunrise to sunset, and t_d is the length of daytime in hour.

3.2 Long-periodtrends of ultraviolet radiation in Beijing

The all-weather empirical model was established and validated with measured data at Beijing. Long-term UV radiation data were computed from Beijing observation station, which is located in the southwest Fourth Ring Roads of Beijing. A data set of R_s values collected from 1958–2005 by the National Meteorological Center (CMA) were used in this study.

The Time series of the annual and seasonal averages of UV radiation for spring (represented by April), summer (represented by July), autumn (represented by Oct.), winter (represented by Jan.) are computed for Beijing and presented in Figure 3. Annual mean UV radiation levels decreased from the early 1960s to late 1990s, but began to increase by the late 1990s. The annual mean daily value of UV radiation from 1958-2005 was 0.46 MJ m-2 at Beijing station. By the latter half of the 20th century, decreases in UV radiation (0.018 MJ m-2 d-1 per decade) have been observed. We observed the decrease trends in the long-term annual mean UV radiation as observed from its value of -0.026 MJ m-2 d-1 per year during the period 1958-1997 and its value of -0.0024 mJ m-2 d-1 per year during the period 1958-2005.

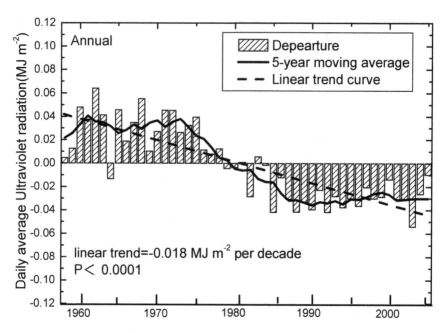

Fig. 5. Long-term variation characteristics of annual average of Ultraviolet radiation for 1958–2005. (Hu et al., 2010 b)

4. Conclusions

The temporal variations in UV radiation using UV radiation and R_s data collected in Beijing over the period January 2005 to December 2010 were studied. The UV radiation levels increased gradually from spring reaching a peak in the summer, and gradually decreased to its lowest levels in winter. Its annual mean daily value was 0.39 ± 0.16 MJ m⁻² d⁻¹, and its lowest and highest daily values were 1.06 and 0.01 MJ m⁻² d⁻¹, respectively. Its highest daily value occurred in May and the lowest value occurred in December. The monthly mean daily value of the ratio, UV/R_s, gradually increased from 2.7% in November to 3.7% in August, after which it gradually decreased. The annual mean daily value of UV/R_s was 3.1%.

The annual mean daily value of UV radiation from 1958-2005 was 0.46 MJ m⁻² d⁻¹. Over the latter half of the 20th century, there have been significant decreases in UV radiation (0.018 MJ m⁻² d⁻¹ per decade).

5. Summary to the chapter

Measurements of total Ultraviolet radiation (UV), broadband global solar radiation (R_S), reflective radiation, net radiation, and Photosynthetically active radiation from 2005 to 2010 in Beijing were used to determine temporal variation characteristics of UV in Beijing. The UV radiation levels increased gradually from spring reaching a peak in the summer, and then gradually declined to its lowest levels in winter. The annual mean daily value of UV radiation was 0.39 ± 0.16 MJ m⁻² d⁻¹, and the lowest and highest daily average UV radiation levels were 1.06 and 0.01 MJ m^{-2} d^{-1}, respectively. The highest daily values of UV radiation occurred in May while the lowest values occurred in December. The monthly mean daily UV/R$_s$ value gradually increased from 2.7% in November to 3.7% in August, after which it gradually decreased. The annual mean daily UV/R$_s$ value was 3.1%.

A simple, efficient, and empirically derived, all-weather, model is proposed to estimate UV from R$_S$. The annual mean daily value of UV radiation from 1958-2005 was 0.46 MJ m^{-2} d^{-1}. Over the latter half of the 20th century, there have been significant decreases in UV radiation (0.018 MJ m^{-2} d^{-1} per decade).

6. References

Caňada,J., Pedrós,G., López,A., Boscá, J.V., 2000, Influences of the clearness index for the whole spectrum and of the relative optical air mass on UV solar irradiance for two locations in the Mediterranean area, Valencia and Cordoba, J. Geophys. Res., 105(D4),4659-4766.

Elhadidy, M. A., Abdel-Nabi, D. Y. , and Kruss, P. D. , 1990, Ultraviolet solar radiation at Dhahran, Saudi Arabia, Sol. Energy, 44, 315-319.

Feister, U., J. Junk, and M. Woldt, 2008: Long-term solar UV radiation reconstructed by Artificial Neural Networks (ANN), Atmos. Chem. Phys. Discuss., 8, 453-488.

Fioletov, V. E., L. McArthur, J. B. Kerr, and D. I. Wardle, 2001: Longterm variations of UV-B irradiance over Canada estimated from Brewer observations and derived from ozone and pyranometer measurements, J. Geophys. Res., 106, 23009-23028.

Geiger, M., L. Diabaté,L. Ménard, L. Wald, 2002: A web service for controlling the quality of measurements of global radiation, Solar energy, 73(6) 475-480.

Grants, R.H., Heisler, G.M., 1997. Obscured overcast sky radiance distributions for ultraviolet and photosynthetically active radiation. J. Appl. Meteorol. ,36, 1336-1345.

Gueymard, C. A., 2000: The sun's total and spectral irradiance for solar energy applications and solar radiation models, Solar Energy, 76, 423-453.

Hu, B., Wang Y., Liu G., 2010 a. Long-term Trends in Photosynthetically Active Radiation in Beijing, Advances in Atmospheric Sciences, 2010,27(6),1380-1388

Hu, B., Wang Y., Liu G., 2010 b. Variation characteristics of ultraviolet radiation derived from measurement and reconstruction in Beijing, China, Tellus, 62B, 100-108

Kaurola, J., P. Taalas, T. Koskela, J. Borkowski, and W. Josefsson,2000: Long-term variations of UV-B doses at three stations in northern Europe, J. Geophys. Res., 105, 20813-20820.

Lindfors, A., J. Kaurola, A. Arola, T. Koskela, K. Lakkala, W. Josefsson, J. A. Olseth, and B. Johnsen,2007: A method for reconstruction of past UV radiation based on radiative transfer modeling: Applied to four stations in northern Europe, J. Geophys. Res., 112, D23201, doi:10.1029/2007JD008454.

McKenzie, R. L., P. V. Johnston, D. Smale, B. A. Barry, and S. Madronich, 2001: Altitude effects on UV spectral irradiance deduced from measurements at Lauder, New Zealand, and at Mauna Loa Observatory, Hawaii. J. Geophys. Res., 106, 22845-22860.

McKenzie, R.L., Matthews, W.A., Johnston, P.V., 1991. The relationship between erythemal UV and ozone, derived from spectral irradiance measurements. Geophys. Res. Lett. ,18, 2262- 2272.

National Radiological Protection Board ,2002, Health effects from ultraviolet radiation: Report of an advisory group on non-ionising radiation,Doc. NRPB 13(1), Chilton.U. K.

Outer, K. , Den, P. N. ,Slaper,H., and Tax,R. B., 2005,UV radiation in the Netherlands: Assessing long-term variability and trends in relation to ozone and clouds, J. Geophys. Res., VOL. 110, D02203, doi:10.1029/2004JD004824.

Piazena, H., 1996: The effect of altitude upon the solar UV-B and UV-A irradiance in the topical Chilean Andes. *Solar Energy*, 57, 133-140.

Seckmeyer, G., B. Mayer, G. Bernhard, A. Albold, R. Erb, H. Jaeger, and W. R. Stockwell, 1997: New Maximum UV Irradiance Levels Observed in Central Europe. *Atmos. Environ.*, 31, 2971-2976.

Slaper,H.,and Koskela,T.,1997,Methodology of intercomparing spectral sky measurements, correcting for wavelength shifts, slit function differences and defining a spectral reference, in The Nordic Intercomparison of Ultraviolet and Total Ozone Instruments at Izana, October 1996, edited by B. Kjeldstad, B. Johnson, and T. Koskela, pp. 89– 108, Finn. Meteorol. Inst., Helsinki.

Su, W. Y., T. P. Charlock, and F. G. Rose, 2005: Deriving surface ultraviolet radiation from CERES surface and atmospheric radiation budget: Methodology. *J. Geophys. Res.*, 110, D14209, doi: 10.1029/2005JD005794

United Nations Environment Programme (UNEP) ,1998, Environmental Effects of Ozone Depletion: 1998 Assessment, 205 pp., Nairobi, Kenya.

7

A New Method to Estimate the Temporal Fraction of Cloud Cover

Esperanza Carrasco[1], Alberto Carramiñana[1], Remy Avila[2],
Leonardo J. Sánchez[3] and Irene Cruz-González[3]
*[1]Instituto Nacional de Astrofísica,
Óptica y Electrónica, Puebla
[2]Centro de Física Aplicada y Tecnología Avanzada
Universidad Nacional Autónoma de México,
Santiago de Querétaro
[3]Instituto de Astronomía,
Universidad Nacional
Autónoma de México, México D.F.
México*

1. Introduction

High altitude astronomical sites are a scarce commodity with increasing demand. A thin atmosphere can make a substantial difference in the performance of scientific research instruments like millimeter-wave telescopes or water Čerenkov observatories. In our planet reaching above an altitude of 4000 m involves confronting highly adverse meteorological conditions. Sierra Negra, the site of the Large Millimeter Telescope (LMT) is exceptional in being one of the highest astronomical sites available with endurable weather conditions.

One of the most important considerations to characterize a ground-based astronomical observatory is cloud cover. Given a site, statistics of daytime cloud cover indicate the usable portion of the time for optical and near-infrared observations and bring key information for the potentiality of that site for millimeter and sub-millimeter astronomy. The relationship between diurnal and nocturnal cloudiness is strongly dependent on the location of the site (1). For several astronomical sites it has been reported (1) that the day versus night variation of the cloud cover is less than 5 %. Therefore, daytime cloud cover statistics is a useful indicator of nighttime cloud conditions.

We developed a model for the radiation that allowed us to estimate the fraction of time when the sky is clear of clouds. It consists of the computation of histograms of solar radiation values measured at the site and corrected for the zenithal angle of the Sun.

The model was applied to estimate the daytime clear fraction for Sierra Negra (2). The results obtained are consistent with values reported by other authors using satellite data (1). The same method was applied to estimate the cloud cover of San Pedro Mártir -another astronomical site (3) . The estimations of the time when the sky is clear of clouds obtained are also consistent with those reported by the same authors (1). The consistency of our results with those obtained applying different and classical techniques shows the great potential of the method developed to estimate cloud cover from in situ measurements.

In this chapter our model will be explained. In §2 the main characteristics of solar radiation through the terrestrial atmosphere are discussed; in §3 our method to estimate the temporal fraction of cloud cover is described using radiation data of the astronomical sites Sierra Negra and San Pedro Mártir. In §4 a brief summary is presented.

2. Solar radiation through the terrestrial atmosphere

2.1 The Sun

The Sun provides energy to the Earth at an average rate of $s_\odot = 1367\,\mathrm{W/m^2}$. This value relates directly to the solar luminosity, $L_\odot = 3.84(4) \times 10^{27}$ Watts, as observed at an average distance of one astronomical unit, $s_\odot = L_\odot / \left(4\pi d_\oplus^2\right)$, with $d_\oplus = 1\,UA \simeq 1.496 \times 10^{11}$ m. The value of $s_\odot$ is stable enough to be often referred as the *Solar constant*. Variations of the solar flux arise from intrinsic variations in the solar luminosity and seasonal variations of the distance between the Earth and the Sun. The eccentricity of the orbit of the Earth around the Sun, $\varepsilon_\oplus = 0.0167$, translates into minimum and maximum distances of $d_\oplus/(1 \pm \varepsilon)$, and hence a yearly modulation ($\propto \varepsilon_\oplus^2$) of $\pm 3.3\%$ in the solar flux over the year. Given a location on the Earth, this modulation is smaller than seasonal variations due to the changes of the apparent trajectory of the Sun in the sky, originated by the inclination of the Earth spin axis relative to the ecliptic. Intrinsic variations of the solar flux due to changes in luminosity, some tentatively related to the 11-year solar activity cycle, are very low, of the order of 0.1%.

The solar radiation is distributed along the infrared to ultraviolet regions of the electromagnetic spectrum. This distribution is shown in terms of apparent magnitudes m_V in standard spectral bands, from the ultraviolet (U) to the infrared (IHJK), in Table 1 and plotted in Fig. 1. The conversion into energy flux F_V is made through the standard formula:

$$F_V = F_V^0 \, 10^{-0.4 m_v}. \tag{1}$$

A comparison of the solar spectrum with a blackbody spectrum can be made defining three temperature measures: the effective temperature; the color temperature; and the brightness temperature:

- the effective temperature T_e is given by the integrated flux F and the angular size of the radiation source, $F = \sigma T_e^4 \delta\theta^2$, with $\delta\theta$ the apparent radius and σ the Stefan Boltzmann constant. For the Sun $T_e \simeq 5770$ K, which corresponds to a maximum emission at a wavelength $\lambda \simeq 0.5\,\mu m$.

- the color temperature is calculated through the best blackbody fit of the spectrum. Fig. 1 shows a blackbody fit to F_V of the form $A\nu^3/(e^{\nu/\nu_c} - 1)$, with best fit parameters

$$A = 1.166 \times 10^{-9} \mathrm{erg\,cm^{-2}s^{-1}}, \quad \text{and} \quad \nu_c = 1.167 \times 10^{14}\,\mathrm{Hz}, \tag{2}$$

which result in a color temperature $T_{col} \simeq 5600$ K. As observed in the plot, a blackbody is a fair fit of the spectral distribution of the solar flux.

- the third temperature indicator of solar conditions is the brightness temperature, defined monocromatically by $I_\nu = B_\nu(T_b)$. It is nearest to the effective temperature in the I band, $\lambda = 0.9\,\mu m$.

Band	Magnitude	λ (μm)	$\Delta\lambda$ (μm)	F_ν^0 Jy	F_ν (10^{-12}W m^{-2}Hz^{-1})	$F_\nu\Delta\nu$ (W m^{-2})
U	-26.03	0.360	0.068	1880	0.485	76.
B	-26.14	0.440	0.098	4650	1.329	202.
V	-26.78	0.550	0.089	3950	2.035	179.
R	-27.12	0.700	0.22	2870	2.022	272.
I	-27.48	0.900	0.24	2240	2.199	195.
J	-27.93	1.26	0.20	1603	2.382	90.
H	-28.26	1.60	0.36	1075	2.165	91.
K	-28.30	2.22	0.52	667	1.394	44.

Table 1. Apparent magnitudes of the Sun in different bands and its conversion into energy fluxes. The integrated sum over bands gives 1150 W/m^2. The magnitudes are from (4).

The radiation we observe from the Sun and the sky is determined by radiation processes occurring in the air, on the ground and on the clouds. This is revised in the following subsections.

2.2 The terrestrial atmosphere

The atmosphere of the Earth is composed of a mixture of gases dominated by molecular nitrogen and oxygen, N_2 and O_2. It can be modeled as an ideal gas of molecular mass $\mu \simeq 29$ in local thermodynamical equilibrium under hydrostatic equilibrium in a gravity field g, thus following

$$P = \frac{\rho kT}{\mu m_H}, \quad \frac{dP}{dz} = -g\rho. \tag{3}$$

The last expression assumes a plane parallel atmosphere, a good approximation when the atmosphere is a layer much thinner than the radius of the planet, as in the Earth. These two equations are insufficient to solve for the three unknowns $\{\rho(z), P(z), T(z)\}$, and a third equation containing the thermodynamical behavior of the gas is needed to close the problem. The simplest atmosphere solutions are for an isothermal gas, $T = $ constant, or an adiabatic gas, $P \propto \rho^\gamma$, where γ is the ratio of specific heat capacities.

- Solving equations 3 for an **isothermal gas** gives an exponential structured atmosphere,

$$T(z) = T, \quad P(z) = P_0 e^{-z/H}, \quad \rho(z) = \rho_0 e^{-z/H}, \tag{4}$$

of characteristic height scale $H = kT/\mu m_H g$. For Earth atmosphere, $\mu = 28.9644$, we get $H \simeq 8781.5$m at $T = 300$K and $H \simeq 7171.6$m at $T = 245$K.

- Using **adiabatic cooling** to equations 3 results in a constant temperature gradient,

$$\frac{dT}{dz} = -\frac{(\gamma - 1)}{\gamma}\frac{T_0}{H},$$

with H the isothermal temperature height and T_0 the ground temperature. The adiabatic temperature gradient for $\gamma = 1.4$, nominal value for a diatomic molecular classical ideal gas, is $dT/dz = -9.8$ K/km.

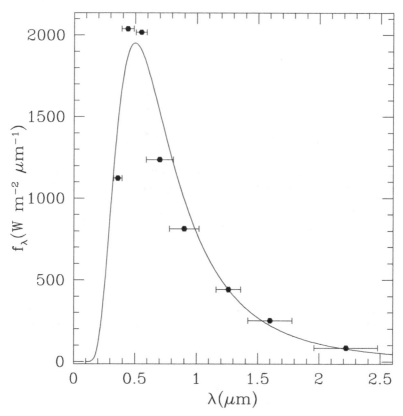

Fig. 1. Solar spectrum and blackbody fit with color temperature $T_c = 5600$ K.

A proper atmospheric solution should consider the transfer of radiation. Still, approximate solutions do provide a fair description of the terrestrial atmosphere. A still relatively simple but more precise description of the Earth's atmosphere is given by the "standard atmosphere" model, which incorporates thermodynamics through defining layers of constant temperature and constant temperature gradient. This is shown in table 2, where boundary conditions refer to the base of the atmosphere, rather than its upper edge. Note that 99 % of the atmosphere is contained in the inner 30 km. The troposphere, which constitutes ~ 80 % of the atmosphere, has an structure analog to adiabatic,

$$T(z) = T_0 - \theta z, \quad P(z) = P_0 \left(1 - \theta z / T_0\right)^{\alpha}, \quad \rho(z) = \rho_0 \left(1 - \theta z / T_0\right)^{\alpha - 1}, \tag{5}$$

but with a lower temperature gradient, $\theta = -6.5$ K/km, and $\alpha = \mu m_H g / k\theta \simeq 5.256$. The weather in Sierra Negra, the high altitude site of the γ-ray observatory HAWC[1] and of the LMT, has been monitored since late 2000. We found its meteorological variables to conform very closely with a standard atmosphere with $T_0 = 304$ K an adequate boundary value, fitting $P(4.1 \text{ km}) \simeq 625.6$ mbar at HAWC, and $P(4.58 \text{ km}) \simeq 569.5$ mbar at LMT (2). One of the

---

[1] High Altitude Water Čerenkov.

Capa	z_{g0} (km)	z_0 (km)	dT/dz K/km	T_0 °C	P_0 Pa
Troposphere	0	0.000	−6.5	+15.0	101 325
Tropopause	11	11.019	0.0	−56.5	22 632
Stratosphere (I)	20	20.063	+1.0	−56.5	5 475
Stratosphere (II)	32	32.162	+2.8	−44.5	868
Stratopause	47	47.350	0.0	−2.5	111
Mesosphere (I)	51	51.413	−2.8	−2.5	67
Mesosfera (II)	71	71.802	−2.0	−58.5	4
Mesopause	84.852	86.000	−−	−86.2	0.37

Table 2. Layers defining the International Standard Atmosphere. z_{g0} y z_0 are the geopotential and geometric heights, respectively, at the base of the layer.

ongoing projects at the Sierra Negra site is the simultaneous measurements of meteorological variables, and hence their gradients, in these two locations.

2.3 The transfer of solar radiation through the atmosphere

One of the most important features of the terrestrial atmosphere is its transparency to visible light, while being opaque to infrared radiation. As a consequence, most of the solar radiation reaches the ground, where it is thermalized and re-emitted as infrared radiation, which is trapped by atmospheric molecules, raising Earth's temperature above its direct equilibrium temperature. This is the basic scenario of our atmosphere; in practice this is complicated by scattering due to small particles suspended in the air, the presence of clouds, the local properties of the ground and the sea, and the non static conditions in the atmosphere (winds) and the sea (currents).

In the absence of an atmosphere, the temperature on a location where the Sun is observed with an angle $\theta_\odot$ would be given by the equilibrium condition

$$(1 - a)s_\odot \cos \theta_\odot = \pi \sigma T^4, \tag{6}$$

where a represent the albedo, or fraction of the radiation reflected by the surface, and the factor π results from integrating the re-emitted radiation ($\propto \cos \theta$) over half a sphere. For an albedo $a = 0.3$, the temperature is $T \simeq 270 \cos \theta_\odot^{1/4}$ K. The radiation emitted by the ground will be emitted in the infrared, at wavelengths of the order of $10\,\mu$m. The resulting simplification is that radiation in the atmosphere can be treated in terms of two separated components, one in the visible ($\lambda \simeq 0.5\,\mu$m) and the other in the infrared ($\lambda \simeq 10\,\mu$m).

A refinement of the previous calculation can be made assuming a *grey* atmosphere, where the term grey indicates the assumption that its radiation properties are independent of the wavelength, at least on a given spectral window. If we now assume the atmosphere absorbs 10 % of the visible light and 80 % of the infrared radiation originated on the ground, we infer that just above 60 % of the solar radiation is trapped by the atmosphere, which acquires a temperature T_a given by the energy density of the radiation captured,

$$u = aT_a^4 = \eta s_\odot/c \quad \Rightarrow T \simeq 245\,\text{K}, \tag{7}$$

with $\eta = 0.604$ and c the speed of light. As seen in the previous subsection the atmosphere is not an homogeneous layer, although most of it can be described as in local thermodynamical equilibrium, with the temperature scale $T(z)$ defined in table 2. The implicit assumption is that radiation absorption and emission rates are nearly equal locally. Although they are less abundant than N_2 and O_2, molecules like water (H_2O), carbon dioxide (CO_2) and methane (CH_4) play important roles in atmospheric radiation transfer processes. Their molecular spectra are rich in electronic, vibrational and rotational transitions. The vibrational components are important in the infrared while the rotational transitions dominate in microwaves.

Relevant to this work are the optical properties of the atmosphere. On the one hand, processes occurring in transparent air and on the other hand, the brightness of clouds. Aerosol particles suspended in the air scatter light, with preferential selection of shorter wavelengths. This process is described in terms of Rayleigh scattering, for which the cross section can be roughly written as $\sigma \approx 5.3 \times 10^{-31}\,m^{-2}\,(\lambda/532\,nm)^{-4}$. The integration of the hydrostatic equilibrium equation (3) shows that the column density of the atmosphere is given by $\mathcal{N} = P/\mu m_H g \simeq 2.1 \times 10^{29}\,m^3$, and a probability $\mathcal{N}\sigma \approx 0.1$ of absorbing a 532 nm photon in one atmosphere, assuming (wrongly) that the density of aerosol particles is proportional to that of air everywhere in the atmosphere. Taking as a benchmark $\lambda = 0.5\,nm$, where the solar emission is maximum, the visible emission of the atmosphere downwards due to scattering amounts to 5 % of the solar flux distributed over an effective solid angle of π steradians, equivalent to 4.3 mag/arcsec^2. This means that even if the direct solar radiation were obstructed, an omnidirectional detector of visible radiation would measure a flux $\sim 0.05 s_\odot$. The fact that Rayleigh scattering is a process whose importance increases at short wavelengths is well known to be the origin of the blue color of diurnal sky.

The most common situation in which direct solar radiation is obstructed is cloudy weather. In cloudy conditions solar light is scatter and reflected by water particles suspended in the clouds. Without entering in details, one can see that a sizable fraction of solar radiation scattered by the clouds does reach the ground. This amount does depend on the actual conditions, but will add to the 5 % grossly estimated to arise from blue sky itself. The results of the studies presented below put the integrated emission of cloudy skies at about 20 % of $s_\odot$.

3. A method to estimate the temporal fraction of cloud cover

3.1 Introduction

In this section a new method to estimate the temporal fraction of cloud cover, based in solar radiation measurements in situ, will be described. The data are compared with the radiation expected given the coordinates of the site and hence the position of the Sun in the sky. It will be illustrated by using real solar radiation data obtained at two astronomical sites: Sierra Negra and San Pedro Mártir (SPM), both in Mexico. First, a brief introduction to the sites and the data sets will be presented. In the next section the solar modulation is explained. In the following two sections the results obtained for Sierra Negra and the statistics of clear time are described. In the two subsequent sections an additional example of the method applied to SPM and the statistics of clear time are discussed. In this case, the model proved to be sensitive enough to determine the presence of other atmospheric phenomena.

Sierra Negra, also known as Tliltepetl, is a 4580 m altitude volcano inside the Parque Nacional Pico de Orizaba, a national park named after the highest mountain of Mexico. With an altitude of 5610 m Pico de Orizaba,[2] also known as Citlaltepetl, is one of the seven most prominent peaks in the world, where prominent is related with the dominance of the mountain over the region. These two peaks are located at the edge of the Mexican plateau which drops at the East to reach the Gulf of Mexico at about 100 km distance, as shown on Fig. 2. Sierra Negra is the site of the LMT, a 50 m antenna for astronomical observations in the 0.8-3 millimeter range. LMT is the largest single-dish millimeter telescope in the world. Its performance depends critically on the amount of water absorbing molecules, mainly H_2O in the atmosphere. The top of Sierra Negra has the coordinates: 97° 18' 51.7'' longitude West, and 18° 59' 08.4'' latitude North. The development of the LMT site led to the installation of further scientific facilities benefiting from its strategic location and basic infrastructure like a solar neutron telescope and cosmic ray detectors, among others. In July 2007 the base of Sierra Negra, about 500 m below the summit, was chosen as the site of HAWC observatory, a $\sim$ 22000 m^2 water Čerenkov observatory for mapping and surveying the high energy γ-ray sky. The HAWC detector incorporates the atmosphere, where particle cascades occur.

Fig. 2. Map of Mexico indicating the locations of Sierra Negra site at the East and San Pedro Mártir at the North West [from Conabio site: www.conabio.gob.mx].

The data presented here were acquired with a Texas Electronics weather station. The radiation sensor was made of a solar panel inside a glass dome. The data are output as time ordered energy fluxes in units of W/m^2. The nominal range is up to 1400 W/m^2 with a resolution of 1 W/m^2 and 5 % accuracy.

The solar radiation data span from April 12, 2002 to March 13, 2008, completing a sample of 990770 minutes. The effective data coverage was 62 %. Coverage was 73 % in 2002, decreases towards 2004, with a 44 % and increases to 81 % in 2005 and 2006. The majority of the data were taken with 1 or 5 minutes sampling. The weather of the site is influenced by the dry weather of the high altitude central Mexican plateau and humid conditions coming from the Gulf of Mexico. Given the weather conditions the data points were divided in two samples:

[2] Instituto Nacional de Estadística, Geografía e Informática (INEGI) official figure.

the *dry season* is the 181 day period from November 1st to April 30; the *wet season* goes from May 1st to October 31st, covering 184 days.

The SPM observatory is located at $31°02'39''$ latitude North , $115°27'49''$ longitude West and at an altitude of 2830 m, inside the Parque Nacional Sierra de San Pedro Mártir. SPM is $\sim$65 km E of the Pacific Coast and $\sim$55 km W to the Gulf of California, as shown in Fig. 2. The largest telescope at the site is an optical 2.1-m Ritchey-Chrétien, operational since 1981. Astroclimatological characterization studies at SPM are reviewed in (5). Other aspects of the site characterization have been reported by several authors e.g. (7; 8). Nevertheless, the first study on the radiation data measured in situ was done by Carrasco and collaborators (3). The data were recorded by the Thirty Meter Telescope (TMT) site-testing team from 2004 to 2008. See (9) for an overview of the TMT project and its main results.

The data presented here consist of records of solar radiation energy fluxes in units of Wm^{-2} acquired with an Monitor automatic weather station (9). The sensor has a spectral response between 400 and 950 nm with an accuracy of 5 %, according to the vendor. The data span from 2005 January 12 to 2008 August 8, with a sampling time of 2 minutes and a 67 % effective coverage of the 3.6 year sample; data exist for 973 out of 1316 days. The complete sample contains 596580 min out of 899520 possible; coverage was 59 % for 2005 and increased to 78 % towards the end of the campaign, in 2008.

3.2 Solar modulation

The method to estimate the temporal fraction of cloud coverage is based on computing the ratio between the expected amount of radiation and that observed. We safely assume that, at least under clear conditions, radiation directly received from the Sun is dominant. This is a term of the form $s_\odot \cos\theta_\odot$, where $\theta_\odot$ is the zenith angle of the Sun as observed from the site under study. The modulation term is removed by simply dividing by $\cos\theta_\odot$. To compute the local zenith angle as a function of time, consider a coordinate system centered on Earth with the $\hat{z}$ axis oriented perpendicular to the ecliptic. By definition, the position of the Sun in this system is restricted to the $x - y$ plane, and is given by $d_\oplus \hat{r}_\odot$, with the direction to the Sun given by the unitary vector

$$\hat{r}_\odot = -\hat{x}\cos\left(\omega_a t\right) - \hat{y}\sin\left(\omega_a t\right), \tag{8}$$

where $\omega_a = 2\pi/\mathrm{year}$ is the angular frequency associated to the yearly modulation. Given a location on Earth of geographical latitude b, the zenith is in the direction given by the unitary vector

$$\hat{n} = -\hat{z}_e \sin b + (\hat{x}_e \cos(\omega_s t) + \hat{y}_e \sin(\omega_s t)) \cos b, \tag{9}$$

where $\omega_s = \omega_a + \omega_d = 2\pi/\mathrm{yr} + 2\pi/\mathrm{day}$, and

$$\hat{x}_e = \hat{x}, \quad \hat{y}_e = \hat{y}\cos\iota - \hat{z}\sin\iota, \quad \hat{z}_e = \hat{y}\sin\iota + \hat{z}\cos\iota, \tag{10}$$

with ι the inclination angle of the Earth axis relative to the $\hat{z}$, the unitary vector normal to the ecliptic plane. With these relations in hand, it follows that the zenith angle of the Sun is given by $\cos\theta_\odot = -\hat{n} \cdot \hat{r}_\odot$. Equations 8 and 9 assume $t = 0$ corresponds to the time of equinox, rather than the beginning of the civil year.

These equations were used to generate solar zenith angles for all years covered by the data. Before the actual data analysis, we verified the proper functioning of the related software, both for Sierra Negra and for San Pedro Mártir. The comparison, shown in Figs. 3 and 4, gives an insight on the differences between sites inside or ouside the tropics: Sierra Negra, at $b \simeq 19°$ is located three degrees south of the tropic of Cancer, while San Pedro Mártir, located in the Baja California at $b \simeq 31°N$ is at a moderately northern location. Fig. 3 shows the daily behavior of $\cos\theta_\odot$ for both sites, with summer months displayed in red and winter months in blue. It is clear that the range of values of $\cos\theta_\odot$ at time of culmination, or simply noon, is larger for San Pedro, going below 0.6 is winter, while these always reach at least 0.7 in Sierra Negra. The effect of latitude is more marked in the behavior of the values of $\theta_\odot$ near noon through the year (fig. 4). Due to its closeness to the equator, the Sun reaches the zenith twice a year at Sierra Negra, while never getting $\theta_\odot < 8°$ for San Pedro Mártir. One can calculate analytically the times of the two passage of the Sun through the zenith, which are given by

$$\cos\phi_\pm = \sin b / \sin\iota,$$

where $t = 0$ is the time of summer solstice and $\phi_\pm$ corresponds to ±35.643 days, or passages through the zenitn on the 17th of May and on the 26th of July, in concordance with Fig. 4.

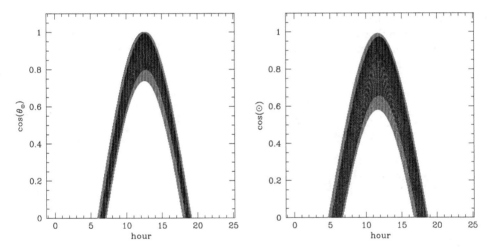

Fig. 3. Cosine of the solar zenith angles as a function of time of day for Sierra Negra (left) and San Pedro Mártir (right). Red color indicates data for June and July while blue indicates December and January. The higher elevation of the Sun as seen from Sierra Negra can be appreciated.

A direct representation of the radiation data can be seen in Fig.5, where the solar radiation is plotted *vs.* the cosine of the Sun zenith angle, $\cos\theta_\odot$. The red line corresponds to the *Solar constant* $s_\odot \times \cos\theta_\odot$. On the left side are the data corresponding to Sierra Negra while on the right side are those to SPM. The dots above the red line are spurious data as the sensor can not received more energy than that provided by the Sun to the Earth, given by $s_\odot$. Note that for Sierra Negra there is a points concentration below the red line corresponding to clear conditions i.e. the radiation measured at the site is very close to that expected. The same effect is more pronounced in the case of SPM. The latter means, as will be shown, that SPM has a larger fraction of clear time.

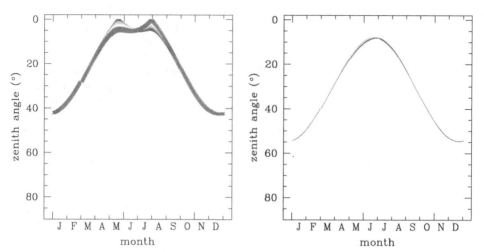

Fig. 4. Minimum zenith angle for the Sun as seen from Sierra Negra (left) and San Pedro Mártir (right). Data shown are for four 10 minute windows at or close to the local time of culmination, or solar noon. The Sun reaches the zenith ($\theta_\odot = 0°$) twice a year in Sierra Negra, while ranging from 55° to 7° at San Pedro.

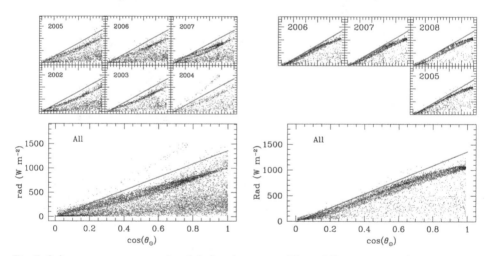

Fig. 5. Solar power versus $\cos\theta_\odot$ global and per year. The red line corresponds to $s_\odot \times \cos\theta_\odot$. Left: for Sierra Negra. Right: for San Pedro Mártir

3.3 The histogram of $\psi(t)$ for Sierra Negra

The radiation flux at ground level is considered, to first approximation, to be given by the *Solar constant* $s_\odot$, modulated by the zenith angle of the Sun and a time variable attenuation factor $\psi(t)$. Knowing the position of the Sun at the site as a function of time, we can estimate the variable ψ, given as

$$\psi(t) = \frac{F(t)}{s_\odot \cos\theta_\odot}. \tag{11}$$

where $F(t)$ is the radiation measured at the site and $\theta_\odot$ is the zenith angle of the Sun. $\psi(t)$ is a time variable factor, nominally below unity, which accounts for the instrumental response (presumed constant), the atmospheric extinction on site and the effects of the cloud coverage.

Knowing the site latitude, the modulation factor $\cos\theta_\odot$ was computed as a function of day, month and local time, minute per minute to study the behavior of the variable ψ and to obtain its distribution. The term z is referred as the airmass, defined as $z = \sec\theta_\odot$; thus $z < 2$ is equivalent to $\theta_\odot < 60°$. Most astronomical observations are carry out at this airmass interval.

The histogram of values of ψ showed a bimodal distribution composed by a broad component for low values of ψ and a narrow peak $\psi \lesssim 1$, as shown in Fig. 6. The narrow component is interpreted as due to direct sunshine while the broad component is originated when solar radiation is partially absorbed by clouds; we then use the relative ratio of these two components to quantify the "clear weather fraction".

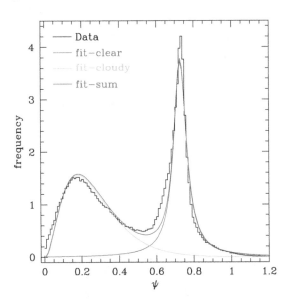

Fig. 6. The distribution of the solar flux divided by the nominal solar flux at the top of the atmosphere, $s_\odot \cos\theta_\odot(t)$, for Sierra Negra. The distribution of the complete sample shows a bimodal behaviour which can be reproduced by a two component fit, shown in solid lines.

The histogram of ψ values can be reproduced by a two component fit. The functional form of the best fit is given by,

$$f(\psi) = A\psi^2 e^{-\beta\psi} + \frac{B}{1 + [(\psi - \psi_0)/\Delta\psi]^2}. \tag{12}$$

The first term on the right hand side is a χ^2 function with six degrees of freedom. It is interpreted as the cloud-cover part of the data, with its integral being the fraction of cloud-covered time. The second term, a Lorentzian function with centre ψ_0 and width $\Delta\psi$, represents the cloud-clear part of the data. A and B provide the normalization and relative weights of both components; β is related to the width and centre of the broad peak. In the appendix the details of the calculation of the fit parameters, including the errors, are discussed.

In Fig. 6 the distribution of ψ for the whole data set is shown in black with the double component fit in red. The first component of the fit corresponding to the cloud cover part of the data is shown in blue while the second one, corresponding to the clear part of the data is shown in green. This bimodal distribution, with a first maximum at around $\psi \sim 0.2$ and a narrow peak at $\psi \sim 0.75$, has a minimum around $\psi \sim 0.55$.

Fig. 7 presents the distribution of ψ for the dry and for the wet seasons. Clearly, in both cases the distribution of ψ is also bimodal. For the dry months when the sky is mostly clear of clouds the narrow peak is higher. In contrast, during the wet season when more clouds are present, the broad peak is more pronounced. Note that, in both cases, the fit given by Eq. 12 reproduces very well the observational data.

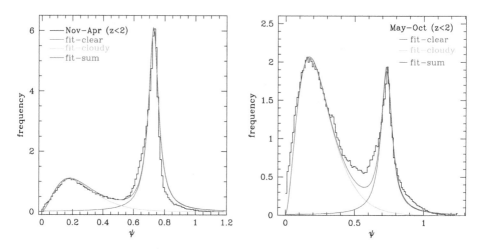

Fig. 7. The distribution of the solar flux divided by the nominal solar flux at the top of the atmosphere, $s_\odot \cos\theta_\odot(t)$, for Sierra Negra. Left: for the dry season. Right: for the wet season. In both cases, the bimodal distribution is well reproduced by a two component fit, show in solid lines.

The data were separated with $\psi \leq \psi_{min}$ as cloudy weather and data with $\psi > \psi_{min}$ as clear weather. ψ_{min} corresponds to the intersection of the two components of the function fitted to the distribution of all the data points. The fraction of clear time f(clear), was computed as clear/(clear+cloudy).

From the global distribution, a clear fraction for the site of 48.4 %, was obtained. This results is consistent with values reported by other authors (1). The authors surveyed cloud cover and water vapor conditions for different sites using observations from the International Satellite Cloud Climatology Project (ISCCP) for the California Extremely Large Telescope (CELT) project. The study period is of 58 months between July 1993 to December 1999 using a methodology that had been tested and successfully applied in previous studies. For Sierra Negra they measured a clear fraction for nighttime of 47 %. They conclude that the day versus night variation of cloud cover is less than 5 %, being clearer at night. Therefore, the results obtained with the method presented here are consistent within 6.4 % with those obtained via a totally independent technique.

3.4 Statistics of clear time for Sierra Negra

The method described above also allow to calculate the statistics of the clear time f(clear). For completness the analysis in this section includes data with airmass less than 10. The fraction of clear time f(clear) was computed for every hour of data. The distribution of the hourly clear fraction is shown in Fig. 8 for the dry (left) and wet (right) seasons. The histograms have a strong modulation. If we consider the semester between November and April $f(\text{clear}) = 1$ has 37.4 % of the data while the $f(\text{clear}) = 0$ peak has 9.0 % of the data. This means that 37.4 % of the time the sky is clear of clouds and 9.0 % of the time the sky is cloudy. During the complementary wet months the $f(\text{clear}) = 1$ peak contains 11.4 % of the data while $f(\text{clear}) = 0$ has 32.5 % of the data. Intermediate conditions prevail around 55 % of the time in both semesters. From the f(clear) histogram of the whole data (not shown) 20.3 % of the hours have $f(\text{clear}) = 0$, while 25.0 % have $f(\text{clear}) = 1$; the remaining have intermediate values.

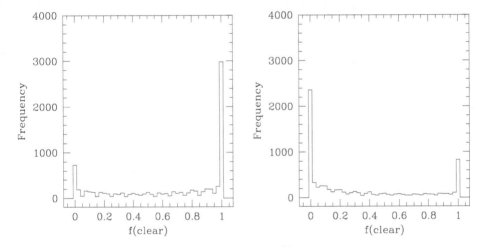

Fig. 8. Distribution of hourly clear fraction for the dry (left) and wet (right) seasons, for Sierra Negra.

The contrast between dry and wet semesters is well illustrated in the left panel of Fig. 9, showing the median and quartile fractions of clear time for successive wet and dry semesters. Semesters are taken continuously, from May to October representing the wet season and November to April of the following year for the dry season. The bars represent the dispersion in the data, measured by the interquartile range. Large fluctuations are observed at any time of the year. The contrast between the clearer dry months, with median daily clear fractions typically above 75 %, and the cloudier wet months, with median clear fractions below 20 %, is evident. The seasonal variation can be seen with more detail in the monthly distribution of the clear weather fraction, combining the data of different years for the same month, shown

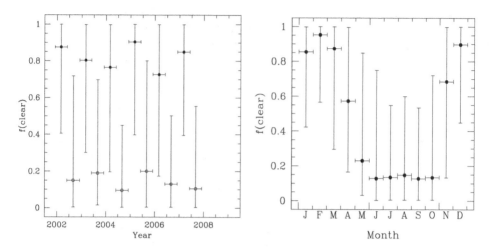

Fig. 9. Left: clear fractions for the different seasons. Points are at median; bars go from 1st to 3rd quartile. Wet season (open dots) is the yearly interval from May to October; dry season (full dots) is from November to April of the following year. Right: the median and quartile values of the fraction of clear weather for the different months of the year.

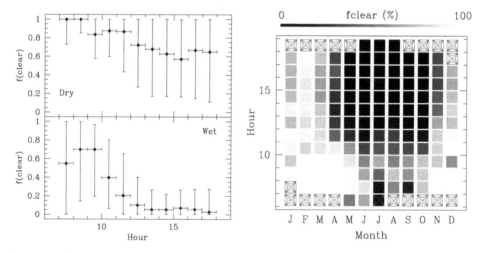

Fig. 10. Left: median and quartile values of the fraction of clear weather fclear, for each hour of day. The lower and upper panels are for wet (MJJASO) and dry (NDJFMA) semesters, respectively. Right: a grey level plot showing the median fraction of clear time f(clear), for each month and hour of day. Squares are drawn when more than 10 h of data are available; crosses indicate less than 10 h of data.

on the right side of the same figure. The skies are clear, f(clear)> 80 %, between December and March, fair in April and November, f(clear)∼ 60 %, and poor between May and October, f(clear)< 30 %. The fluctuations in the data are such that clear fractions above 55 % can be found 25 % of the time, even in the worst observing months.

Fig. 10 (left) shows the median and quartile clear fractions as function of hour of day for the wet/dry subsets. The interquartile range practically covers the (0-1) interval at most times. We note that good conditions are more common in the mornings of the dry semesters, while the worst conditions prevail in the afternoon of the wet season, dominated by Monsoon rain storms. The trend in our results for daytime is consistent with that obtained by applying different methods (1). By analysing the clear fraction during day and nighttime these authors found that the clear fraction is highest before noon, has a minimum in the afternoon and increases during nighttime.

On the right panel of Fig. 10 a grey level plot of the median percentage of clear time for a given combination of month and hour of day, is shown. Dark squares show cloudy weather, clearly dominant in the afternoons of the rainy months. These are known to be the times of stormy weather in the near-equator. Clear conditions are present in the colder and drier months. This plot is similar to that of humidity. In fact, when relative humidity decreases, the fraction of clear time increases, as shown in Fig. 11.

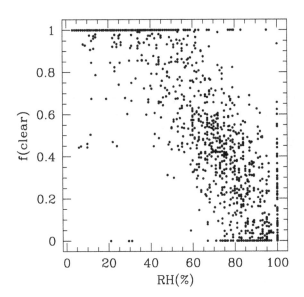

Fig. 11. Relative humidity (RH) vs. fraction of clear time, for Sierra Negra. It is apparent that there is a trend: for low values of RH, f(clear) is larger.

3.5 The histogram of $\psi(t)$ for San Pedro Mártir

The same analysis was carried out for SPM radiation data. The normalized histogram of ψ for all the data is shown in Fig. 12. It presents a double peak in the clear component, not fully consistent with the standard narrow component fit function, and the cloud component with maximum at $\psi \lesssim 0.3$. We applied the double component fit of Eq. 12, shown in the same figure. The fit for the clear component is drawn in blue, for the cloud component in green and for the sum in red. The coefficients of the fit and associated errors are presented in Table 3. Fit errors were obtained through a bootstrap analysis using 10000 samples. The fit agrees with the data within the statistics, except in the wings of the clear peak. Still, the Lorentzian function proved to fit much better the data than a Gaussian. The fit can be better appreciated in a logarithm version shown on the right side of the same figure.

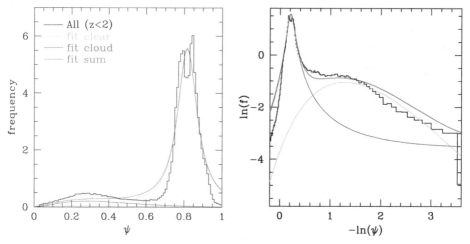

Fig. 12. *Left*: the normalized observed distribution of ψ for all the data and the corresponding fits for SPM. The blue line is the fit to clear weather; the green one to cloudy weather and the red line to the sum. *Right* : the logarithm of the normalized observed distribution of ψ and of the corresponding fits.

Parameter	Global $z < 2$	Bootstrap	errors	relative error (10^{-3})
A	40.8	40.766	± 0.612	15.0
β	7.19	7.175	± 0.044	6.2
B	6.03	6.035	± 0.055	9.1
ψ_0	0.815	0.8151	± 0.0002	0.3
$\Delta\psi$	0.063	0.0629	± 0.0005	7.5
f(clear)	0.824	0.8238	± 0.0009	1.1
f(cloud)	0.176	0.1762	± 0.0009	5.0

Table 3. Parameters of the fit shown in red Fig. 12.

We considered data with $\psi \leq \psi_{min}$, where $\psi_{min} = 0.58$, as cloudy weather and data with $\psi > \psi_{min}$ as clear weather. The value $\psi_{min} = 0.58$ corresponds to the intersection of the two components of the function fitted to the distribution of all data points. As mentioned, the fraction of clear time f(clear), was computed as clear/(clear+cloudy). From the global histogram we obtained for SPM a clear fraction for the site of 82.4 %. The errors in the determination of f(clear) and f(cloud), were also obtained by generating 10000 bootstrap samples; they are shown in Table 3.

The 82.4 % of clear fraction obtained from the global distribution shown in Fig. 12 is similar to that reported using satellite data (1). These authors estimated that the usable fractions of nightime at SPM was 81 %. Their definition of usable time includes conditions with high cirrus. For the case of SPM, they conclude that the day versus night variation of cloud cover is less than 5 %, being clearer at night. Therefore, the results presented here are consistent within 6.4 % with those reported in the literature (1).

Another estimation of the useful observing time at SPM by Tapia (10) reports a 20 yr statistics of the fractional number of nights with totally clear, partially clear and mostly cloudy based in the observing log file of the 2.1m telescope night assistants. The author reports a total fraction of useful observing time of 80.8 % and compares his results with those from other authors (1); he concludes that the monthly results from both studies agree within 5 % while for the yearly fraction, the discrepancies are lower than 2.5 %. Therefore, our results in this case, are also consistent with those obtained with completely different techniques.

Futhermore, when analyzing the fits per month we realized that the Lorentzian fits for the clear weather peak were better than that of the complete dataset: to study the seasonal variation of ψ we created histograms and the corresponding fits per month. Consider the histogram and corresponding fit for July and November shown Fig. 13. It can be appreciated that the fits reproduce the distribution of ψ very well. The narrow clear component is consistent with prevailing clear sky conditions, for which the solar radiation reaches the site with only the attenuation of the atmosphere. The coefficients of the fits presented in Fig. 13, according to the functional form of $f(\psi)$, Eq. 12, are shown in Table 4. The fits can be better valued in the logarithm displays of Fig. 13, presented in Fig. 14. The fits to the complete data (red line), to clear weather (blue line) and to cloudy weather (green line) are indicated.

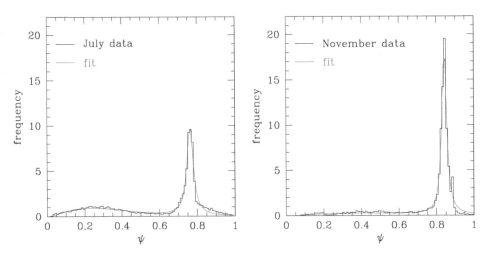

Fig. 13. The observed distribution of ψ and the two component fit for July (*left*) and November (*right*). Comparing both plots a shift in the centre of the narrow component is clearly appreciated.

We studied the position of the centre of the peak corresponding to the clear fraction as a function of the month of each observed year. We found that for every year there is a cyclic

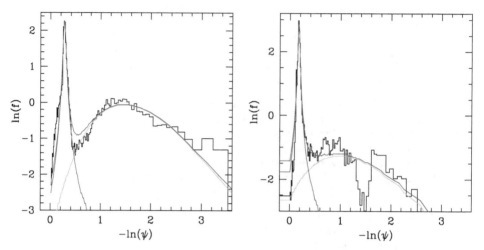

Fig. 14. The logarithm plots of the observed distribution of ψ for July (*left*) and November (*right*) shown in Fig. 13. Note that even in the case of low statistics the fit proves to be good.

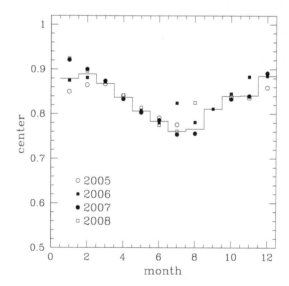

Fig. 15. The centre of the narrow component of ψ for each month. The monthly values for all the data are indicated by the histogram, while the dots mark individual months of different years. The position of the narrow peak component is not constant during the year, reaching a minimum in July. The error bars are smaller than the symbols.

effect: the centre of the peak is 0.880 in January, reaches a maximum around 0.889 in February, decreases to a minimum of 0.761 in July and increases towards the end of the year to 0.885 in

Sample	A	β	B	ψ_0	$\Delta\psi_0$
July	135.2	8.68	9.24	0.761	0.021
	±1.5	±0.112	±0.141	±0.0005	±0.001
November	11.7	3.83	19.07	0.841	0.015
	±2.6	±0.189	±0.236	±0.0009	±0.002

Table 4. Coefficients of the fits shown in Fig. 13.

December. Errors in the statistical determination of ψ_0 are $\lesssim 0.001$. Fig. 15 shows the position of the centre of the narrow component obtained using the whole data set for each month with a solid line. The corresponding values of the individual months for different years are indicated by distinct symbols.

The solar radiation sensor accuracy is $\pm5\,\%$. However, considering N ($\sim$20000) data points per bin, the position of the peaks are statistical variables determined with an accuracy $\propto 1/\sqrt{N}$ times the individual measurement error i.e. much better than 5 %. Hence, the variations in the position of the centres observed with an amplitude of up to 14 % are statistically robust. Still, the amount of radiation corresponding to the clear peak in July is higher by 277 Wm^{-2} than that received in November.

The variation trend in the centre of the clear peak shown in Fig. 15 can be interpreted in terms of seasonal variations of the atmospheric transmission: there are more aerosols during spring (maximum) and summer than in the rest of the year (6). This is also consistent with the seasonal variation of the Precipitable Water Vapor (PWV) at 210 GHz reported by different authors (11), (12), (13; 14), as the maximum PWV values occur during the Summer. From these results we concluded that the double peak in the global distribution of ψ is a real effect due to absorption variations in the atmosphere.

The larger value of the centers of the clear peak for July 2006 and August 2008 relative to the same months of the other years, shown in Fig. 15, suggest that the atmosphere was more transparent. We analyzed the aerosol optical thickness reported by (6). The larger value of the centre for July 2006 is consistent with smaller values of the aerosol optical thickness for July 2005 and 2007 but marginally for July 2008. The bigger value of the centre for August 2008 is also consistent with smaller values of the aerosol optical depth for August 2007 and marginally for August 2006 while for August 2005 there is not data available.

3.6 Statistics of clear time for SPM

First, the fraction of clear time obtained for every hour of data, accumulating 7828 h is shown in Fig. 16. We note that it behaves in a rather unimodal fashion: 78.6 % have $f(\text{clear}) = 1$ while 9.5 % of the hours have $f(\text{clear}) = 0$. The remaining fraction of data (12.5%) have intermediate values.

The solar radiation data observed at airmass lower than 2 is a subset of that observed below 10. For completeness, in this analysis we considered data with airmass less than 10.

The contrast between summer and the other seasons is well illustrated on the right side of Fig. 16, showing the median and quartile fractions of clear time for successive years. The bars represent the dispersion in the data measured by the interquartile range. The quartiles are indicative of the fluctuations and therefore more representative than averages. Large variations are observed mainly during the summer months for the whole period.

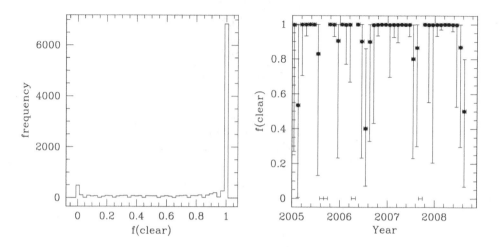

Fig. 16. Left: distribution of hourly clear fraction for the 7828 datapoints available. Right: clear fractions for the different months. Points are at median; bars go from 1st to 3rd quartile. The annual cycle can be appreciated.

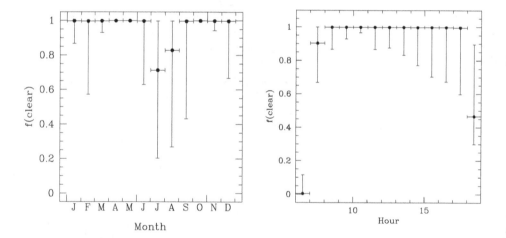

Fig. 17. Left: median and quartile values of the monthly clear fraction. Right: median and quartile values of the hourly clear fraction

Considerable fluctuations are also present for 2005 in January, February and December. The latter is not reproduced in 2006 but in 2007 there is also a large fluctuation in December. The contrast between the spring and autumn months, with median daily clear fractions typically above 98 %, and the cloudier months with median clear fractions below 80 % is evident.

The seasonal variation can be seen with more detail in the monthly distribution of the clear weather fraction, combining the data of different years for the same month, shown on the left side of Fig. 17. The skies are clear, f(clear)> 99 %, between March and May, relatively poor between June and September with a minimum median value of f(clear)< 72 % and fair between December and February when in the worst case 25 % of the time f(clear)< 57 %.

The right side of Fig. 17 shows the median and quartile clear fractions as function of hour of day. Good conditions are more common in the mornings. The trend in our results for daytime is consistent with that obtained by (1). By analysing the clear fraction during day and nighttime they found that the clear fraction is highest before noon, has a minimum in the afternoon and increases during nighttime. The authors associated the afternoon maximum in cloudiness with lifting of the inversion and cloud layer because the site is high enough to be located above the inversion layer at night and in the mornings.

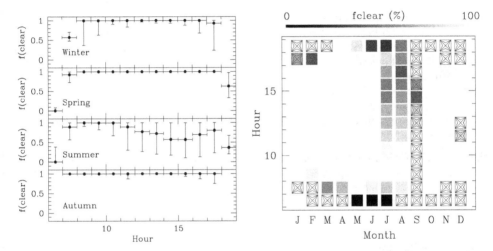

Fig. 18. Left: graph showing the median and quartile values of the fraction of clear weather for each hour of day for each season. Right: a grey level plot showing the median fraction of clear time f(clear), for each month and hour of day. Squares are drawn when more than 10 h of data are available; crosses indicate less than 10 h of data.

The left panel of Fig. 18 presents the median and quartiles of clear fraction as a function of hour of day for the seasons subsets. Seasons were considered as follows, winter: January, February and March; spring: Abril, May and June; summer: July, August and September and autumn: October, November and December. It is clear that during the summer the conditions are more variable than at any other epoch of the year. In the other seasons the conditions are very stable.

The right side of Fig. 18 shows a grey level plot of the median percentage of clear time for a given combination of month and hour of day. Squares are drawn when more than 10 h of data are available; crosses indicate less than 10 h of data. Clear conditions are present in the colder and drier months, from October to June. Dark squares show cloudy weather, clearly dominant in the afternoons of the summer months, from July to September.

We repeated the analysis for airmass lower than 2. An equivalent histogram to that shown in Fig. 16, was created by computing the fraction of f(clear) for every hour, adding 5211 h. As expected, it also has an almost unimodal distribution: 82.5 % have f(clear) = 1 while 6.7 %

of hours f(clear) = 0. The remaining fraction of data have intermediate values. The values of f(clear) obtained for the periodicities presented in this section are very similar but with less dispersion. In fact, in the analysis per hour the difference in median values are within 0.1 %. For the analysis per month the differences are also in that range except for July and August with differences between 0.3 to 13 %, with a maximum of 20 % for July 2006. The lower values obtained for the global distribution and for different periods can be explained by the presence of clouds formed at airmass $2 < z < 10$. The equivalent grey level plot of Fig. 18 for airmass less than 2 (not shown) does not include the contribution of clouds formed in the early morning and late afternoon hours, specially during the summer months.

3.7 Summary

We have presented a method to estimate the temporal fraction of cloud cover based on calculating the ratio between the expected amount of radiation and that observed. We described the equations to compute the solar modulation given the latitude of the site under study. These equations were applied to two different sites: Sierra Negra, at $b \simeq 19°$ located three degrees south of the tropic of Cancer and San Pedro Mártir, located in the Baja California at $b \simeq 31°N$. Knowing the position of the Sun at the site as a function of time, we computed the variable ψ, given by $\psi(t) = F(t)/s_\odot \cos\theta_\odot$, where F(t) is the solar flux measured and $s_\odot \cos\theta_\odot(t)$ is the nominal solar flux at the top of the atmosphere. From the global normalized observed distribution of ψ, we calculated the fraction of time when the sky is clear of clouds.

The fit to the histograms of ψ developed for Sierra Negra (2) also reproduced the SPM data (3), showing that this method might be generalized to other sites. Furthermore, the consistency of our results with those obtained by other authors shows the great potential of our method as cloud cover is a crucial parameter for characterization of any site and can be estimated from i*in situ* measurements.

4. Acknowledgments

The authors acknowledge the kindness of the TMT site-testing group. The authors specially thank G. Sanders, G. Djorgovski, A. Walker and M. Schöck for their permission to use the results from the report by Erasmus & Van Staden (1) for Sierra Negra and SPM. The authors also thank Jorge Reyes for his help with the images.

5. Apendix: fit and errors

The radiation flux $F(t)$ is normalized to a function $\psi = F(t)/s_\odot \cos\theta_\odot(t)$. The observed distribution of ψ is well fitted by the function

$$f(x) = Ax^2 e^{-\beta x} + \frac{B}{1 + ((x - x_0)/\Delta x)^2}.$$

- The first term on the right hand side is a χ^2 function with six degrees of freedom. The function has maximum at $x_{max} = 2/\beta \simeq 0.28$ for our data. It is interpreted as the "cloud-cover" part of the data, with its integral over the unitary interval,

$$\int_0^1 Ax^2 e^{-\beta x} dx = \frac{2A}{\beta^3}\left[1 - \left(1 + \beta + \beta^2/2\right)e^{-\beta}\right]$$

$$\rightarrow 2A/\beta^3 \quad \text{for} \quad \beta \gg 1,$$

the fraction of "cloud-covered" time. For the SPM($z < 2$) sample we have $A = 40.8, \beta = 7.19$ and $2A/\beta^3 \left[1 - \left(1 + \beta + \beta^2/2\right) e^{-\beta}\right] = 0.2195(0.974) = 0.2139$. The correct unitary normalization is for

$$A^* = \frac{\beta^3/2}{1 - (1 + \beta + \beta^2/2)e^{-\beta}} = 189.74.$$

- The second term, a Lorentzian function of center x_0 and width Δx, represents the "cloud-clear" part of the data. The fitting function is normalized around x_0 such that

$$\int_{x_0-\eta_0\Delta x}^{x_0+\eta_1\Delta x} \frac{B\,dx}{1 + (x - x_0/\Delta x)^2} = B\Delta x\left(\arctan(\eta_1) + \arctan(\eta_0)\right)$$

$$\rightarrow B\pi\Delta x \quad \text{for} \quad \eta_{0,1} \gg 1,$$

the term $B\Delta x(\arctan(\eta_1) + \arctan(\eta_0))$ represents the "clear" fraction. For the SPM($z < 2$) data $B = 6.03$, $x_0 = 0.8151$, $\Delta x = 0.063$, I take $\eta_0 = x_0/\Delta x = 12.938$ and $\eta_1 = (1 - x_0)/\Delta x = 2.935$ so $\arctan \eta_1 + \arctan \eta_0 = 0.8709\pi$ and the correct normalization factor should be $B^* = 5.801$.

The fit requires determining the five parameters through residual minimization. The determination of $\{\beta, x_0, \Delta x\}$, determining the shape of the distribution, is numerical; that of A, B is analytical through solving

$$\left\langle yx^2e^{-\beta x}\right\rangle = A\left\langle \left(x^2e^{-\beta x}\right)^2\right\rangle + B\left\langle \frac{x^2e^{-\beta x}}{1 + ((x - x_0)/\Delta x)^2}\right\rangle$$

$$\left\langle \frac{y}{1 + ((x - x_0)/\Delta x)^2}\right\rangle = A\left\langle \frac{x^2e^{-\beta x}}{1 + ((x - x_0)/\Delta x)^2}\right\rangle + B\left\langle \left(\frac{1}{1 + ((x - x_0)/\Delta x)^2}\right)^2\right\rangle$$

Strictly speaking, we should have $A = (1 - w)A^*$ and $B = wB^*$ with w defining the "clear fraction".

Error determination can, in principle, be done with the process of residual minimization, through a parabolic fit to the function describing the figure of merit. Given the nature of the fitting functions this is not practical; we proceeded through a bootstrap analysis of the ($z < 2$) sample containing $N \simeq 180,000$ points, obtaining the results shown in tables 3 and 4. To determine the errors in subsamples of size N_s we assume errors scale as $\sqrt{N/N_s}$.

6. References

[1] Erasmus A, & Van Staden C. A., 2002, "A satellite survey of cloud cover and water vapor in the western USA and Northen Mexico. A study conducted for the CELT project.", internal report

[2] Carrasco, E., Carramiñana, A., Avila, A., Guitérrez, C., Avilés, J.L., Reyes, J., Meza, J. & Yam, O., (2009), MNRAS, 398, 407

[3] Carrasco, E., Carramiñana, A., Sánchez, L. J., Avila, R. & Cruz-González, I., (2012), MNRAS, 420, 1273-1280

[4] http://mips.as.arizona.edu/~cnaw/sun.html

[5] Tapia M., Hiriart D., Richer M. & Cruz-González, I.,(2007), Rev. Mex. AA (SC), 31, 47

[6] Araiza M.R. & Cruz-González I., (2011) Rev. Mex. AA 47, 409

[7] Cruz-González I., Avila R. & Tapia M., eds, (2003), Rev. Mex. AA (SC), 19.

[8] Cruz-González I., Echevarría J. & Hiriart D., eds, (2007), Rev. Mex. AA (SC), 31

[9] Schöck M. et al., (2009), Publ. Astr. Soc. Pac., 121, 384

[10] Tapia M., (2003), Rev. Mex. AA (SC), 19, 75

[11] Hiriart D. et al., (1997), Rev. Mex. AA, 33, 59

[12] Hiriart D. et al., (2003), Rev. Mex. AA (SC), 19, 90

[13] Otárola A. et al., (2009), Rev. Mex. AA, 45, 161

[14] Otárola A. et al., (2010), Publ. Astr. Soc. Pac., 122, 470

8

Correlation and Persistence in Global Solar Radiation

Isabel Tamara Pedron
Paraná Western State University
Brazil

1. Introduction

This work is focused on the investigation of correlations and memory effects in daily global solar radiation data series and in the capture of underlying multifractality. It is well known that the behaviour of the climatological variables affect directly crucial aspects of the people's daily lives. Typically measurements of these variables are a sequence of values that constitute a time series, and time series analysis tools can contribute effectively to the study of such variables. An interesting investigation that can be done on the series is to identify the occurrence of correlation in the records of the sequence and to detect an effect of long-term memory in this data set over time. One possible approach is to estimate how a particular measure of fluctuations in the series scales with the size s of the time window considered. A specific method for this analysis is the Detrended Fluctuation Analysis – DFA, a well-established method for the detection of long-range correlations. Usually, trends may mask the effect of correlations. DFA can systematically eliminate trends of polynomial of different orders. This method was proposed in (Peng et al., 1994) and has successfully been applied to many different fields, and particularly in the study of data series of variables associated with the weather and climate. The fluctuation function behaves as a power law with the values chosen for s. The exponent can be identified as the Hurst exponent (H). The DFA method gives the Hurst exponent, and estimating such exponent from a given data set is an effective way to determine the nature of correlation in it. Values of H in the range (0, 0.5) characterize anti-persistence, whereas those in the range (0.5,1) characterize persistence, long-range correlations. The value $H=0.5$ is associated with uncorrelated noise. Temperature and precipitation have characteristic values of H (Koscielny-Bunde et al., 1998; Bunde & Havlin, 2002) although some claim that the scaling exponent is not universal for temperature data (Király & Jánosi, 2005; Rybski et al. 2008). Relative humidity shows stronger persistence (Chen et al., 2007; Lin et al., 2007), and wind speed also exhibit behaviour with long-range correlation (Govindan & Kantz, 2004; Kavasseri & Nagarajan, 2005; Koçak, 2009; Feng et al., 2009). One can be sure of the universality of the correlations in climatological time series but its exponents can be related to local patterns.

The variation of the Hurst exponent in time for a given series indicates the existence of non-stationary fluctuations, pointing to a multi-fractal. Thus, when the series points to the existence of more than one exponent for its characterization we are dealing with multifractal behaviour. Multifractal signals are far more complex than monofractal signals and require more exponents (theoretically infinite) to characterize their scaling properties. In this work multifractality in time series data of global solar radiation is studied by applying the Multifractal Detrended Fluctuation Analysis (MF-DFA) proposed in (Kantelhardt *et al.*, 2002). It is a modified version of DFA to detect multifractal properties of time series and provides a systematic tool to identify and quantify the multiple scaling exponents in the data. This method was applied in several cases, in particular in climatological data series as presented in (Kantelhardt *et al.* 2003; Alvarez-Ramirez *et al.*, 2008; Pedron, 2010; Zhang, 2010).

2. Data set and method

Series of (daily) global solar radiation data were studied and all data set is within the period (Jul 97 – Apr 2011). Data were provided by Technological Institute SIMEPAR-Brazil from 18 meteorological stations in Paraná State (PR) (Fig. 1) located in Southern Brazil (Fig. 2). The coordinates of each station are in Table 1. Data cover a region with 199,314.9 km² and the climate is classified as subtropical. Mean temperature oscillates between 4 °C and 21 °C. The weather conditions in Paraná are usually associated with incursions of polar air toward the Equator and areas of convective clouds associated with extra tropical cold fronts. It also shows areas of local instability associated with mesoscale convective complexes, squall lines, convective clouds, heavy precipitation and lightning.

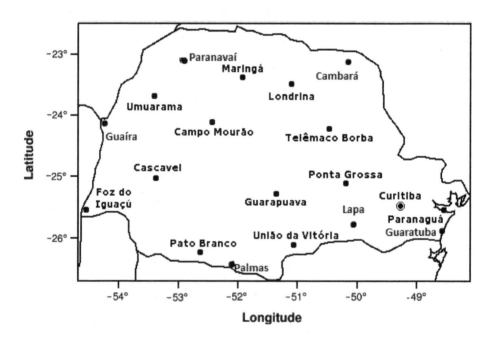

Fig. 1. Meteorological stations in Paraná state. Adapted from: <http://www.simepar.br>.

Fig. 2. The Paraná (PR) state location in Southern Brazil.

Data series provide the incidence of direct solar radiation measured on the surface. There is no correction carried out regarding the presence of clouds.

The MF-DFA method is a generalization of the standard DFA, being based on identification of the scaling of the *mth*-order moments of the time series which may be non-stationary. The modified MF-DFA procedure consists of a sequence of steps and detailed information about computation can be found in Kantelhardt et al. (2002). The steps are essentially identical to the conventional DFA procedure. First we construct the profile $X(i)$ as $X(i) = \sum_{k}^{i} x'_k$ .The index i counts the data points in the record, i.e., $i=1,2,...,$ N. For eliminating the periodic seasonal trends daily differences $x'_i = x_i - \overline{x}_i$ were computed, where $\overline{x}_i$ represents the average value of radiation for each calendar date i. The profile is then divided into $N_s = int(N/s)$ non-overlapping segments of length s, where s represents time intervals measured in days. To accommodate the fact that some of the data points may be left out, the procedure is repeated from other end of the data set. The local trend is determined by using the least-squared fit to each segment v and we obtain the detrended time series $X_s(i) = X(i) - p_v(i)$ where $p_v(i)$ is the polynomial fit to the vth segment. In this work it is used as linear fit (DFA1). The variance of the detrended time series is calculated for each segment as

$$F^2(v,s) = \frac{1}{s} \left\{ \sum_{i=1}^{s} X_s^2[(v\text{-}1)s+i] \right\}. \tag{1}$$

Averaging over all segments the *mth* order fluctuation can be obtained and

$$F_m(s) = \left\{ \frac{1}{2N_s} \sum_{v=1}^{2N_s} [\, F^2(v,s)\,]^{m/2} \right\}^{1/m}. \tag{2}$$

For $m=2$ the standard DFA procedure is retrieved. If the original series are long-range power-law correlated, the fluctuation function will vary as $F_m(s) \propto s^{h(m)}$ (for large values of s). Note that $h(m)$ is the generalized Hurst exponent and $h(2)$ is the usual exponent H previously mentioned. A multifractal description can also be obtained from considering partitions functions $Z_m(s) = \sum_{v=1}^{N_s} \left| X_{vs} - x_{(v-1)s} \right|^m \propto s^{\tau(m)}$ where $\tau(m)$ is the Renyi exponent (Barabasi & Vicsek, 1991). A linear scaling of $\tau(m)$ with m is characteristic of a monofractal data set, whereas a nonlinear scaling is indicative of multifractal behaviour. The exponent $h(m)$ is related to the Renyi exponent $\tau(m)$ by

$$\tau(m) = m\,h(m) - 1. \tag{3}$$

It is also possible to verify the multifractality degree by defining the ratio $H_< / H_>$ where $H_<$ is the slope of the function $\tau(m)$ *versus* m for $m < 0$ and $H_>$ is the equivalent for $m > 0$. By definition such relation is equal to unity in monofractal signals and a deviation from this value indicates multifractal properties.

3. Results and discussion

The mean global solar radiation at each station is showed in (Fig. 3).

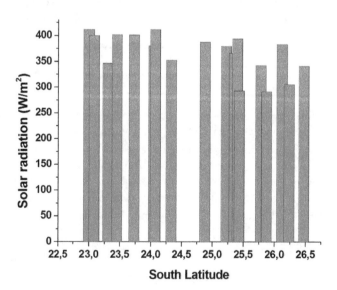

Fig. 3. Latitude dependence of the mean global solar radiation in the Paraná state.

In general, the mean intensity of solar radiation measured on the surface (global solar radiation) at each station decreases with increasing latitude, it is expected (Table 1). The lowest values were recorded for the stations closest to the coast, Curitiba and Guaratuba, not necessarily being of higher latitude. Local characteristics of clouds can affect actual radiation on the surface.

The typical distribution of radiation during the year is shown in Fig. 4. This seasonal behaviour implies a natural correlation, the consequent periodic trends are eliminated being the daily differences, previously discussed in the DFA method. On the other hand, the phenomena ElNiño/LaNiña affect the region, particularly in precipitation and temperatures. The radiation itself is not affected by the phenomenon, but a greater or lesser distribution of clouds would affect the radiation values measured directly on the surface. In a first approach, it is assumed that the presence of clouds have a random effect in the data stream and does not represent an actual correlation.

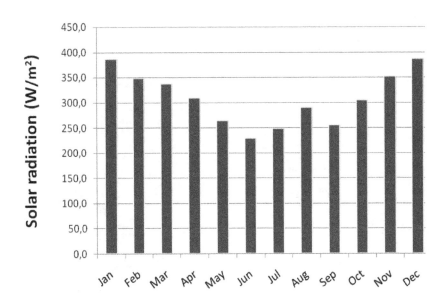

Fig. 4. Distribution of mean global solar radiation during the year, typical of the South Hemisphere.

Fig. 5 shows the global solar radiation series and the profile of cumulative series $X(i)$ for the Curitiba station. To perform the DFA method the lower and upper limits for s values for the time windows were chosen as 4 and $N/4$, where N corresponds to the number of records. The result for Hurst exponent is presented in Fig. 6 for both original and shuffled series. For randomic data series is expected $H=0.5$.

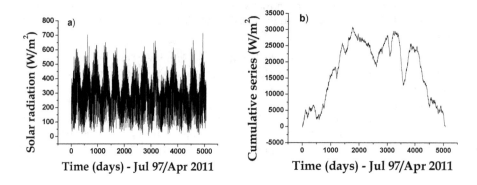

Fig. 5. a) Solar radiation measured on the surface at Curitiba station. b) Cumulative series of daily deviations for the same station.

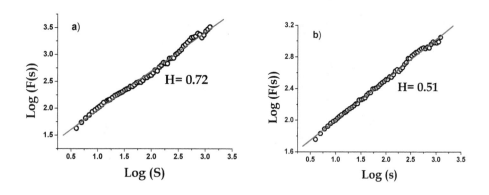

Fig. 6. a) Hurst exponent for Curitiba station. b) The exponent H for the shuffled series.

The H exponent for all studied data series presented value in the range 0.53 to 0.76 with mean value 0.65 (Table 1). Remembering that $H > 0.5$ is related with correlation and persistence the values for it in solar radiation is not surprising. It is reasonable to suppose that the incidence of solar radiation in the next day is not too different from the previous day. No correlation, in a statistical sense, it was found with the exponent H and latitude, altitude and average solar radiation, respectively.

Station	LAT (S)	LONG (W)	ALT (m)	Solar Rad (Wm^{-2})	H	$H_</H_>$
Cambará	23°00′	50°02′	450	352.6	0.66	0.97
Campo Mourão	24°03′	52°22′	601	319	0.68	1.15
Cascavel	24°53′	53°33′	719	331.6	0.65	0.98
Curitiba	25°26′	49°16′	935	256.9	0.72	1.32
Foz do Iguaçu	25°24′	54°37′	232	327.3	0.64	1.38
Guarapuava	25°21′	51°30′	1070	347.8	0.64	1.14
Guaratuba	25°52′	48°34′	0	311.9	0.55	0.88
Guaíra	24°04′	54°15′	227	252.4	0.53	1.23
Lapa	25°46′	49°46′	909	296.8	0.57	0.94
Londrina	23°22′	51°10′	585	281.8	0.76	0.97
Maringá	23°25′	51°59′	570	332.5	0.63	1.24
Palmas	26°29′	51°59′	1100	314.1	0.73	0.80
Pato Branco	26°07′	52°41′	721	341.9	0.68	0.96
Paranavaí	23°04′	52°27′	480	327.9	0.63	1.48
Ponta Grossa	25°05′	50°09′	885	328.1	0.61	0.84
Telêmaco Borba	23°44′	53°17′	768	321.4	0.65	1.35
Umuarama	23°45′	53°19′	480	342.8	0.66	1.24
União da Vitória	26°14′	51°04′	756	255.4	0.76	0.80

Table 1. Location of the meteorological stations, mean value of global solar radiation, Hurst exponent H and the relation $H_</H_>$, the multifractal degree.

The values grouped by frequency and intensity of Hurst exponent occurrences are presented in Fig. 7. Our results present contrasting value with those presented by (Harrouni &, Guessoum, 2009). They indicate high degree of anti-persistence. When the method is applied to temperature data series, the H value obtained is closer to a universal value, however this may depend of geographical position. It is expected that global solar radiation has similar performance regarding the persistence, in this case depending on the distribution of clouds or other elements in the local atmosphere.

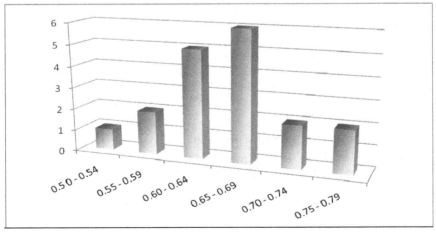

Fig. 7. Frequency of occurrence of Hurst exponents H.

In Table 1 it is also possible to observe the ratio $H_</H_>$ which indicates the deviation from monofractal behaviour. Monofractal signals are characterized by unity in this relation. In this sense, a representative graphic is presented in Fig. 8 were the perfomance of the exponent $\tau(m)$ with different values of m is showed. In general the values for all stations indicate weak multifractality (Table 1). In this sense radiation time series present stationarity. Intrinsic correlations of time series represent the behavior of global solar radiation and, despite the presence of clouds, it does not demand the need for multiple Hurst exponents to describe the time series. Note that negative values of m emphasize on the parts with small fluctuations. For positive values of m the focus is on the parts with large fluctuations. Small deviations of the ratio above or below from unity are related with this characteristic.

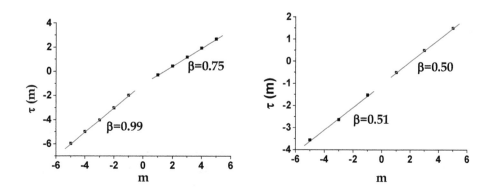

Fig. 8. The multifractal exponent versus several moments m. Different slopes indicate multifractality. At the left is the result for original series from Curitiba station. In the right is the result for the shuffled series.

As described in (Kantelhardt et al., 2002) two different types of multifractality in time series can be identified. In the first case the multifractality of a time series which can be due to the shape of probability density function and the second case, the multifractality which can also be due to different long-range correlations for small and large fluctuations. It is possible to distinguish between these two types of multifractality. To achieve this the corresponding randomly shuffled series was analyzed. The correlations are destroyed by shuffling procedure but dependence of broad distribution function remains, and consequently the multiplicity of exponents is maintained.

Applying the method in the shuffled series of all stations, the ratio $H_</H_>$ approaches unity, indicating a tendency toward monofractal nature. This demonstrates that the multifractality present in the series of solar radiation are due to different long-range correlations of the small and large fluctuations. In Fig. 8 the right graph presents the result for the shuffled series from Curitiba station.

It is interesting to note here the role of clouds or the presence of other particles or aerosols in the atmosphere. The method could be applied to time series related with solar radiation at the top of the atmosphere, for each location, and the series of solar radiation measured on the surface. In this case it would be possible to obtain information about the behaviour of clouds or other elements in the atmosphere at each location. On the other hand, variations of radiation incident on the planet due to periodic variations in solar activity could be detected in longer series.

4. Conclusion

In this work the DFA method was applied to detect long range correlation in data series of daily global solar radiation, measured on the surface, from 18 climatological stations, in Southern Brazil. The Hurst exponent presented mean value 0.65. Results indicate that series exhibit correlation of persistent character. The MF-DFA method was applied to the data set to analyze their scaling properties. Results indicate that the series are multifractal, but in a weak degree. These characteristics means that the processes may be governed by more than one scaling exponent to capture the complex dynamics inherent in the data. However, the low degree of multifractality of these series indicates small effect and points to a possible stationarity in the time series. Furthermore, the shuffled series present monofractal nature, indicating that the multifractal behaviour in the original ones arises from long-range (time) correlations. Both methods are powerful to analyze climatological time series from fluctuations and their statistical behaviour to obtain important information about the nature of the phenomena from its historical data.

5. References

Alvarez-Ramirez, J; Alvarez, J.; Dagdug L.; Rodriguez, E., Echeverria, J. C. (2008). Long-term memory dynamics of continental and oceanic monthly temperatures in the recent 125 years. *Physica A*, Vol. 387, No. 14, pp. 3629-3640

Barabasi, A. L. & Vicsek, T. (1991). Multifractality of self-affine fractals. *Phys. Rev. A*, Vol. 44, pp. 2730-2733

Bunde, A. & Havlin, S. (2002). Power-law persistence in the atmosphere and in the oceans. *Physica A* , Vol. 314, pp. 15-24

Chen X, Lin, G. & Fu, Z. (2007). Long-range correlations in daily relative humidity fluctuations: A new index to characterize the climate regions over China. *Geophys. Res. Lett.*, Vol. 34, L07804

Feng, T.; Fu, Z.; Deng, X. & Mao, J. (2009). A brief description to different multi-fractal behaviors of daily wind speed records over China. *Physics Letters A* , Vol. 373, pp. 4134-4141

Govindan, R. B. & Kantz, H. (2004). Long-term correlations and multifractality in surface wind speed. *Europhys. Lett.* , Vol. 68, 184

Harrouni, S. & Guessoum, A. (2009). Using fractal dimension to quantify long-range persistence in global solar radiation. *Chaos, Solitons & Fractals*, Vol. 41, pp. 1520-1530

Kantelhardt, J. W.; Rybski, D.; Zschiegner, S. A. et al. (2003). Multifractality of river runoff and precipitation: comparison of fluctuation analysis and wavelet methods. *Physica A*. Vol. 330, pp. 240 – 245

Kantelhardt, J. W.; Zschiegner, S. A.; Koscielny-Bunde, E.; Bunde, A.; Havlin, S. & Stanley, H. E. (2002). A Multifractal detrended fluctuation analysis of nonstationary time series. *Physica A* , Vol. 316, pp. 87-114

Kavasseri, R. G. & Nagarajan, R. (2005). A multifractal description of wind speed records. *Chaos, Solitons & Fractals,* Vol. 24, pp. 165-173

Király, A. & Jánosi, I. M. (2005). Detrended fluctuation analysis of daily temperature records: Geographic dependence over Australia. *Meteorol. Atmos. Phys.*, Vol. 88, pp. 119-128

Koçak, K. (2009) Examination of persistence properties of wind speed records using detrended fluctuation analysis. *Energy* , Vol. 34., No. 11, pp. 1980-1985

Koscielny-Bunde, E.; Bunde, A.; Havlin, S.; Roman, H. E., Goldreich, Y. & Schellnhuber, H. J. (1998). Indication of a Universal Persistence Law Governing Atmospheric Variability. *Phys. Rev. Lett.* , Vol. 81, 729-732

Lin, G.; Chen, X. & Fu, Z. (2007). Temporal–spatial diversities of long-range correlation for relative humidity over China. *Physica A*, Vol. 383, No. 2, pp. 585-594

Pedron, I. T. (2010). Correlation and multifractality in climatological time series. *Journal of Physics: Conference Series.* Vol. 246, 012034

Peng, C.-K.; Buldyrev, S. V.; Havlin, S.; Simons, M.; Stanley, H. E. & Goldberger, A. L.(1994). Mosaic organization of DNA nucleotides. *Phys. Rev. E* , Vol. 49, pp. 1685-1689

Rybski, D.; Bunde, A. & von Storch, H. (2008). Long-term memory in 1000-year simulated temperature records. *J. Geophys. Res.* , Vol. 113, D02106

Zhang, Q .; Yu, Z.G.; Xu, C.Y.; Anh, V. (2010). Multifractal analysis of measure representation of flood/drought grade series in the Yangtze Delta, China, during the past millennium and their fractal model simulation. *International Journal of Climatology.* Vol. 30, No. 3, pp. 450-457

9

Surface Albedo Estimation and Variation Characteristics at a Tropical Station

E. B. Babatunde
Covenant University, Ota, Ogun State
Nigeria

1. Introduction

1.1 Albedo

Reflection of radiation is one of the mechanisms by which solar radiation is depleted in the atmosphere and it is mostly done by clouds. By its definition, reflected radiation is lost to space completely(i.e.100%).

Albedo is related to reflection of solar radiation at a surface and therefore defined in terms of it, as the ratio of the reflected solar radiation to the incident solar radiation at the surface, i.e., H_r/H_o,in this chapter. It is assumed however that the reflected radiation, H_r, is both diffuse and specular in nature, that is, it is diffuse if the reflected radiation is uniform or isotropic in all angular directions, and specular if the surface of reflection is smooth with respect to the wavelength of the incident radiation such that the laws of reflection are satisfied (Igbal, 1983). It was said by Gutman (1988) that the observed albedo assumed that the radiation field is isotropic. The extraterrestrial radiation, H_o at the edge of the atmosphere, from the sun, is considered the incident solar radiation.

Albedo is also known as reflectance or reflectivity of a surface; by this, the surface albedo of the Earth is regarded the same as planetary albedo by many scientists (Igbal, 1983). Albedo, as a property of a surface, therefore, can be used to determine the brightness of a surface. According to Prado et al (2005), materials with high albedo and emittance attain low temperature when exposed to solar radiation, and therefore reduce transference of heat to their surroundings. Thus albedo is an important input parameter or quantity in evaluating the total insolation on a building or a solar energy collector. It is also important in the studies dealing with thermal balance in the atmosphere.

Several other definitions of albedo are given based on the source of the reflected radiation; only some are mentioned here.

The reflected radiation measured at several portions of the electromagnetic radiation is used to estimate the spectral surface albedos (Gutman et al,1989); the linear combination of them constitutes the broad band surface albedo (Wydick et al, 1987; Brest et al, 1987; Saunders et al, 1990). The broadband or total wavelengths surface albedo is simply surface albedo. Prado et al (2005), however, gave an encompassing definition of albedo as the specular and diffuse reflectance, integrated over 290 and 2500nm wavelengths range, which corresponds approximately to 95% of the solar radiation that reaches the Earth's surface. The albedos of the individual surfaces on the Earth, such as water, vegetation, snow, sand, surfaces of buildings, dry soil, that of the atmosphere, etc, all constitute the surface or planetry albedo.

We estimated the surface albedo of the earth's surface at Ilorin by using equ.3 to simulate the daily and monthly averages of the shortwave solar radiation reflection, H_r and reflection coefficient, H_r/H_o, and studied the daily and seasonal variation characteristics of H_r/H_o used to define the albedo. This is the objective of this chapter, which is a report from BSRN station in Ilorin, Nigeria.

2. Determination of solar radiation reflection coefficient, H_r/H_o (albedo)

H_r/H_o is a ratio of short wave reflected radiation H_r, towards the space, to the extraterrestrial radiation H_o

incident on the surface of the earth at the edge of the earth' s atmosphere. Here at Ilorin, the location of this work, H_r , the reflected radiation is not measured nor is there a formula in literatures by which it may be predicted or estimated. The apparatus to measure surface albedo or reflected solar radiation is not available here nor in many other under-developed countries. It is therefore determined to produce its data by estimating or simulating it. Therefore the work done on short wave energy balance at the edge of the atmosphere becomes relevant, as it provides a means by which the short wave reflected solar radiation back to space could be estimated. Once the reflected radiation flux is obtained, the solar radiation or short wave radiation reflection co-efficient is easily obtained. It is reasonable to want to know the fraction of the incident radiation H_o is returned back to space on daily, seasonal, and annual basis. Therefore the knowledge of reflection co-efficient, H_r/H_o, used to define albedo, is desirable, and is a very important and relevant radiation parameter in radiative transfer in the atmosphere.

In estimating and studying the characteristics of albedo, global (total) solar radiation H, and diffuse solar radiation H_d, of wavelengths range, mostly from 0.2 to 4.0μm, were used to simulate solar radiation reflection, H_r and the reflection coefficient, H_r/H_o. The radiation fluxes were obtained from the BSRN station, Physics Department, University of Ilorin. The extraterrestrial radiation, H_o, at the top of the atmosphere at Ilorin, computed for year 2000, were used.

The global (total) radiation was measured by Eppley Pyranometer, PSP, with calibration constant of 8.2 x 10⁻⁶ V/Wm⁻² , while the diffuse radiation H_d was measured by the Black and White Eppley Pyranometer model 8-48 with calibration constant 9.18 x 10⁻⁶ V/Wm⁻².

From the measured and computed radiation fluxes, the daily and monthly averages of the fluxes, and the ratios H/H_o, H_d/H and H_r/H_o were computed. Thus, the sw- solar radiation reflection, and total wavelengths reflection co-efficient or reflectance simulated using the data of year 2000 at Ilorin were used for the study. In compliance with the world WRR, sampling rate of 1-second duration of the radiation fluxes was done every minute with integration time of 3-minutes maintained for averaging and recording.

In the work on shortwave solar energy balancing at the edge of the atmosphere carried out in 2003 by Babatunde (2003; 2003), the relation

$$H/H_o + H_a/H_o + H_r/H_o = 1 \tag{1}$$

was used to establish the sw-solar radiation energy balance at the edge of the Earth's atmosphere. H/H_o is the fraction of the extraterrestrial radiation, H_o, transmitted through the atmosphere to the ground surface, and called clearness index (Babatunde and Aro,1995; Udo,2000), H_a/H_o is the fraction absorbed, called the absorption co-efficient or absorbance, and H_r/H_o is the fraction reflected back to space called the reflection co-efficient or reflectance (Babatunde,2003). Further in the work, H_a/H_o was found to be very small in value compared with the other ratios and therefore negligible, i.e. $H_a / H_o << 1$

Hence equation (1) becomes

$$H/H_o + H_r/H_o \approx 1 \qquad (2)$$

From this, an expression for estimating H_r / H_o was obtained as

$$H_r/H_o = 1 - H/H_o \qquad (3)$$

A similar equation was obtained by Babatunde and Aro (1995) for cloudiness index, H_d/H, i.e.

$$H_d/H = 1 - H/H_o \qquad (4)$$

Thus, can these two parameters be said to be twins of the same physical quantity?

Both H_r and H_r/H_o could be and were estimated using eqn.3.

The sw- reflected solar radiation, H_r, is understood to include the reflected radiation from the Earth's surface, reflected radiation back to space by clouds, and the scattered radiation back to space by atmospheric particles and clouds. Reflection, with regards to solar radiation, is redirection of radiation by $180°$ after striking a surface or any atmospheric particle; it is a lost radiation to the space. The fraction,

H_r / H_o, called total wavelengths $(0.2 - 4\mu m)$ reflection co-efficient or reflectance from all the surfaces on the Earth's surface defines generally the Earth's surface albedo (Igbal,1983).The monthly averages of H_r and H_r/H_o produced are shown in columns 8 and 7 respectively of Table 1.

3. Characteristic variation and atmospheric effects on albedo

3.1 Daily variations of H/H_o, H_d/H, H_r/H_o

It is instructive and informative to compare the variations and characteristics of the reflection coefficient H_r/H_o, cloudiness index H_d/H and clearness index H/H_o as done in the graphs in Figs. 1 – 4 for February, April, August and November, months representing four periods of different atmospheric or sky conditions in the year. By the graphs the atmospheric conditions causing the variations could be discerned. The graphs represent daily and unequal fluctuations of the parameters in those months as shown in the figures. The fluctuations in the values of the parameters in turn indicate daily changes in the atmospheric conditions causing the variations.

In the graphs the reflectance, H_r/H_o and the cloudiness index, H_d/H have the same characteristics but show slight differences in magnitudes, while they both have opposite characteristics to H/H_o. Reflection coefficient, H_r/H_o, from this observation, may therefore be interpreted to be a sort of cloudiness index as H_d/H is (Prado et al, 2005), and confirmed by Eqns. 3 and 4.

The magnitude of the cloudiness index could be interpreted to mean the degree of cloudiness or turbidity in the sky and to imply the magnitude of the diffuse radiation in the global, while the magnitude of reflection co-efficient would indicate the degree of brightness of the surface and the amount of reflected radiation back to space.

It could be said by this, that if the sky was relatively cloudless, albedo, or reflection coefficient, H_r/H_o would be relatively small, thus, more radiation would be available to solar energy devices on the earth. On the other hand the variation of H_d/H which was simultaneously significant, was observed to be high in magnitude more than those of the others for the same changes in atmospheric conditions. This implies that it is more sensitive to the atmospheric condition changes responsible for the variations than the others.

Discussing specifically the variation of the parameters in each of the sampled months, and since H and H_r are each a fraction of the same quantity, H_o, it is plausible that, H/H_o and H_r/H_o are compared. In Fig.1, representing variation in February, the daily fluctuation of H/H_o and H_r/H_o were observed not to be significantly big, however the values of H/H_o were bigger than those of H_r/H_o practically throughout the month. This implies that since H is toward the ground surface and H_r is toward the space, the global radiation H, received on the Earth's surface was more than the reflected radiation, H_r lost to space in February at Ilorin.

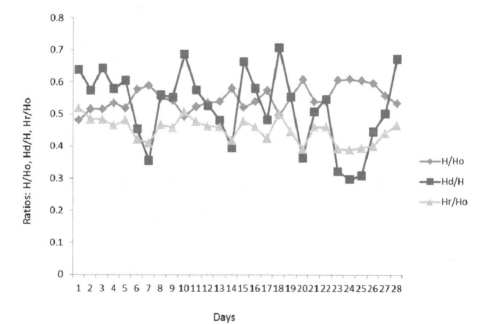

Fig. 1. H/H_o, H_d/H, H_r/H_o for February 2000

Fig.2, presenting the variations or the fluctuations of the parameters in April, indicates very significant variations of the parameters. The high and frequent fluctuations of the parameters could indicate corresponding high, dynamic changes in the atmospheric conditions in the month. Again, H/H_o was higher than H_r/H_o on many days in the month, indicating that more radiation was available on the ground surface than lost to space in reflection. There is however a rise in the value of albedo, i.e. H_r/H_o, observed in this month. This could be due to the presence of some clouds in the sky and heavy hygroscopic particles replacing the harmattan dust particles in the sky.

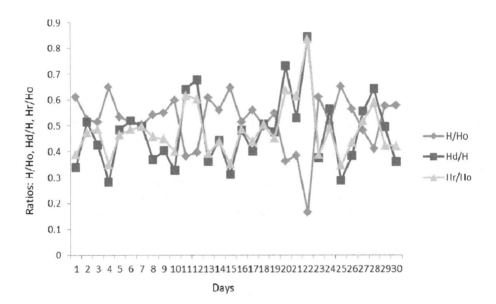

Fig. 2. H/H_o, H_d/H, H_r/H_o Graphs for April 2000

In August, in fig.3, the variations of the parameters were very significant with bigger values of fluctuation. However H_r/H_o and H_d/H were much bigger than H/H_o for almost all the days in the month. This is a reversal of the case in February and November, and which could only imply that more radiation was reflected back to space than received on the Earth's surface. The high values of H_d/H at the period would indicate that the little global radiation received was mostly diffuse radiation. The high values of H_r/H_o, the reflectance, or albedo, would imply high brightness of the Earth's surface toward the space, and low surface temperature of the Earth in this month.

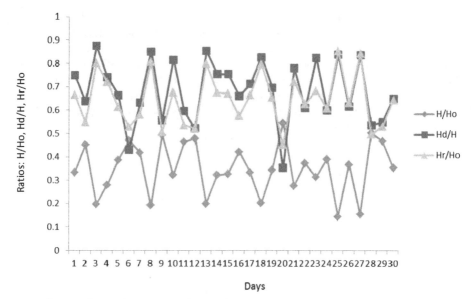

Fig. 3. H/H_o, H_d/H, H_r/H_o Graphs for August 2000

But in November (in Fig.4), the values of H/H_o were much higher than those of H_r/H_o and H_d/H for almost all the days in the month. The low values of H_r/H_o imply that little amount of radiation was reflected back to space, and large amount of radiation was received on the ground surface. They also imply low values of albedo, and therefore less brightness of the surface of the Earth but high surface temperature. All these, and the very low values of H_d/H could indicate that very little amount of diffuse radiation, little or no clouds and little or no dust particles in the sky are the characteristics of the November month in the year.

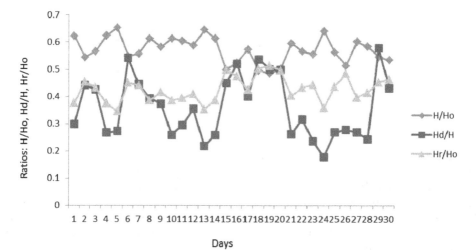

Fig. 4. H/H_o, H_d/H, H_r/H_o Graphs for November 2000

3.2 Monthly average variations of H/H$_o$, H$_d$/H, H$_r$/H$_o$ examined

The graphs of the monthly averages of H$_r$/H$_o$, H$_d$/H and H/H$_o$ in Fig.5, indicate high values of H$_r$/H$_o$ in July, August and September with the highest in August, and relatively low values in October, November, December and January with the lowest in November. Similarly the graphs indicate high values of H$_d$/H in June, July and August with the highest in August, and low values of it in November, December and January with the lowest in November. The result, that the parameters, H$_d$/H and H$_r$/H$_o$, have their highest and lowest values occurring in the same months respectively confirm that the two parameters are twins of the same physical quantity, cloudiness and turbidity, but in the opposite directions, thus answering the question raised earlier on.

The results further show that the high values of H$_r$/H$_o$ in July, August and September, with the highest in August, the peak of rainy season and a predominantly cloudy month, confirm the more, that the reflection of solar radiation by the planet Earth, in this region, is mostly by clouds. The lowest value of the parameter in November confirms also that November is relatively cloudless and dustless, that is, relatively clear and clean (Babatunde and Aro, 1990). A high value of reflectance observed in February, though a relatively cloudless month, would indicate that reflection of radiation at this period is by the dust particles in the atmosphere, and indicates that the atmosphere in February of that year was heavily laden with harmattan dust.

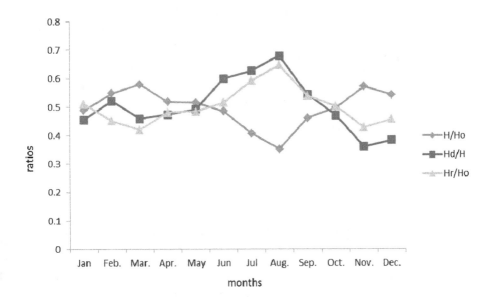

Fig. 5. Monthly average variation of H/H$_0$, H$_d$/H, H$_r$/H$_0$ for year 2000

Mon	H_o	H	Hd	H/H_o	Hd/H	Hr/H_o	Hr
Jan	32.61	16.21	7.24	0.489	0.456	0.511	16.664
Feb	34.85	19.14	9.83	0.549	0.522	0.451	15.717
Mar	37.03	21.44	9.74	0.58	0.459	0.42	15.553
Apr	37.75	18.85	8.79	0.519	0.474	0.481	18.158
May	37.16	19.17	8.90	0.516	0.492	0.484	17.985
Jun	36.51	17.72	8.34	0.485	0.6	0.515	18.803
Jul	36.66	14.95	8.83	0.408	0.627	0.592	21.703
Aug	37.29	13.14	8.46	0.353	0.68	0.647	24.127
Sep	37.11	17.10	9.05	0.461	0.545	0.539	20.002
Oct.	35.44	17.67	8.01	0.498	0.47	0.502	17.791
Nov.	33.11	18.96	6.73	0.572	0.361	0.428	14.171
Dec.	31.18	17.26	6.47	0.543	0.384	0.457	14.537

Table 1. The Monthly Average of the Radiation coefficients and Fluxes in (MJ/m^2 day) for year 2000

3.3 Seasonal variation and sky conditions by H_r/H_o, the albedo

Table 1 and Fig.5 above, present the monthly average values of reflectivity, H_r/H_o, of the Earth and its atmosphere at this location. The reflectance or reflectivity of radiation or albedo property of the contemporary atmosphere is seen to vary from month to month at this location as in any other location on the earth's surface. The seasonal values can be inferred from the monthly average values. An interesting implication of this is that reflectance or albedo could be used as a radiation or atmospheric parameter to determine the sky conditions of a location or region. It may also be used to estimate the surface temperature of the Earth at the location. The following expression, though not very accurate, relates the surface temperature, T of the Earth to its albedo, i.e.

$$T = [(1- a)S/4\sigma]^{1/4} \text{ (McIlveen,1992)}$$

where **a** is the albedo, **S** is the solar constant and σ is the universal Stefan-Boltzman constant. The expression indicates that the temperature T would decrease as albedo increases.

Nigeria, the case study, being in the tropics, experiences two main seasons: dry season and rainy season. While temporal demarcation between them is not rigid, the dry season is from about November to April and the rainy season is from about May to October. The two seasons may be divided into sub- seasons or periods with slightly different weather or atmospheric conditions (Falaiye et al, 2003). For the purpose of determining the sky conditions using seasonal variations of albedo in this work, the two seasons were sub-divided into four divisions. For each period, the representative value of the albedo or reflectivity of the Earth was computed. The sub-divisions are presented in Table 2 below.

Period	Albedo(H_r/H_o)	Sky Conditions
December–March	0.447 ± 0.049	Dry, Cloudless, Dusty
April–May	0.465 ± 0.001	Transition period: dry to rain, small cloudiness, dust clearing.
June–September	0.559 ± 0.065	Rains, cloudiness with low thick clouds, no dust.
October–November	0.404 ± 0.049	Transition period: rain to dry, little clouds, very little or no dust.

Table 2. Sub- Seasons with Albedo Values in Year 2000

However for the two main seasons, the sky conditions parameters are summarized as follow in Table 3.

Season	H/H_0(clearness index)	H_r/H_0(albedo)	H_d/H(cloudiness index)
*[1]DS(Nov.–Apr.)	0.542 ± 0.031	0.435 ± 0.054	0.443 ± 0.055
*[2]RS(May–Oct)	0.454 ± 0.056	0.525 ± 0.071	0.569 ± 0.074

*[1]- Dry Season, *[2] - Rainy Season

Table 3. Seasonal sky conditions parameters for year 2000

For the dry season, the sky is generally and relatively cloudless as indicated by the relatively low average value of albedo, as seen in Table 3. More solar radiation is therefore expected to be available at the Earth's surface at this period, while in the rainy season the albedo is relatively high, and this is attributed to high cloudiness at this period, see Table 3. Hence, relatively little amount of radiation is expected on the Earth's surface, and the surface of the Earth-Atmosphere is expected to be brighter and cooler. A further analysis of these results shows that the sums of the ratios $H/H_o + H_d/H$ and $H/H_o + H_r/H_o$ are each approximately equal to unity, a deduction that these quantities are compliments of each other, in the two seasons. This confirms further that H_d/H and H_r/H_0 are mirror images of one another. They are the same atmospheric or sky condition parameter, cloudiness index.

4. Discussions

The results obtained confirm that the atmospheric constituents responsible for reflecting solar radiation back to space are clouds, aerosols and dust particles of different sizes, of which cloud is the chief (McIlveen,1992). When therefore an atmosphere is clear and clean, that is, cloudless and dustless, the values of H_r/H_o are expected to be relatively small and that of H/H_o to be relatively large. The implication of this is that, when the value of H/H_o is large, most of the radiation on such days is expected to reach the ground surface not deviated and not scattered, and the reflection of radiation to space on such days is expected to be small and mostly from the surface of the Earth, because the atmosphere is cloudless. In general, it can be said that reflection of radiation back to space by the planet would be mostly that of clouds and aerosol in the atmosphere. That is, the shortwave radiation reflection by the Earth's surface alone is comparatively small to that by its atmosphere. Thus it can be safely said that the atmospheric conditions that influence reflection of shortwave radiation back to space most are clouds and aerosol particles, particularly those of molecular size.

High values of reflectivity or reflectance indicate period of low altitude and thick clouds, and rains, dominating the sky. The large albedo values, therefore, in June to September must be mainly due to clouds. The implication of this is that there will be the possibility of poor performance of the solar energy systems, particularly solar concentrating devices, poor fruition of crops and plants and low surface temperature of the Earth during this period as most of the sunlight is sent back to space by reflection. According to the value of the albedo of this period, about 60% of the sunlight that strike the Earth-Atmosphere surface is reflected back, and was not available to solar energy devices for operation.

October to November is a transition period between rainy and dry seasons; it had the lowest average value of albedo of 0.404 (Table 2 above). This indicates about 40% of sunlight being reflected away back to space. This does indicate a period of little or no clouds to reflect radiation, little or no dust to scatter radiation back to space but enhances more sunlight reaching the ground surface; hence performance of solar energy devices is expected to be high, fruition of crops and plants to be enhanced and the Earth's surface temperature is expected to rise (Babatunde et al, 2009).

April-May period is another transition period between the dry and rainy seasons. Changes in the sky conditions were dynamic during this period as the variations of all the parameters were significantly high and frequent. It is therefore relatively cloudy and contained more of hygroscopic particles than dust. The next highest average value of albedo of 0.465 (Table 2 above) was recorded in this period. This value indicates less than half or about half of the sunlight being reflected back. The albedo of the period was however higher than that of the period termed, very dry, cloudless and with high concentration of the harmattan dust, this period is known to be, December to March, a period, with albedo of 0.447(Table 2 above).

This analysis indicates that an atmosphere with low altitude and thick clouds will reflect more radiation than the scattering one, even with large dust concentration.

Since it is possible to use equation 3 to estimate the reflectance of a surface at a location, the values of it, obtainable at Maceio, Brazil (9^0 40'S, 35^0 42'W), of coordinates almost similar to that of Ilorin (8^0 32'N, 4^0 34'E), are compared. It has a value of 0.53 for H/H_o in the rainy season and 0.59 in the dry season. These correspond to, by computation, reflectance or albedo of 0.47 and 0.41 for the rainy and dry seasons respectively (De Sonsa et al, 2005). Brazil is covered with thick rain forest and also in the tropics, with clouds cover most of the time. These albedos are comparable to the ones obtained here at Ilorin. Hence this method of estimating albedo, though simple, may give a reasonable estimation of it at other locations.

5. Conclusion

The daily and monthly variation patterns of the simulated shortwave reflection co-efficient, H_r/H_o, known as albedo, a surface phenomenon, were compared with the corresponding cloudiness index, H_d/H, an atmosphere phenomenon. They were both found to be mirror images of one another. While the shortwave diffuse radiation is toward the Earth's surface, its mirror image, the shortwave reflection radiation is back toward space.

The shortwave solar radiation reflection co-efficient was used to define the Earth's surface albedo which was found to vary daily and monthly in accordance with changes in the atmospheric conditions causing the variations.

The surface albedo according to the analysis, therefore, at Ilorin in year 2000 was found to range between 0.4 and 0.6. These values seem high when compared with the average value of 0.3 obtained for the Earth's surface albedo, but would be considered consistent with the acceptable ones since the values fall within the many possible values averaged statistically to obtain the quoted value for the Earth. The values of albedo can vary from 0 for no reflection to 1 for complete reflection of light striking the surface. For a spot like Ilorin on the surface of the Earth to have values of albedo between 0.4 and 0.6, is not unexpected. Ilorin, in Nigeria, is in the tropical region which is cloudy with different types of clouds most of the time in the year. The atmospheric factors which influence radiation reflection and scattering in the Earth-Atmosphere system most, are clouds and particles, clouds being chief. The Earth and its atmosphere, in this regard, were found most reflective of radiation in August and least reflective in November at Ilorin in year 2000.

6. Summary of the chapter

The expression, $H_r/ H_o = 1 - H/H_o$, developed by Babatunde,(2003) and Babatunde et al,(2003), at Ilorin (8° 30' N, 4° 34' E), Nigeria was used to simulate short-wave (SW) reflected radiation H_r, and reflection coefficient, H_r/H_o. H_r/H_o was used to define total wavelengths surface albedo. The temporal variations of the simulated reflectance, H_r/H_o, the clearness index, H/H_o, and cloudiness index, H_d/H obtained for year 2000 were studied to establish any inter-relationship between them. It was observed, in the relationship between them, that the clearness of the atmosphere characteristically, is diametrically opposite to that of reflectivity of the atmosphere, while the cloudiness and reflectivity of the atmosphere have the same characteristics. It is thus observed, in the effects on solar radiation, while high value of clearness index will enhance the performance of solar energy devices on earth, high value of reflectance will adversely affect it.

The highest reflectance recorded was 0.644 at the peak period of cloud activity in August, and the lowest was 0.361 in November when it was relatively cloudless and dustless. It was deduced that, characteristically, shortwave solar radiation reflection is a mirror image of shortwave diffuse solar radiation, and that reflectance is a sort of cloudiness index.

The albedo deduced from the study, for the Earth–Atmosphere at Ilorin in 2000, ranged between 0.4 and 0.6. These values were consistent with the possible values of albedo of different surfaces on the Earth's surface. The above equation therefore, may be found suitable for estimating surface albedo at any other place on the Earth's surface.

7. References

Babatunde E.B., C.O. Akoshile, O.A. Falaiye, A.A. Willoughby, T.B. Ajibola, I.A. Adimula and T.O. Aro (2009), Observation Bio-Effect of SW-Global Solar Radiation in Ilorin in the tropics. J.Advances in Space Research 43; pp 990-994

Babatunde, E. B (2003): Some solar radiation ratios and their interpretations with regards to radiation transfer in the atmosphere. Nig. J. of Pure and Appl.Sc. In press. Vol. 4

Babatunde, E. B and T. O. Aro, (1990): Characteristic variations of global (total) solar radiation at Ilorin, Nigeria. Nig. J. solar energy, 9,157-173.

Babatunde, E.B and Aro, T.O (1995) Relation between clearness index and cloudiness index. Renewable Energy, 6, 7 801-805.

Babatunde, E.B, Falaiye, A. O. and Afolabi A. B. (2003): Solar Radiation energy balancing at the edge of the Earth's atmosphere. Proc. of regional conference on climate change of Nigerian Meteorological Society (NMS)at Oshodi, Lagos, 11-14 November.

Brest, C and Goward, S.N.(1987) Deriving surface albedo measurements from narrow band satellite data. International Journal of remote sensing, 8, 351-367

De sonsa, J.L, Nicacio, M. and Monra, M.A.L (2005) solar radiation measurements in Maceio, Brazil.Renewable Energy,30, 1203-1220.

Falaiye, O.A., Afolabi, B.A. and Babatunde, E. B(2003)Investigating the effect of ambient temperature on clearness index (K_c) and cloudiness index (K_d).Accepted for publication in the Nig. Met. Journ. (NMS)

Gutman, G., A. Gruber, D. Tarpley, and Taylor. R.(1989):Application of angular models to AVHRR data. for determination of clear-sky planetary albedo over land surfaces. Journal of Geophysical Research 94:99959-9970

Gutman,G. (1988): A simple method for estimating monthly mean albedo of land surfaces from AVHRR data. Journal of Applied Meterology.27 27:973-984

Igbal, M., (1983): An Introduction to solar radiation. Academic Press Toronto. 607-610.

McIlveen,R., (1992):Fundamental of weather and Climate. McGraw Hill

Prado, R.T.A and Ferreira, F.L(2005):Management of albedo and analysis of its influence on the surface temperature of building materials. Energy and Buildings 37, 4,295-301

Sanders, R.W.(1990): The determination of broadband surface albedo from AVHRR visible and near-infrared radiances. International Journal of remote Sensing 11, 49-67

Udo,S.O.(2000): Sky conditions at Ilorin as charcterized by clearness and relative sunshine. Solar Energy. V.69,1,45-53.

Wydick, J.E., Davis and Gruber. A. (1987). Estimation of broad-band planetary albedo from operational narrowband satellite measurements. NOAA Technical report NESDIS 27, U.S. Dept. of Commerce,32 pp

Section 3

Agricultural Application – Bioeffect

10

Solar Radiation Effect on Crop Production

Carlos Campillo, Rafael Fortes and Maria del Henar Prieto
Centro de Investigación finca la Orden-Valdesequera
Spain

1. Introduction

Solar radiation is the set of electromagnetic radiation emitted by the Sun. The Sun behaves almost like a black body which emits energy according to Planck's law at a temperature of 6000 K. The solar radiation ranges goes from infrared to ultraviolet. Not all the radiation reaches Earth's surface, because the ultraviolet wavelengths, that are the shorter wavelengths, are absorbed by gases in the atmosphere, primarily by ozone.

The atmosphere acts as a filter to the bands of solar spectrum, and at its different layers as solar radiation passes through it to the Earth's surface, so that only a fraction of it reaches the surface. The atmosphere absorbs part of the radiation reflects and scatters the rest some directly back to space, and some to the Earth, and then it is irradiated. All of this produce a thermal balance, resulting in radiant equilibrium cycle (figure 1).

EARTH'S ENERGY BUDGET

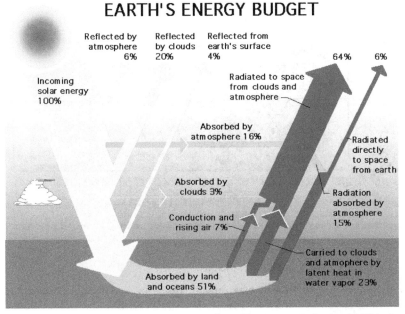

Fig. 1. Effects of clouds on the Earth's Energy Budget. This image is from a NASA site

Depending on the type of radiation, it is known that the 324 Wm⁻² reaching the Earth in the upper atmosphere (1400 Wm⁻² is the solar constant), 236 Wm⁻² are reissued into space infrared radiation, 86 Wm⁻² are reflected by the clouds and 20 Wm⁻² are reflected by the ground as short-wave radiation. But part of the re-emitted energy is absorbed by the atmosphere and returned to the earth surface, causing the "greenhouse effect".

The average energy that reaches the outside edge of the atmosphere from the sun is a fixed amount, called solar constant. The energy contains between the 200 and 4000 nm wavelengths and it is divided into ultraviolet radiation, visible light and infrared radiation.

Ultraviolet radiation: Consists of the shorter wavelengths band (360 nm), it has a lot of energy and interacts with the molecular bonds. These waves are absorbed by the upper atmosphere, especially by the ozone layer.

Visible Light: This radiation band corresponds to the visible area with wavelengths between 360 nm (violet) and 760 nm (red), it has a great influence on living beings.

Infrared radiation: Consists of wavelengths between 760 and 4000 nm, it corresponds to the longer wavelengths and it has little energy associated with it. Its absorption increases molecular agitation, causing the increase of temperature.

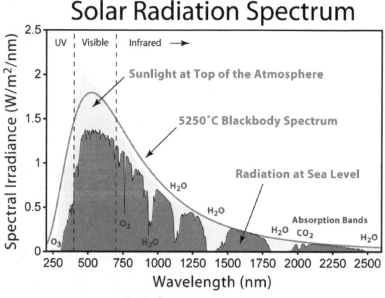

Fig. 2. Spectrum of solar radiation above the atmosphere and sea level. prepared by Robert A. Rohde as part of the Global Warming Art project

Solar radiation on the earth can be classified as:

Direct radiation: This radiation comes directly from the sun without any change in its direction. This type of radiation is characterized by projecting defined shadow onto the objects that intersect.

Diffuse radiation: This radiation comes from all over the atmosphere as a result of reflection and scattering by clouds, particles in the atmosphere, dust, mountains, trees, buildings, the ground itself, and so on. Global radiation: Is the total radiation. It is the sum of the two radiations above. On a clear day with a clear sky, the direct radiation is predominant above the diffuse radiation.

Animals with thermoregulatory abilities and mobility can seek or avoid certain features of current weather. In contrast, terrestrial plants are rooted in place and must accept that the rates of their metabolic processes are determined by the ambient conditions.

Crop communities exert a strong influence over their local microenvironment. Nearly all cropping practices are geared toward, or have the effect of, modifying chemical and physical aspects of that environment (aerial and soils properties).

One of the most important factors that influences plants development is the solar radiation intercepted by the crop. The solar radiation brings energy to the metabolic process of the plants. The principal process is the photosynthetic assimilation that makes synthesize vegetal components from water, CO_2 and the light energy possible. A part of this, energy is used in the evaporation process inside the different organs of the plants, and also in the transpiration through the stomas.

Photosynthesis is a chemical process that converts carbon dioxide into organic compounds, especially sugars, using the energy from sunlight. Depending on how carbon dioxide is fixed the plants can be grouped into three types: C3, C4, and CAM. The C3 plants are the more usual superior plants, which are the temperate weather crops (wheat, barley and sunflower, etc); the C4 category are species from arid weathers or hotter or tropical weathers (corn, sugar or sorghum). The C3 type are generally considered less productive than C4 (figure 3).

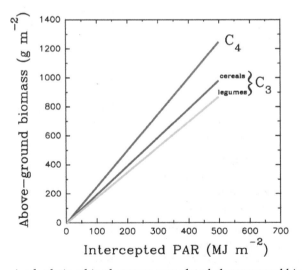

Fig. 3. Typical theorized relationships between cumulated aboveground biomass and cumulated intercepted solar radiation for C4 and C3 species. From Gosse et al. 1986.

One difference lies in the fact that photorespiration is very active in C3 plants. The photorespiration makes plants increase the oxygen consumption when they are illuminated by the sun, and this is very important for agriculture in temperate zones. In a hot day with no wind, the CO_2 concentration in the plant decreases considerably for photosynthesis consumption, therefore, the relationship between carbon and oxygen decreases, and the CO_2 fixation increases the photorespiration.

2. Interception of radiation

In the interception of light (LI) by a canopy, difference between the solar incident radiation and reflected radiation by the soil surface (Villalobos et al., 2002), is a determining factor in crop development and provides the energy needed for fundamental physiological processes such as photosynthesis and transpiration.

Plants intercept direct and diffuse sunlight. The upper leaves receive both types of radiation, while the lower leaves intercept a small portion of direct radiation. Diffuse radiation therefore, becomes more significant in the lower leaves due to radiation transmitted and reflected from the leaves and the soil surface. Solar radiation transmitted by the leaves is predominantly infrared. From a practical point of view, the solar radiation spectrum is divided into regions, each with its own characteristic properties. Appropriate procedures and sensors must be chosen according to the specific objectives of the radiation measurements. Visible radiation, between the wavelengths of 400 and 700 nm, is the most important type from an ecophysiological viewpoint, as it relates to photosynthetically active radiation (PAR). Only 50% of the incident radiation is employed by the plant to perform photosynthesis (Varlet-Gancher et al, 1993). The quantity of radiation intercepted by plant cover is influenced by a series of factors such as leaf angle, the properties of the leaf surface affecting light reflection, the thickness and chlorophyll concentration, which affect the light transmission, the size and shape of the leaf phyllotaxis and vertical stratification, and the elevation of the sun and distribution of direct and diffuse solar radiation. Of the 100% total energy received by the leaf only 5% is converted into carbohydrates for biomass production later. Losses of energy are: by non-absorbed wavelengths: 60%. Reflection and transmission: 8%.Heat dissipation: 8%. Metabolism: 19%.

Of the global radiation incident on the plant canopy only a proportion is used to carry out photosynthesis: PAR (photosynthetic active radiation). The plant's response differs with different wavelengths. Chlorophyll is the main pigment that absorbs the light, other accessory pigments are the b-carotene, red isoprenoid compound which is the precursor of vitamin A in animals and the xanthophyll, a yellow carotenoid.

Essentially the entire visible light is capable of promoting photosynthesis, but the regions from 400 to 500 and 600 to 700 nm are the most effective (figure 4). In addition, pure chlorophyll has a very weak absorption, between 500 and 600 nm. The accessory pigments complement the absorption of light in this region, supplementing the chlorophylls.

-   620-700 nm (red): A greater absorption bands of chlorophyll.
-   510-620 nm (orange, yellow- green); Low photosynthetic activity.
-   380-510 nm (purple, blue and green): Is the most energetic. Strong absorption by chlorophyll.
-   < 380 nm (ultraviolet). Germicides effects, even lethal < 260 nm.

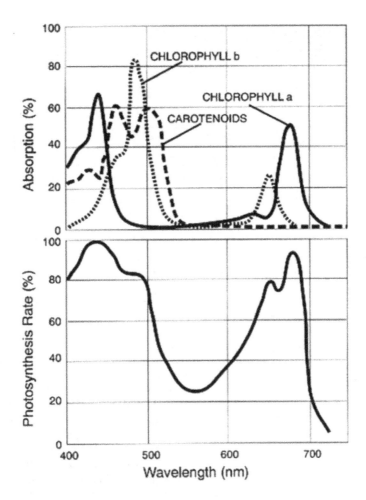

Fig. 4. Typical PAR action spectrum, shown beside absorption spectra for chlorophyll-A, chlorophyll-B, and carotenoids. From Whitmarsh and Govindjee, 1999.

3. Leaf area index

For an efficient use of solar radiation by crop, the great part of the radiation must be absorbed by the photosynthetic tissues. Leaf is the principal photosynthetic functional unit, therefore its efficiency on the capture and use of solar energy determines the vegetable productivity. The area and arrangement of foliage (the canopy architecture), determine the interception of solar radiation (LI) by a crop and the distribution of irradiance among individual leaves (Loomis and Connor, 2002). Leaf area and arrangement change during the life of a crop and, by leaf movement, even during the course of a single day. Maximum crop production requires complete capture of incident solar radiation and can only be achieved with supporting levels of water and nutrients (Loomis and Connor, 2002).

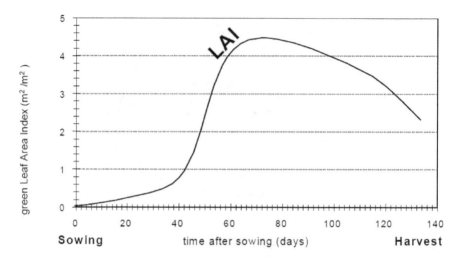

Fig. 5. Typical presentation of the variation in the active (green) Leaf Area Index over the growing season for a maize crop. From Allen et al., 1998

The leaf area index (LAI) is other concept for estimate the crop's ability to capture the light energy. LAI is often treated as a core element of ecological field and modeling studies. LAI is broadly defined as the amount of leaf area (m^2) in a canopy per unit ground area (m^2) Watson (1947). Because it is a dimensionless quantity, LAI can be measured, analyzed and modeled across a range of spatial scales, from individual tree crowns or clusters to whole regions or continents. As a result, LAI has become a central and basic descriptor of vegetation condition in a wide variety of physiological, climatological, and biogeochemical studies. LAI is a key vegetation characteristic needed by the global change research community. For example, LAI is required for scaling between leaf and canopy measurements of water vapour and CO_2 conductance and flux, and for estimates of these variables across the global biosphere–atmosphere interface. Because solar radiation covers the entire surface of the ground, the LAI is a robust measure of leaf area per unit of solar radiation available.

4. Effect of intercepted radiation and leaf area index on growth and crop production

The productivity of a crop depends on the ability of plant cover to intercept the incident radiation, which is a function of leaf area available, the architecture of vegetation cover and conversion efficiency of the energy captured by the plant into biomass. Most production strategies are directed towards maximizing the interception of solar radiation. In the case of crops, this implies adapting agricultural practices in such a way as to obtain complete canopy cover as soon as possible. Deficiencies in water and nutrient inputs may reduce the rate of leaf growth, reducing yield below optimum levels due to insufficient energy capture (Gardner et al., 1985).

The efficiency of interception of PAR depends on the leaf area of the plant population (Varlet-Grancher et al., 1989) as well as on the leaf shape and inclination to the canopy. Gallo & Daughtry (1986) observed that the difference between the intercepted and absorbed PAR, along the maize crop cycle, was lower than 3.5%. According to this, Müller (2001) showed that maize leaves absorb 92% of the intercepted radiation by the canopy. The efficiency of interception of a canopy corresponds to the capacity of the plant population in intercepting the incident solar radiation, which is the main factor influencing the photosynthesis and the transpiration processes (Thorpe, 1978). The efficient crops tend to spend their early growth to expand their leaf area; they make a better use of solar radiation. Agronomic practices, such as fertilization boot, high stocking densities and better spatial arrangement of plants (eg narrow rows) are used to accelerate ground cover and increase light interception.

Solar radiation also has an important role in the processes of evaporation and transpiration. Evaporation takes place mainly from the soil surface and transpiration is the evaporation that occurs across different plant organs, mainly leaves. Because both processes are closely linked, they are often considered together (evapotranspiration); water consumption account, linked to the crop itself, is considered "crop water needs" and is a fundamental aspect in the planning and designing irrigation strategies. Apart from the availability of water in the surface horizons, the cultivated soil evaporation is determined mainly by the fraction of solar radiation reaching the soil surface. This fraction decreases over the growing season, and at the same time the crop canopy cover grows (figure 6). The development of a crop can be divided into four stages (Allen et al., 1998):

Initial Stage: The early growth of individual plants, with little plant-plant competition is very fast. As the LAI develops, there is a shade of lower leaves, so that descriptions of crop growth are based on leaf area depending on the soil surface (Gardner et al., 1985). The water lost during this phase is mainly due to direct soil evaporation.

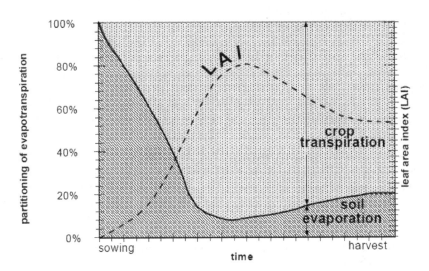

Fig. 6. The partitioning of evapotranspiration into evaporation and transpiration over the growing period for an annual field crop. From Allen et al., 1998

Crop Development Stage: LAI grows exponentially, changing the dominant component of evapotranspiration, predominating evaporation in the initial period and the plant transpiration at the end of the stage. As the leaf area grows, the radiation intercepted by leaves increases. At flowering time, leaf area development ends, with the goal of cultural practices to maximize crop photosynthesis intercepting virtually all of the incoming solar radiation.

Mid-season stage: The late season stage runs from the start of maturity to harvest or full senescence. In the vegetative period radiation interception does not increase, starting from fruit ripening to leaf senescence. (Late season stage).

From the point of view of optimizing the use of irrigation water, it is important to have an accurate estimate of the needs of the plant at any time. All of this will be determined in the development stage, which affect the distribution of solar energy in the process that occurs in the water consumption. Crop conditions (cultural practices, climate, soil, etc) that modify the development of vegetation cover along the life cycle change the water needs of the plant, which would imply a change in the watering schedule when the goal is meet in those needs. There are different procedures to determine the needs of the crop (ETc): the most popular is that proposed by (Doorenbos & Pruitt, 1977) ETc = ETo x Kc[1]

Where ETo is the evapotranspiration reference, (Kc) is the crop coefficient, which varies with the state of crop development and is adapted as the reference evapotranspiration (ETo) for each crop. It is related directly to the LI or the PGC, since it determines the distribution of energy available from plant surfaces and bare soil.

Because the leaf surface is the main photosynthetic organ of the plant, it is sometimes convenient to express the growth per unit leaf area. The rate of accumulation of dry matter per unit leaf area and per unit time is called net assimilation rate (NAR) and is usually expressed in g/m^2 (leaf area) day. The NAR is a measure of average photosynthetic efficiency of leaves in a population. This is high when the plants are small and most of the leaves are exposed to direct sunlight. As the plant grows and the leaf area index increases, the leaves begin to shade, causing a decrease in NAR. For covers with a high LAI, the young leaves at the top take the highest proportion of absorbed radiation, thus having a high rate of CO2 assimilation and also assimilate many other parts translocated. In contrast, the older leaves at the bottom of the cover, which are shaded, have a low rate of assimilation of CO_2 and provide a small assimilation to other parts of the plant.

Under no-stressed environmental conditions, the amount of dry matter produced by a crop is linearly related to the amount of solar radiation, specifically photo synthetically active radiation (PAR), intercepted by the crop. The slope of the regression between biomass and cumulative radiation intercepted by a crop has been used to determine the radiation use efficiency (RUE), which is calculated as the ratio of the biological yield (Kg/ha) to the intercepted PAR (MJ) by the crop plants. Monteith (1977), demonstrated that cumulative seasonal light interception for several crops grown with adequate soil water supply was closely related to biomass production. He formalized and fully established the experimental and theoretical grounds for the relationship (RUE) between accumulated crop dry-matter and solar radiation, arguing that this approach is robust and theoretically appropriate to describe crop growth. RUE is highly dependent on the photosynthetic performance of crop canopies and can be influenced by several factors, namely, extremes temperature, water, and nutrient status. This is indicated by the variation reported in RUE among and within crop species and across locations and growing environments (Subbarao et al 2005). The literature reported quite a large number of RUE values for different crops and locations (Gallagher & Biscoe, 1978; Gosse et al., 1986; Kiniry et al., 1989). Stockle & Kemanian (2009) at intervals showed the value of RUE in g / MJ for large groups of plants: C3 Annuals (1.2-1.7), C4 Annuals (1.7-2.0), C3 Oil crops (1.3-1.6), Legumes (1.0-1.2) and Tuber and root (1.6-1.9). Moreover, the radiation use efficiency (RUE) approach that relates dry mass accumulation to the amount of intercepted PAR (Monteith, 1994; Kiniry, 1999) is widely used to estimate biomass accumulation in horticultural crops, fruit trees and forest (Landsberg & Hingston, 1996; Kiniry et al., 1998; Mariscal et al., 2000).

The efficiency of radiation interception is also influenced by the levels of nutrients in plants, mainly by nitrogen (Dewar, 1996; Scott Green et al., 2003). High crop RUE is directly dependent on obtaining the maximum leaf photosynthetic rate (Sinclair and Horie, 1989; Hammer and Wright, 1993). Nearly 70% of the soluble protein in leaf is concentrated in the carboxylation enzymes (i.e., Rubisco). A positive relationship between leaf nitrogen content per unit area (specific leaf nitrogen) and photosynthetic rates has been reported for a number of crops including wheat, maize, sorghum, rice, soybean, potato, sunflower, peanut, and sugarcane (Muchow & Sinclair, 1994; Sinclair & Shiraiwa, 1993; Sinclair & Horie, 1989; Hammer and Wright, 1993; Evans, 1983; Marshall and Vos, 1991; Giminez, et al 1994; Anten, et al, 1995; Peng, et al, 1994 and Vos & Van Der Putten, 1998 as cited in Subbarao et al 2005). The quantum yield of CO_2 assimilation, which is one of the major determinants of the photosynthetic efficiency of crop canopies, reportedly decreases under N deficiency Meinzer and Zhu, 1998). Levels of photoinhibition also increase under N deficiency (Henley et al., 1991). Thus, a favorable crop nitrogen status appears to be necessary for the realization/expression of maximum RUE in a given crop species. Several studies have reported a positive response of RUE to N fertilization in a number of crops (Muchow & Sinclair, 1994; Hall et al., 1995; Green, 1987). Nitrogen deficiency should decrease the range where there is a linear response between PAR and increased light and thus the range of maximum RUE (Sinclair, 1990; Muchow, 1988). A substantial decrease in RUE under nitrogen stress has been reported for maize (Muchow & Davis, 1988; Muchow, 1994), sorghum (Muchow, 1988), kenaf (Muchow, 1992), wheat (Green, 1987), sunflower (Hall et al., 1995 and Bange et al., 1997), and peanut (Wright et al., 1993). Uhart & Andrade (1995) showed the differences in RUE produced in a crop of corn with nitrogen and without nitrogen, the latter being 40% lower (Figure 7).

The water deficit reduces the interception of solar radiation due to rolling up the leaves (Müller, 2001). If the water deficit is prolonged, the number and size of leaves may be reduced or the total leaf area may decrease, reducing as a result, the interception of radiation (Collinson et al., 1999). Soil water and the resulting plant water status play a key role in determining stomata conductance and canopy photosynthesis. Soil water deficit results in plant water deficits that lead to stomata closure and reduced photosynthesis, and results in loss of photosynthetic efficiency of the canopy and thus to a decrease in RUE (Monteith, 1977). Plants have developed a number of adaptive mechanisms to cope with water deficits to minimize the impact on their productivity (Subbarao et al 1995 and Tunner, 1997).

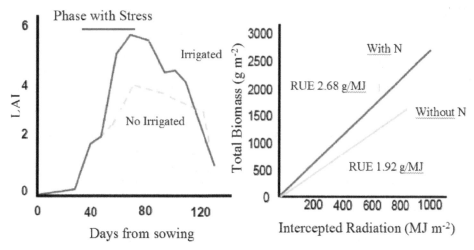

Fig. 7. Effect of water stress and nutrition in two trials in corn, adapted from Otegui (1992) and Uhart & Andrade (1995)

Nearly a 70% decline in RUE due to drought stress was observed in a number of grain legumes (Subbarao et al 2005). Though RUE of C4 crop species is generally higher than that of C3 crop species, the photosynthetic advantage disappears as the water stress increases (Subbarao et al., 2005). When drought stress is imposed from flowering until physiological maturity, a 25% decline in RUE occurred in pigeonpea (Nam et al., 1998). The growth of many field crops can be slowed down or even stopped by a relatively moderate water stress (Boyer, 1970). Stress of this magnitude develops following only a few days without rain, resulting in stomata closure, thus limiting photosynthesis (Sheehy et al., 1975). For rice, wheat, maize, sorghum, and pearl millet, drought stress has been reported to decrease RUE (Gallagher and Biscoe, 1978; Lecoeur & Ney, 1991; Inthapan & Fukai, 1988; Muchow, 1989; Whitfield & Smith, 1989; Robertson and Giunta, 1994; Jamieson et al, 1995 as cited in Subbarao et al 2005). A variety of mechanisms that include leaf movements (that can reduce the radiation load on the canopy when exposed to water deficits) and osmotic adjustment, and root attributes (that can maintain water supply during drought spells) play a major role in maintaining high levels of RUE during water stress (Subbarao et al., 2005). Otegui (1992) compared two maize crops under irrigation and no irrigation during a particular time of cycle, LAI experienced a decrease in cultivation without irrigation (figure 7).

Figure 8 shows a test conducted by the author (unpublished data), processing tomato crop irrigated with two doses. According to crop requirements (T100) and a deficit treatment of 75% of crop needs throughout all crop cycle (T75), LAI measurement and the evolution of dry biomass (aerial biomass) deficit treatment has a lower accumulation of biomass and LAI throughout the crop cycle. This aspect affected the final crop production. Reductions in RUE due to water deficits have been reported by Hughes and Keatinge (1983) and Singh and Sri Rama (1989) in grain legumes. Tesfaye et al., (2006) indicated that dry matter production in grain legumes is highly associated with the fraction of PAR intercepted, which in turn is highly associated with LAI. Li et al. (2008) showed that furrow planting pattern should be used in combination with deficit irrigation to increase the RUE and grain yield of winter wheat in North China. Miralles and Slafer (1997) indicated that post-anthesis RUE appeared to be closely and positively associated and with the number of grains set per unit biomass at anthesis in winter wheat, and Uhart & Andrade (1995) found that stresses reduced the leaf photosynthetic rate and could result in lowering RUE. Whitfield and Smith (1989), Chen et al. (2003), and Li et al. (2008) showed that crop yield was positively related to RUE in winter wheat.

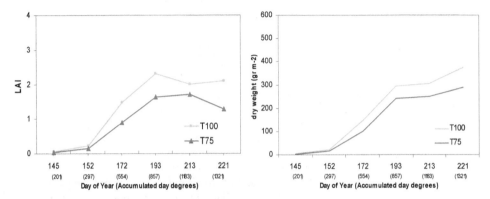

Fig. 8. Effect of water stress in processing tomato crop irrigated with two doses. According to crop requirements (T100) and a deficit treatment 75% of crop needs (T75)

In the modern agricultural research one of the methods in analyzing the crop production along the growth season is simulation by means of the crop production model (Aquacrop, Cropsys, CERES,…); mathematical crop simulation models can quantify the different processes that lead to the yield formation. Once calibrated and validated for a zone, a theoretical harvest with different types of soil management and certain climatic conditions is obtained. The predictive ability of these models can be significantly improved by adjusting the model input data on biomass generated at certain stages of crop development. (Baret et al., 1989; Chistensen & Goudriaan, 1993). Water stress and nutrition reduces LAI for a smaller size and greater leaf senescence. The smaller size of LAI agrees with light capture and thus crop growth, decreasing the efficiency of radiation.

The measurement of the radiation intercepted by a crop for the formation of leaf area is an important factor in monitoring crops, water relations studies, nutrition and crop simulation models. A good measurement of both parameters will be important in studying the effects of solar radiation on crops.

5. Intercepted radiation and leaf area index measurement methods

5.1 Intercepted radiation measurement

Quantifying the intercepted radiation (LI) is therefore an important consideration when studying the different agricultural or environmental factors on yield; it is the main source of data in the most widely used methods for estimating crop water needs.

The LI measurement methods are not necessarily destructive, since the provision of plants on the ground plays a key role. However, there are differences between the different methods in terms of the changes introduced into the covers to make measurements, direct methods and indirect methods.

5.1.1 LI measurement with direct methods

A direct method for determining the percentage of intercepted radiation (LI) is to measure PAR both above and below the canopy at noon on completely cloudless days (Board et al., 1992; Purcell, 2000; Reta-Sánchez y Fowler, 2002):

$$LI = \left| 1 - \left(\frac{PAR \; below \; canopy}{PAR \; above \; canopy} \right) \right| \tag{1}$$

Commercially available lineal PAR sensors are used to take these measurements which are based on PAR values registered by the sensor. These measurements can be taken either by locating sensors perpendicular to the crop rows (Egli, 1994) or by taking multiple measurements parallel to them (Board et al., 1992). The latter method can be costly, according to the number of measurements needed to characterize the study area, especially in the case of low-lying crops, where it may be necessary to remove vegetation in order to place sensors under it, which also has the drawback of introducing alterations during data collection. Using the percentage of shaded soil at solar noon or the percentage of ground cover (PGC) to estimate LI, is an easier and more economical way to obtain the required data. It is generally assumed that the shaded area at soil level corresponds to the fraction of incident radiation which has been intercepted by the crop. This is an approximation that is valid as long as the percentage of light transmission through the leaves is small in comparison to its absorption. The precision with which PGC estimates LI will therefore depend on how well the shaded area is defined and on the capacity of the canopy to capture all of the radiation within the shaded area. In this second case, estimates could be improved by taking complementary measurements of radiation at a sufficient number of points within the shaded area to characterize the radiation traversing the canopy (Lang et al., 1985).

Some of the methods used to determine PGC involve visual estimates (Olmstead et al., 2004; Ortega-Farias et al., 2004). Methods such as the "interception line" (Gallo y Daughtry 1986; Mohillo y Moran, 1991), the analysis the intersection of shadows on metric strips and paper drawings of the sampling areas were used to determine PGC in a non-destructive way (García et al., 2001). However, to apply these last three methods, cloudless days are needed, as a sufficient number of measurements at different orientations are needed to allow a reliable characterization of the area (Ewing y Horton, 1999). The precision of the visual estimation method varies, because it depends on the skill of the operator; results will not be comparable when several people are involved (Olmstead et al., 2004). Furthermore, it has been shown that coverage values tend to be overestimated (Olmstead et al., 2004). In the cases of the interception line and metric strip methods, similar problems are encountered as those associated with the use of PAR bars in the case of low-lying crops and it is difficult to take measurements below the canopy. Finally, making paper drawings is very costly when working under field conditions and when a relatively large area must be characterized.

5.1.2 LI measurement with indirect methods

In indirect methods, different apparatus for estimating the different components of radiation, such as direct radiation, diffuse, land, atmospheric... are used which consider the net radiation balance in order to know how much available radiation reaching the surface.

The inherent difficulties in measuring PAR throughout a canopy and advances in radiometric techniques have led to the development of methods for remotely sensing radiation capture. Radiometric methods rely on differences in the spectral reflectance of vegetation and soil. Vegetative indices based on reflectance in broad wavebands have provided good estimates of radiation capture and yield in crop plants (Gallo et al., 1985; Hatfield et al., 1984). Vegetation indices have also provided good estimates of fractional groundcover (Boissard et al., 1992; White et al., 2000). More recently, spectroradiometers capable of measuring narrow band radiation have been used to monitor plant stress (Elvidge & Chen, 1995). Radiometric satellite data are now available for the evaluation of large areas, and small portable radiometers are becoming less expensive as the technology progresses. In this respect, good results have been obtained with measurements using digital photographic images to determine crop cover and radiation interception in soybean (Purcell, 2000) and lettuce (Klassen et al., 2003), crop cover in turfgrass (Richardson et al., 2001), and canopy and soil cover with straw mulch (Bennet et al., 2000; Beverly, 1996; Olmstead et al., 2004). Other important points are that the area of soil exposed to the sun can be differentiated from that covered by leaves while the angle of the camera is close to that of the sun (Purcell, 2000). With regards to differentiating between the green parts of the crop and the soil surface, results could vary in the case of soils of different colors as a result of their different behavior with respect to the reflection and absorption of radiation; this is particularly the case for different kinds of mulches. In this case, the validity of the method will largely depend on the capacity of the software to discriminate between parts of the crop's green canopy. In the presence of weeds or green cover, it may be necessary to prescreen images.

Digital images offer a series of additional advantages over other methods for estimating LI, assuming that the soil background can be distinguished from leaves, light transmission of leaves is small relative to light absorption, and that the angle of the camera to the horizon approximates the solar angle (Purcell, 2000) such as the direct treatment of images by computers. Moreover, a graphic record of the crop is generated in the case of studies of canopy evolution. This can be used for phonological monitoring (Shelton et al., 1988) to determine differences in color and fertility in maize (Ewing and Horton, 1999) and to study the incidence of pests and diseases.

Automated methods of digital image analysis are indirect methods of LI measurement. Initially they were not widely used because they generally require complex and expensive instrumentation, as well as making mistakes with the changing colors of soil and plant (Hayes & Han, 1993; Van Henten & Bontsema, 1995; Beverly 1996). However, no alteration of vegetation cover and the automation of image analysis has allowed the elimination of many subjective decisions of the observer.

Recent advances in high-resolution digital cameras and associated image manipulation software provide enhanced methods of visual discrimination and computer thresholding that are user-friendly and inexpensive. Three recent studies have demonstrated the accuracy of digital imaging analysis for monitoring plant growth. Paruelo et al. (2000) described a method for estimating aboveground biomass in semiarid grasslands using digitized photographs and a DOS-based program they developed. Purcell (2000) described a method for measuring canopy coverage and light interception in soybean fields using a digital camera and standard imaging software. Richardson et al. (2001) described a digital method for quantifying turfgrass cover following a modified version of Purcell (2000). Klassen 2002 used standard methods of measurement of radiation for comparison with the analysis of vegetation cover as with digital photography using the analysis software Adobe Photoshop 6.0 image. Olmstead et al. (2004) analyzed vegetation cover in grapevine crop through the analysis of digital images, using Sigma Scan Pro 5.0 compared with estimated visualization measures. Other authors used the measurement by digital photography analysis for other uses, Adamsen et al. (1999) to measure maize senescence.

A seemingly key advantage of using digital cameras is that they allow for continuous monitoring of vegetation (White et al., 2000), in the case of low-lying horticultural crops. These measures do not alter the disposition of the crop. Replacing standard procedures, such as the width of cultivation, direct quantification of the shadows or linear PAR sensors, are subjective and costly, and often inaccurate (Campillo et al., 2008).

Taking advantage of the latest developments in digital technology, it is now possible to measure the evolution of vegetation cover through digital photography and to determine the PGC using image interpretation techniques (Campillo et al., 2008; Rodríguez et al., 2000).

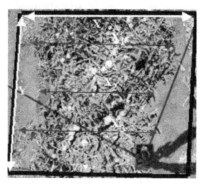

Fig. 9a. Digital images of processing tomato measure with a area method.

Campillo et al (2008), compared LI methodology (PAR) with various methods of PGC measurement. They used three methodologies to measure PGC in two low-lying crops, a winter crop (cauliflower) and a summer crop (processing tomato) in two consecutive years (2005 and 2006) and (2005) in cauliflower crop.

Area method (SA): In this method, crop row width was estimated by simulation based on measurements taken at three points within the marked area using a metric strip. The data were then used to estimate average row width and the PGC (Adams & Arkin, 1977; Giménez, 1985). Both row and frame width were determined in pixels using the measuring tool (IMAGE J 1.33). The sampling area was delimited by the width (X) and length (Y) of the reference frame (Fig. 9a) and the three measurements of row width were: x1, x2, x3. PGC was calculated using the expression:

$$PGC = \left(\frac{\left(\frac{(x_1 + x_2 + x_3)}{3} \right) * Y}{(X * Y)} \right) * 100 \qquad (2)$$

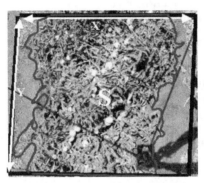

Fig. 9b. Digital images of processing tomato measure with a contour method.

Contour method (SC): In this method, the technique of drawing the crop's shade contour on paper and the subsequent measurement of the area in question is simulated (Kvet & Marshall, 1971). Figure 9b shows the processing of the digital image. To measure the area, the crop's contour was previously delimited using the IMAGE J 1.33 program. Areas with no vegetation cover that were within the canopy were measured and omitted from the surface area count. The crop surface area (S) was measured in pixels using the same program. This area was then related to the sampling area to estimate the PGC according to the following expression:

$$PGC = \left(\frac{S^2}{(X * Y)} \right) * 100 \qquad (3)$$

Fig. 9c. Digital images of processing tomato measure with a reclassification method.

Reclassification method (SR). With this method (Fig. 9c), the crop area (S) is determined by classifying the image according to the range of radiation levels shown on an RGB image of the crop (0 to 255 colors); this was done using a RGB max reclassification tool (GIMP 2.2). After the classification process, it is possible to measure the surface area occupied by green parts (crop) and to differentiate them from the soil or plastic. In contrast to the other two methods, here the crop must be subjected to homogeneous lighting conditions, because the presence of shadows may reduce a crop's color and impede subsequent color reclassification. PGC was calculated according to formula [3].

PGC measurements were compared with measurements made with a LI PAR bar.

Intercepted radiation: LI measurements were made using a 100-cm linear PAR sensor (LICOR Li-190; LI-COR, Lincoln, NE). They were made at solar noon, perpendicular to the crop row, in the same area in which the photographs had been taken. Samples taken from below the crop were compared with reference measurements taken above the crop row (ref). Percentages of LI were calculated by applying Eq. [4], in which it was necessary to know the percentage of radiation that was not intercepted by the crop (RP) as a quotient of the PAR measurements taken both above and below the canopy. According to the degree of plant development two situations for measurement of RP were proposed:

1. When the crop row width was less than 100 cm , RP was calculated by applying Eq. [5] as the average of five measurements taken under the crop (r1, r2, r3 , r4, r5). Measurements were taken every 20 cm using the total length of the PAR bar (100 cm) and adding 50 cm to the reference measurement to include the total width of crop (150 cm). In this situation, ref was measured using the total length of the PAR bar.
2. When the crop row width was greater than 100 cm the maximum length of the PAR sensor, RP was calculated applying Eq. [6] as the average of three measurements taken beneath the crop on each side of the crop row (r1, r2, r3 left side and r4, r5, r6 right side). Measurements were taken at 20-cm intervals using a half-length PAR bar (50 cm).

The sensor was covered with a material that blocks light and average measurements were taken in the center of the row (r7, r8), also using a half-length PAR bar (50 cm). This was done in a way that included the total width of culture (150 cm). In this situation, ref was measured using a half-length PAR bar (50 cm).

$$LI = (1 - RP) * 100 \tag{4}$$

$$RP = \left(\frac{\left(mean(r_1; r_2; r_3; r_4; r_5) + 0.5 * mean(r_{ef}) \right)}{\left(mean(r_{ef}) * 1.5 \right)} \right) \tag{5}$$

$$RP = \left(\frac{\left(mean(r_1; r_2; r_3) + \left(mean(r_4; r_5; r_6) + mean(r_7; r_8) \right) \right)}{\left(mean(r_{ef}) * 3 \right)} \right) \tag{6}$$

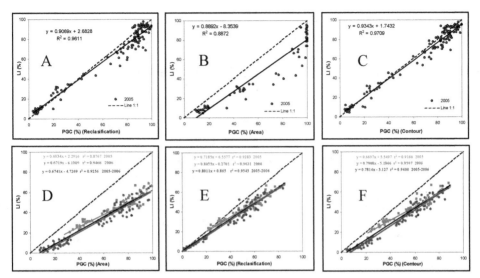

Fig. 10. Relationship between canopy percent light interception (LI) and percentage of groundcover (PGC) determined by the different methods of analysis: area (A,D), contour (B,E), and reclassification (C,F) for the cauliflower in 2005 (A,B,C) and tomato crop in 2005 and 2006 (D,E,F). Values with different letters differ (P < 0.05) between years. From (Campillo et al., 2008)

Figure 10 show that there was a close relationship between the fraction of light intercepted by the canopy at solar noon and estimated PGC for all three methodologies in both years and crops. In all cases, there was a linear adjustment with a significant correlation coefficient (P < 0.01) and an r^2 greater than 0.87. This indicates that any of the described methods would have been valid for estimating the amount of radiation intercepted by the crop.

However, the adjustment was different according to the method used. The adjustment with LI was narrower when using the SR method to estimate PGC (r^2 = 0.92 and 0.96) followed by SC (r^2 = 0.91 and 0.95) and SA (r^2 = 0.87 and 0.94) for 2005 and 2006, respectively in processing tomato and (0.97, 0.96 and 0.89). The relationship between LI and PGC was somewhat stronger when the SR and SC methods were used, whereas the SA method produced greater errors in estimation.

Finally, the most accurate estimate of PGC for both crops was obtained with SR, because color discrimination made it easier to differentiate between vegetation and soil (Fig. 10c). Although the results obtained with SR and SC were similar, it should be borne in mind that the SR method was cheaper to apply because all image processing was performed by software without the need for human definition of the area to be measured. The use of processed digital images with the SR method supposed a considerable improvement with respect to the other two methods and also in the gathering of data directly from the crop. This method was economical and easy to apply. It also eliminated the subjectivity associated with operators having to define areas or points of measurement. Slight but significant differences were found between years applying the same methodology with the adjustment being better in 2006.

5.2 Leaf area index measurement

Determination of LAI is often the most expensive in a field study, because direct measurement (destructive methods) is time-consuming. We can classify in the same way as with LI, in direct methods (which can be destructive or non destructive) and indirect, based on properties of vegetation cover, being non-destructive.

5.2.1 LAI measurement with direct methods

Direct methods for determining leaf area have so far been restricted to the use of an automatic area-integrating meter. Tracing, shadow graphing, and the use of a planimeter to measure the total leaf area attached to shoots are all time-consuming and are tedious approaches; furthermore, in some experiments, there is not enough time to make such measurements (Manivel & Weaver, 1974). All direct methods are similar in that they are difficult, extremely labor-intensive, require many replicates to account for spatial variability in the canopy, and are therefore costly in terms of time and money and also destructive.

5.2.2 LAI measurement with indirect methods

Many indirect methods for measuring LAI have been developed.

Methods based on empirical relationships between leaf area and easily obtainable parameters such as the size of the leaves are available. In any case, the empirical relationship should always be check with direct action as they may vary during the crop cycle and some other varieties. Some used are S = A * L * I and S = A * LB, where S is the area, L the length and I the maximum width plant element, A and B are empirical elements. Also can estimate the leaf area through relationships with the weight. A first group of methods is based on the S = M / σ where M is the leaf weight in grams and σ is the specific weight (g/m2) (Patón et al., 1998). Techniques based on gap-fraction analysis assume that leaf area can be calculated from the canopy transmittance (the fraction of direct solar radiation which penetrates the canopy) (Ford, 1997). Optical methods are indirect, non-contact, and are commonly implemented. They are based on the measurement of light transmission through canopies (Jonckheere et al., 2004). These methods apply the Beer-Lambert law, taking into account the fact that the total amount of radiation intercepted by a canopy layer depends on the incident irradiance, the canopy structure, and its optical properties (Breda, 2003). Monsi & Saeki (1953) expanded the Beer-Lambert extinction law to apply it to plant canopies. The Beer-Lambert law expresses the attenuation of radiation in a homogenous turbid medium. In such a medium, the flux is absorbed in proportion to the optical distance. The LAI is related to the incident solar radiation intercepted by the crop (LI) and extinction coefficient (K), which describes the angle of the blades in relation to the sun, through the formula proposed by Monsi and Saeki (1953):

$$LI = 1 - e^{\left(-(K*LAI)\right)} \tag{7}$$

This approach could also be used to estimate LAI using Eq. [7]; however, we would need to know the extinction coefficient for each crop and variety (Campbell, 1986). Several authors have discussed how to determine k (Hassika et al., 1997; Ledent, 1977; Smith, 1993; Vose et al., 1995) and the accuracy of methodology to be applied (Nel & Wessman, 1993). It is also important to consider that the extinction coefficient also depends on stand structure and canopy architecture (Smith et al., 1991; Turton, 1985) and that the canopy extinction coefficient is a function of wavelength (Jones, 1992), radiation type, and direction (Berbigier & Bonnefond, 1995). It is also important to maximize spatial integration by using large, linear and/or mobile sensors. Extinction coefficient, which varies with species, season and environmental conditions (Hay & Walter, 1989), take values in terms of leaf angles: spherical (0.5-0.7), conical (1), vertical or erectofila (0.3-0.7). The distributions of leaf angles have agronomic and ecological implications. Horizontal distribution implies a high k, allowing for increased intercepted radiation by small plants. the disadvantage is that when the LAI is high the light distribution is very unequal, the lower leaves receive little light, which tends to accelerate senescence. In the opposite, erectofila distribution can be advantageous to intercept radiation when the zenith angle is large (winter, high latitudes) and represents a more homogeneous distribution of radiation when the LAI is high.

This method involves ground-based measurements of total, direct, and/or diffuse radiation transmittance to the forest floor and it makes use of line quantum sensors or radiometers (Pierce and Running, 1988), laser point quadrats (Wilson, 1963), and capacitance sensors (Vickery et al., 1980). These instruments have already proven their value in estimations of LAI for coniferous (Marshall and Waring, 1986; Pierce and Running, 1988) as well as broad-leafed (Chason et al., 1991) stands. In comparison with allometric methods, the approach provides more accurate LAI estimates (Smith et al., 1991). However, the light measurements required to calculate LAI require cloudless skies, and there is generally a need to incorporate a light extinction coefficient that is both site- and species-specific as a result of leaf angle, leaf form, and leaf clumping, etc. (Vose et al., 1995). Measurements can be taken either by locating the sensors perpendicular to the crop rows (Egli, 1994) or by taking multiple measurements parallel to them (Board et al., 1992). This determination can, however, be costly; it depends on the number of measurements needed to characterize the study area, especially in low-lying crops, where vegetation must be moved to place sensors under it, which implies introducing alterations during data collection. There are several commercial systems available to measure indirectly the structure of vegetation and LAI, based on the Beer-Lambert law, including analyzer plant canopy (plant canopy analyzer LiCor LAI-2000) (Li-Cor, 1989); (Cintra et al., 2001; Malone, 2002). El LiCor LAI-2000 has an optical sensor and a control box easily manipulated by an operator. The LAI is estimated according to a model developed by Miller (1967), based on gap-fraction analysis (Barclay et al., 2000). Similar instruments is the CI-100 (Digital plant canopy imager). It consists of a digital camera with a lens of "fish eye" with a 180 degrees field of view.

The analysis of remote estimation methods, provides a temporal and spatial information. The new technologies, provide LAI data from digital cameras (Adamsen et al., 1999), video images (Beverly, 1996), multispectral digital sensors (Bellairs et al., 1996; Shanahan et al., 2001), aerial imagery (Blackmer et al., 1996; Flowers et al., 2001) and satellite images (Wiegand et al., 1979; Thenkabail et al., 1992; Green et al., 1997). One of the remote methods most used is Normalized Difference Vegetation Index (NDVI). These spectral reflectances are themselves ratios of the reflected to the incoming radiation in each spectral band individually, hence they take on values between 0.0 and 1.0. By design, the NDVI itself thus varies between -1.0 and +1.0. It should be noted that NDVI is functionally, but not linearly, equivalent to the simple infrared/red ratio (NIR/VIS). The advantage of NDVI over a simple infrared/red ratio is therefore generally limited to any possible linearity of its functional relationship with vegetation properties (e.g. biomass). This method is sensitive to background soil and weather conditions (Gilabert et al., 1997). There are different satellite sources where one can get the values of NDVI with different resolutions; AVHRR (Advanced Very High Resolution Radiometer), MODIS (Moderate Resolution Imaging Spectroradiometer), SPOT. Vegetation indices are widely used for the calculation of biomass and LAI (Blazquez et al., 1981; Serrano et al., 2000; Wanjura and Hatfield, 1986) and D'Urso, et al (2010) (figure 11)

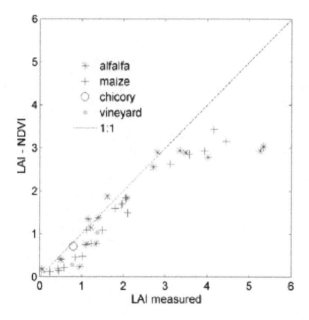

Fig. 11. Correlation between measured LAI using LAI-2000 instrument and estimated LAI from NDVI; RMSE= 0.71, Italian case study (13% of ground measured data used for calibration). From D'Urso, et al 2010.

Campillo et al 2010, compared in two low-lying crops, a winter crop (cauliflower) and a summer crop (processing tomato) in two consecutive years (2005 and 2006) measurements of PGC (non-destructive method) and LAI (destructive method). The objective was relations of two parameters and the possibility of using a PGC methodology as a LAI measurement in vegetable crops.

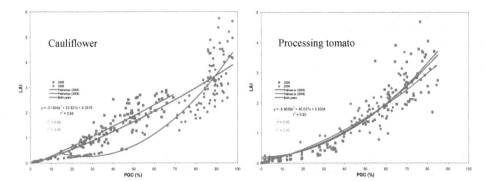

Fig. 12. Relationship between leaf area index (LAI) estimated through the destructive sampling of biomass and the percentage of shaded ground measured (PGC) by the reclassification method for cauliflower and processing tomato crops in 2005 and 2006.

A polynomial relationship ($r^2 > 0.88$) was observed between the two variables in both crops. PGC increased with leaf area development in a curve-linear pattern composed of an initial linear phase followed by a saturating phase, which approached a maximum asymptotic value at full groundcover.

In cauliflower, significant differences were observed between the curves obtained for each year ($r^2 = 0.89$ and 0.95 for the first and second years, respectively, Figure 12). This difference was the result of a significant change in the prevailing weather conditions during the crop cycle that affected the morphology of the leaves. Temperatures in the first year were lower than in the second and frequent frosts caused the curling of leaf margins, resulting in a lower PGC for the same LAI. The PGC–LAI curve adjustments for the tomato crop were significant (Figure 12). The curves coincided for both years, although with differences in the adjustment (0.89 and 0.93 for 2005 and 2006, respectively). In this case, a single curve would have enabled us to estimate the LAI by nondestructive methods using digital images. Although, in principle, the factors that can modify the arrangement of leaves could alter this relationship, this trial included treatments with different water statuses that could have induced changes in plant leaf angle, but this aspect did not affect the goodness of fit. It still remains to be seen how this equation would be influenced by morphological (plant height and leaf type) differences between varieties. The PGC of a crop depends on the leaf area development and on the distribution of the plant leaves on the space (plant architecture). PGC is therefore the dependent variable in the relationship between LAI and PGC. The equations obtained for these two crops are highly significant with a narrow adjustment; they therefore provide a method for estimating LAI based on known PGC values.

From the data obtained when comparing LAI values with those of PGC, and from that obtained by Campillo et al. (2008) relating to PGC as a good estimator of LI, from Eq. [7], we obtained extinction coefficients for growing tomatoes and cauliflower with values ranging between 0.75 and 0.85 and 0.60 and 0.70, respectively. These data are consistent with the value of 0.75 obtained by Heuvelink et al. (2005) for the cultivation of tomato and of 0.55 for growing cauliflower proposed by Olesen & Grevsen (1997). Tei et al. (1996) obtained similar extinction coefficients for other horticultural crops with morphological similarities to cauliflower such as beets and lettuce (0.68 and 0.60, respectively). Campbell (1986) made an overall estimate of extinction coefficients for various crops based on the angle distribution of their leaves; considering average values for crops with leaf angles that were mainly almost horizontal, the values obtained ranged between 0.50 and 0.70.

The method of estimation of LAI was applied on four crops: Tobacco, pepper, soybean and eggplant. The crops chosen for evaluation sought validation of the method in species with very different morphological architectures, both in distribution and area occupied by the plant, as well as the height of it. It turned to be an exponential relationship between the results derived from photography method and leaf area calculated with the planimeter. Linear correlation coefficients obtained for the various crops were 0.89 for eggplant, 0.91 for pepper, soya and 0.87 to 0.88 for tobacco (Figure 13). The correlations obtained in this evaluation were quite heterogeneous despite the morphological disparity and architecture of the different species tested, yielding correlations above 0.87 in all cases. In the case of crops of eggplant, pepper and tobacco, the growth of the canopy crop growth and leaf area occurred at the same time, the highest percentage agreeing with the largest canopy leaf area indices. This dynamic growth was not followed in the case of soybean, where the plant develops its canopy in great haste once covered. The development was apical exponential until reaching a constant height. Therefore, the feasibility of this approach is restricted to the early stages of soybean development, until it reaches the highest percentage of land shaded.

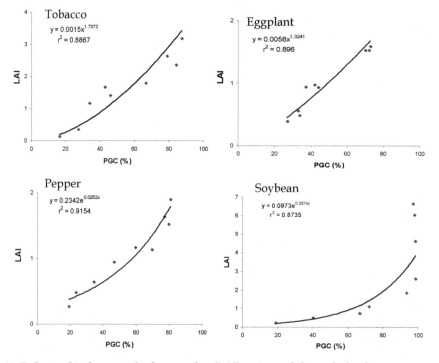

Fig. 13. Relationship between leaf area index (LAI) estimated through the destructive sampling of biomass and the percentage of shaded ground measured (PGC) by the reclassification method for Tobacco, Eggplant, Pepper and Soybean.

6. Conclusions

The productivity of a crop depends on the ability of plant cover to intercept the incident radiation, which is a function of the leaf area available, the architecture of vegetation cover and conversion efficiency of the energy captured by the plant in biomass.

Water stress and nutrition reduce LAI to a smaller size and greater leaf senescence. The smaller size of LAI agrees with light capture and thus crop growth, decreasing the efficiency of radiation.

The measurement of radiation intercepted by a crop for formation of leaf area is an important factor in monitoring crops, water relations studies, and nutrition and in crop simulation models

Measurements taken from digital images exhibit practical advantages with respect to the PAR bar, which must be used at solar noon. In contrast, measurements obtained with a digital camera can be taken at any time of the day and full sunshine is not necessary.

The data obtained by digital photography PGC allow rapid estimation of leaf area using a camera and free software obtaining LAI values in the simplest way than when measured with a planimeter carried out using destructive measurement and individual leaf analysis.

7. Acknowledgment

- INIA proyect RTA04-060-C6-03
- Junta de Extremadura proyect PRI-08A069 and GRU 10130
- Cofinanced by FEDER

8. References

Adams, J.E. & G.F.Arkin. (1977).Alight interception method for measuring row crop ground cover. Soil Sci. Soc. Amer. J. 41:789–792.

Adamsen, FJ; Pinter, PJJ; Barnes, EM; LaMorte, RL; Wall, GW; Leavitt, SW & Kimball, BA. (1999). Measuring wheat senescence with a digital camera. *Crop Science* 39:719-724.

Allen, RG; Pereira, LS; Raes, D & Smith, M. (1998). Crop evapotranspiration - Guidelines for computing crop water requirements. FAO Irrigation and drainage paper 56.

Bange, MP ; Hammer, GL & Rickert, KG. (1997). Effect of specific leaf nitrogen on radiation use efficiency and growth of sunflower. *Crop Sci.*; 37:1201 – 1207.

Barclay, HJ; Trofymow, JA & Leach RI. (2000). Assessing bias from boles in calculating leaf area index in immature Douglas-fir with the LI-COR canopy analyzer. *Agricultural and Forest Meteorology* 100:255-260.

Baret, F.; Guyot, G. & Major, D.J. (1989). Crop biomass evaluation using radiometric measurements. Photogrammetría (PRS), 43: 241- 256.

Becker, P; Erhart, DW & Smith, AP. (1989). Analysis of forest light environments. Part I. Computerized estimación of solar radiation from hemispherical canopy photographs. *Agriculture Forest meteorology* 44(3-4):217-232.

Bellairs, SM; Turner, NC; Hick, PT & Smith, RCG. (1996). Plant and soil influences on estimating biomass of wheat in plant breeding plots using field spectral radiometers. *Australian Journal Agriculture Research* 47:1017-1034.

Beverly, RB. (1996). Video image analysis as nondestructive measure of plant vigor for precision agriculture. *Soil Science and Plant analysis* 27:607-614.

Blackmer, TM; Schepers, JS; Varvel, GE & Meyer GE. (1996). Analysis of aerial photography for nitrogen stress within corn fields. *Agronomy Journal* 88:729-733.

Blazquez, CH; Elliot, RA & Edwards GJ. (1981). Vegetable crop management with remote sensing. *Photogrammetric Engineering and Remote Sensing* 47:543-547.

Boissard, P; Pointel, J & Tranchefort J. (1992). Estimation of the ground cover ratio of a wheat canopy using radiometry. *International Journal Remote Sensing* 13(9):1681-1692.

Boyer JS. (1970). Leaf enlargement and metabolic rates. *Plant Physiol.* 46:233 – 235.

Brougham, RW. (1957). Interception of light by the foliage of pure and mixed stands of pasture plants. *Australian Journal Agriculture Research* 9(1):39-52.

Campbell, GS. (1986). Extinction coefficients for radiation in plant canopies calculated using an ellipsoidal inclination angle distribution. *Agriculture Forest meteorology 36:317-321*

Campillo, C. ; García, M.I. ; Daza, C. & Prieto M.H.(2010). Study of a non-destructive method for estimating the leaf area index in vegetable crops using digital images. Hortscience 45 (10): 1459-1463.

Campillo, C. ; Prieto M.H. ; Daza, C. ; Moñino, M.J. & García, M.I. (2008). Using digital images to characterizing the canopy coverage and light interception in a processing tomato crop. Hortscience 43 (6): 1780-1786.

Cintra, WJ; Ribeiro, FX; Resende, R & Costa LZ. (2001). Comparison of Two Methods for Estimating Leaf Area Index on Common Bean. *Agronomy Journal* 93:989-991.

Coombs, J; Hall, D.O.; Long, S.P. & Scurlock; M.O. (1985). Techniques in bioproductivity and photosyntesis.21-25.

Chen, Y.H.; Yu S.L. & Yu Z.W. (2003): Relationship between amount or distribution of PAR interception and grain output of wheat communities. *Acta Agron. Sin., 29*: 730-734.

Christensen, S. & Goudriaan, J. (1993). Deriving light interception and biomass from spectral reflectance ratio. Remote Sensing of Environment, 43: 87-95.

D'urso, G.; Richter, K.; Calera A.; Osann M.A.; Escadafal R.; Garatuzapajan J.; Hanich I.; Perdigão A.; Tapia J.B. & Vuolo F. (2010). Earth observation products for operational irrigation management in the context of the pleiades project, *Agricultural water management*, 98, 271-282

Doorenbos, J & Pruitt WO. (1977). Guidelines for predicting crop water requirements. FAO Irrigation and Drainage; Rome. p 144.

Egli, D.B. (1994). Mechanisms responsible for soybean yield response to equidistant planting patterns. Agron. J. 86:1046-1049. Evans LT. (1975). *Crop Physilogy.* 1-22 p.

Evans JR. (1983). Nitrogen and photosynthesis in the flag leaf of wheat (Triticum aestivum L.). *Plant Physiol.* 72:297 – 302.

Ewing, RP & Horton R. (1999). Quantitative Color Image Analysis of Agronomic Images. *Agronomy Journal* 91:148-153

Flowers, M; Heiniger, R & Weisz R. (2001). Remote sensing of winter wheat tiller density for early nitrogen application decisions. *Agronomy Journal* 93:783-789.

Ford, DR. (1997). Minimizing errors in LAI estimates from laser-probe inclined-point quadrats. *Field Crops Research* 51:231-240.

Frazer, GW; Fournier, RA; Trofymow, JA & Hall, RJ. (2001). A comparison of digital and film fisheye photography for analysis of forest canopy structure and gap light transmission. *Agricultural and Forest Meteorology* 109:249-263.

Gallagher, JN & Biscoe, PV. (1978). Radiation absorption, growth and yield of cereals. J. *Agric. Sci. Camb.* 91:47 - 60.

Gallo, KP & Daughtry CST. (1986). Techniques for measuring intercepted and absorbed photosynthetically active radiation in corn canopies. *Agronomy Journal* 78:752-756.

Gardner, FP; Parce; R. & Mitchel; R.L. (1985). Carbon fixation by crop canopies. *Physiology of Crop Plants*:31-57.

Gilabert, MA; Gozález-Piqueras, J & García-Haro J. (1997). Acerca de los índices de vegetación. *Revista de teledetección* 8:1-10.

Giménez, C. 1985. Resistencia a sequía de cultivares de girasol bajo condiciones de campo. Tesis Escuela Superior De Ingenieros Agrónomos. Universidad de Cordoba.

Gosse, G ; Varlet-Grancher, C ; Bonhomme, R ; Chartier, M ; Allirand, JM & Lemaire G (1986). Maximum dry matter production and solar radiation intercepted by a canopy. *Agronomie*, 6: 47-56.

Goward, SN; Markham, B; Dye, DG; Dulaney, W & Yang J. (1991). Normalized difference vegetation index measurements from the Advanced Very High Resolution Radiometer. *Remote Sensing Environment* 35:257-277.

Grattan, SR; Bowers, W; Dong, A; Snyder, RL; Carroll, JJ & William G. (1998). New crop coefficients estimate water use of vegetables; row crops. *California Agriculture* 52(1):16-20.

Green, EP; Mumby, PJ; Edwards, AJ; Clark, CD &Ellis AC. (1997). Estimating leaf area index of mangroves from satellite data. *Aquatic Botany* 58:11-19.

Green, CF. (1987). Nitrogen nutrition and wheat in growth in relation to absorbed solar radiation. *Agric. For. Meteorol.* 41:207 - 248.

Han, H.; Li, H.; Ning, T.; Zhang, X.; Shan, Y & Bai, M. (2008). Radiation use efficiency and yield of winter wheat under deficit irrigation in North China. *Plant soil environ.*, 54, 2008 (7): 313-319.

Hall, AJ; Connor, DJ & Sadras VO. (1995). Radiation use efficiency of sunflower crops: effects of specific leaf nitrogen and ontogeny. *Field Crops Res.* 41:65 - 77.

Hay, R; Walker, A. (1989). An introduction to the Physiology of Crop Yield. 292 p.

Hayes, JC & Han, YJ. (1993). Comparison of crop-cover measuring systems. Transactions of the ASAE 36:1727-1732.

Henley, WJ; Levavasseur, G; Franklin, LA; Osmond, CB & Ramus, J. (1991). Photoacclimation and photoinhibition in Ulva rotundata as influenced by nitrogen availability. *Planta* 184:235 - 243.

Inthapan, P & Fukai, S. (1988). Growth and yield of rice cultivars under sprinkler irrigation in South Eastern Queensland. 2. Comparison with maize and grain sorghum under wet and dry conditions. *Aust. J. Exp. Agric.* 28:243-248.

Jamieson, PD; Martin, RJ; Francis, GS & Wilson, DR. (1995). Drought effects on biomass production and radiation use efficiency in barley. *Field Crops Res.* 43: 77–86.

Kiniry, J.R. (1999). Response to questions raised by Sinclair and Muchow. *Field Crops Research* 62:245-247.

Kiniry, J.R.; Landivar, J.A.; Witt, M.; Gerik, T.J.; Cavero, J. & Wade, L.J. (1998). Radiation-use efficiency response to vapor pressure deficit for maize and sorghum. *Field Crops Research* 56:265-270.

Klassen, SP; Ritchie, G; Frantz, JM; Pinnock, D & Bugbee B. (2001). Real-time imaging of ground cover: Relationships with radiation capture; canopy photosynthesis; and daily growth rate. Crop Science Society of America Special Publication on Digital Technologies;19.

Kvet, J. & J.K. Marshall. (1971). Assessments of leaf area and other assimilating plant surfaces. Z.sesta´k jc, and p.g. jarvis, The Hague, The Netherlands. p. 517–574.

Lang, ARG; Yuequin, X & Norman JM. (1985). Crop structure and the penetration of direct sunlight. *Agriculture Forest Meteorology* 35:83-101.

Landsberg, J.J. & F.J. Hingston. (1996). Evaluating a simple radiation/dry matter conversion model using data from Eucalyptus globulus plantations in Western Australia. *Tree Physiology* 16:801-808.

Lecoeur, J & Ney B. (2003). Change with time in potential radiation use efficiency in field pea. *Eur. J. Agron.* 19:91-105.

Li-Cor. (1989). LAI-2000 Plant canopy analizer operating manual. Li-cor; Lincoln;NE. 180 p.

Li, Q.Q.; Chen, Y.H.; Liu, M.Y.; Zhou, X.B; Yu S.L. & Dong B.D. (2008): Effects of irrigation and planting patterns on radiation use efficiency and yield of winter wheat in North China. *Agric. Water Manage.*, 95: 469-476.

Loomis, RS & Connor DJ. (2002). Crop Ecology: Productivity and management in agricultural system. Cambridge University press.

Malone, S; Herbert, D. A.Jr. & Holshouser D.L. (2002). Evaluation of the LAI-2000 Plant Canopy Analyzer to Estimate Leaf Area in Manually Defoliated Soybean. *Agronomy Journal* 94:1012-1019.

Mariscal, M.J.; Orgaz, F. & Villalobos, F.J. (2000). Radiation-use efficiency and dry matter partitioning of a young olive (Olea europaea) orchard. *Tree Physiology* 20:65-72.

Meliá, J & Gilabert MA. (1990). La signatura espectral en teledetección. Aplicaciones a problematicas vegetales; Universidad del País Vasco.

Meinzer, FC & Zhu, J. (1998). Nitrogen stress reduces the efficiency of the C_4 CO_2 concentrating system and, therefore, the quantum yield, in Saccharum (sugarcane) species. *J. Exp. Bot.* 49:1227 – 1234.

Miralles, D.J. & Slafer, G.A. (1997): Radiation interception and radiation use efficiency of near-isogenic wheat lines with different height. *Euphytica, 97*: 201-208.

Monteith, JL. (1977). Climate and the efficiency of crop production in Britain. Philos. *Trans. R. Soc. Lond. Ser.* B 281:277-294.

Monteith, J.L. (1994). Validity of the correlation between intercepted radiation and biomass. *Agricultural and Forest Meteorology* 68:213-220.

Molloy, JM & Moran, CJ. (1991). Compiling a field manual from overhead photographs for estimating crop residue cover. *Soil Use and Management* 7(4):177-182.

Muchow, RC. (1992). Effects of water and nitrogen supply on radiation interception and biomass accumulation of kenaf (Hibiscus cannabinus) in a semi-arid tropical environment. *Field Crops Res.* 28:281 – 293.

Muchow, RC & Davis R. (1988). Effect of nitrogen supply on the comparative productivity of maize and sorghum in a semi-arid tropical environment. II. Radiation interception and biomass accumulation. *Field Crops Res.* 18:17 – 30.

Muchow, RC. (1994). Effect of nitrogen on yield determination in irrigated maize in tropical and subtropical environments. *Field Crops Res.* 38:1 - 13.

Muchow, RC. (1989). Comparative productivity of maize, sorghum and pearl millet in a semiarid tropical environment. 1. Yield potential. *Field Crops Res.* 20:191 - 205.

Nam, NH; Subbarao, GV; Chauhan, YS & Johansen C. (1998). Importance of canopy attributes in determining dry matter accumulation of pigeonpea under contrasting moisture regimes. *Crop Sci.* 38:955 - 961.

Olmstead, MA; Wample, R; Greene, S & Tarara J. (2004). Nondestructive measurement of vegetative cover using digital image analysis. Hortscience 39(1):55-59.

Olmstead, MA; Lang, ARG & Grove GG. (2001). Assessment of severity of powdery mildew infection of sweet cherry leaves by digital image analysis. Hortscience 36(1):107-109.

Ortega-Farías, S; Calderón, R; Martelli, N & Antonioletti R. (2004). Evaluación de un modelo para estimar la radiación neta sobre un cultivo de tomate industrial. *Agricultura técnica* 64:1-11.

Otegui, ME (1992). Incidencia de una sequía alrededor de antesis en el cultivo de maíz. Consumo de agua, producción de materia seca y determinación del rendimiento. Tesis de Magister Scientiae Thesis, UNMdP, Balcarce, Buenos Aires, Argentina, 93 pp.

Paruelo, JM; Lauenroth, WK & Roset PA. (2000). Estimating aboveground plant biomass using a photographic technique. *Journal Rangeland Management* 53(2):190-193.

Patón, D; Núñez-Trujillo, J; Muñoz, A & Tovar J. (1998). Determinación de la fitomasa forrajera de cinco especies del género Cistus procedentes del parque natural de Monfragüe mediante regresiones múltiples. *Archivos de Zootecnía* 47(177):95-105.

Purcell, LC. (2000). Soybean canopy coverage and light interception measurements using digital imagery. *Crop Science* 40:834-837.

Richardson, MD; Karcher DE & Purcell LC. (2001). Quantifying Turfgrass Cover Using Digital Image Analysis. *Crop Science* 41:1884-1888.

Robertson, MJ & Giunta, F. (1994). Responses of spring wheat exposed to pre-anthesis water stress. *Aust. J. Agric. Res.* 45:19 - 35.

Rodríguez, A.; de la Casa, A.; Accietto, R.; Bressanini, L. & Ovando, G. (2000). Determinación del área foliar en papa (Solanumtuberosum L., var. Spunta) por medio de fotografíias digitales conociendo la relación entre el número de pixeles y la altura de adquisición. *Revista brasileira de agrometeorologia* 8:215–221.

Röhrig, M; Stützel, H; Alt, C. (1999). A three-dimensional approach to modeling light interception in heterogeneous canopies. *Agronomy Journal* 91:1024-1032.

Sheehy, J; Green, R & Robson M. (1975). The influence of water stress on the photosynthesis of a simulated sward of perennial ryegrass. *Ann. Bot.* 39:387–401.

Serrano, L; Filella, I & Penuelas J. (2000). Remote sensing of biomass and yield of winter wheat under different nitrogen supplies. *Crop Science* 40:723-731.

Shanahan, JF; Schepers, JS; Francis, D; Varvel, GE; Wilhelm, WW; Tringe, JM; Schlemmer, MR & Major DJ. (2001). Use of remote-sensing imagery to estimate corn grain yield. *Agronomy Journal* 93:583-589.

Sinclair, TR. (1990). Nitrogen influence on the physiology of crop yield. In: Rabbinge R, Goudriaan J, van Keulen H, Penning de Vries T, van Laar HH, eds. Theoretical Production Ecology: Reflections and Prospects. Wageningen: PUDOC, 41 - 55.

Stöckle, C.O. & A.R. Kemanian. (2009). Crop radiation capture and use efficiency: A framework for crop growth analysis. In Crop Physiology (V. Sadras and D. Calderini Eds). Academic Press, Elsevier Inc. p 145-170.

Subbarao, GV; Johansen, C; Slinkard, AE; Nageswara; Rao, RC; Saxena, NP & Chauhan, YS. (1995). Strategies for improving drought resistance in grain legumes. Crit. Rev. Plant Sci. 14:469 - 523.

Subbarao, G.V.; Ito, O. & Berry, W. (2005). Crop Radiation Use Efficiency and Photosynthate Formation-Avenues for Genetic Improvement, M. Pessarakli, Editor, Handbook of photosynthesis, (2nd ed.), Taylor and Francis, New York.

Thenkabail, P; Ward, A; Lyon, J & Van Deventer, P. (1992). Landsat thematic mapper indices for evaluating management and growth caracteristics of soybean and corn. transactions of the ASAE 35:1441-1448.

Turner, NC. (1997). Further progress in crop water relations. Adv. Agron. 58:293 - 337.

Uhart, S.A. & Andrade, F.H. (1995). Nitrogen deficiency in maize. I. Effects on crop growth, development, dry matter partitioning and kernel set. Crop Sci., 35: 1376-1383.

Van Henten, EJ & Bontsema J. (1995). Non-destructive crop measurements by image processing for crop growth control. Journal Agric Engng Res 61:97-105.

Varlet-Grancher, C; Bonhomme, R. & Sinoquet, H. (1993). Crop structure and light microclimate. 518 p.

Villalobos, FJ; Mateos, L; Orgaz, F & Fereres E. (2002). Fitotecnia: Bases y tecnologías de la producción agrícola. Mundi-Prensa; editor. 496 p.

Wanjura, DF & Hatfield JL. (1986). Sensitivity of spectral vegetative indices to crop biomass. transactions of the ASAE 30:810-816.

Watson DJ. 1947. Comparative physiological studies on the growth of field crops: I. Variation in net assimilation rate and leaf area between species and varieties, and with and between years. Annals of Botany. 11:41-76

Welles, JM & Norman JM. (1991). Instrument for indirect measurement of canopy architecture. Agronomy Journal 83:818-825.

White, MA; Asner, GP; Nemani, RR; Privette, JL & Running SW. (2000). Measuring fractional cover and leaf area index in arid ecosystems: digital camera; radiatuon transmittace and laser altimetry methods. Remote Sensing Environment 74(45-57).

Wiegand, CL; Richardson, AJ & Kanemasu ET. (1979). Leaf area index estimates for wheat from LANDSAT and their implications for evapotranspiration and crop modeling. Agronomy Journal 71:336-342.

Whitfield, D.M. & Smith C.J. (1989). Effect of irrigation and nitrogen on growth, light interception and efficiency of light conversion in wheat. Field Crops Res. 20: 279-295.

Whitmarsh & Govindjee. (1999). The photosynthetic process. In: "Concepts in Photobiology: Photosynthesis and Photomorphogenesis", Edited by Singhal, Renger, SK Sopory, K-D Irrgang and Govindjee, Narosa Publishers and Kluwer Academic, pp. 11-51.

Wright, GC; Bell, MJ & Hammer GL. (1993). Leaf nitrogen content and minimum temperate interactions affect radiation-use efficiency in peanut. Crop Sci.33:476-481.

Yonekawa, S; Sakai, N & Kitani O. (1996). Identification of idealized leaf types using simple dimensionless shape factors by image analysis. transactions of the ASAE 39(4):1525-1533.

11

Solar Radiation in Tidal Flat

M. Azizul Moqsud
Kyushu University
Japan

1. Introduction

The Ariake Sea, which is located in the north-western part of Kyushu Island, is one of the best-known semi-closed shallow seas in Japan. Many rivers flow into the eastern coast area of the Ariake Sea and carry 4.4×10^8 kg of sediments per year (Azad et al. 2005).Coarse sediments accumulate in the eastern coast, and fine grains brought by the residual current accumulate in the bay head to form vast tidal flats with fine sediments (Kato and Seguchi 2001). The vast tidal flat mud of the Ariake Sea, which is almost 40% of the total tidal flat area of Japan, is famous for its rich fishery products and *Porphyra* sp. (sea weed) cultivation. Different types of shells like *Sinonovacula constricta, Atrina pectinata* and *Crassostrea gigas* are important creatures in the Ariake tidal mud. However, a dramatic decrease in the catch of these shells is observed in the tidal flat area. From Fig. 1 it is seen that the catch of *Crassostrea gigas* usually living in the near surface mud, dropped from 7.99×10^5 kg in 1976 to only 1.26 $\times 10^5$ kg in 1999; that of *Atrina pectinata,* living in the upper 0.10-0.15 m of the mud, declined from 1.3395×10^7 kg in 1976 to 7.9×10^4 kg in 1999, and the situation in the case of *Sinonovacula constricta,* living in the depth of 0-0.7 m of the mud, was even worse: 1.7×10^5 kg catch in 1976 dropped to practically nil by 1992.

The acid treatment practice for *Porphyra* sp. cultivation is one of the major causes for this declination of the shells as this practice has made the geo-environmental condition of the Ariake tidal mud unfavorable for the living creatures of the tidal mud (Hayashi and Du, 2005, Moqsud et al. 2007). During the period of the cultivation (December -March), the acid (which is mainly organic chemicals) is used as the disinfectant acid to treat the *Porphyra* sp. cultivated in the sea and also to provide some nutrient phosphorus to it.

This organic acid provides ample of foods for the sulphate reducing bacteria living in the mud and consequently increase the sulfide content in the mud. The generation of sulfide is also influenced by the seasonal temperature and shows a higher value during the summer and the late autumn as bacteria becomes more active in the higher temperature. The higher sulfide content created by acid treatment practice is the main reason for the unfavorable condition for the benthos in the Ariake Sea. Moreover, the activities of the benthos depend strongly on the thermal environment near the sediment surface. Photosynthetic capacity of micro phytobenthos on an intertidal flat was strongly influenced by mud surface temperature (Blanchard el. Al, 1997). The filtration rate of bivalves was dependent on the water temperature (Hosokawa et al., 1996). As a result, to evaluate geo-thermal environment is important especially for the acid contaminated Ariake Sea. Thermal properties dictate the storage and movement of heat in soils and as such influence the temperature and heat flux in soils as a function of time and depth (Anandkumar et. al, 2001). In recent years, considerable efforts have gone into developing techniques to determine these properties (Ochsner et al, 2001). The propagation of heat in a soil is governed by its thermal characteristics (De Vries, 1963). Main factors influencing soil thermal properties are mineralogical composition, the organic content and water content (De Vries, 1952, Wierenga et. al, 1969). No study has been carried out before to get the information about the thermal properties as well as thermal environment of the Ariake sea mud. So the objective of this study is to assess the thermal environment of the tidal mud by getting the information of the temperature distribution in different depths and find a diurnal and seasonal profile of it in the tidal flat region, and finally thermal properties variation with respect to depth for the temperature distribution in different seasons. The thermal properties of the Ariake Sea mud collected from both tidal flat and inside the deep sea of the Ariake Sea were conducted as a part of thermal environmental studies of the Ariake Sea.

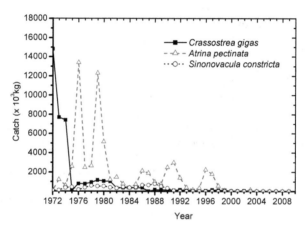

Fig. 1. The graph of catch vs year

2. Sampling sites

Two sampling sites from tidal flat areas, sample 1(S1) and sample 2 (S2) and three sampling sites (sample 3 (S3), sample 4 (S4) and sample 5 (S5)) inside the Ariake Sea were selected as the study areas. Figure 2 shows the locations of the two tidal flat areas (Higashiyoka and Iida) and the three different areas inside the sea, along with the two types (pillar type and float type) of *Porphyra* sp. cultivation areas.

The tidal currents sweep into the sea and move northwards along the eastern shoreline and create a counterclockwise water movement. This would sweep the finer suspended particles delivered by rivers on the east side towards the inland end, where sedimentation would occur. Sediments in the Ariake Sea tidal flats are medium sand to silty mud. Medium sand, which accounts for 71% of the total tidal flats, is located mainly in the east and south coast areas (Azad et al. 2005). The silty mud is mainly in the bay head. Higashiyoka tidal flat located in the bay head was chosen as a study area (S1) which is near to Chikugo River (the biggest river in Kyushu Island), Okinohota River as well as other rivers and thought to be affected by the river waters. Another study area in tidal flat was Iida (S2), which seems to be the most affected by the acid treatment practice. The other three study areas are chosen inside the Ariake Sea where all the time they are under water. The sample 1 and sample 2 (Higashiyoka and Iida) were collected during the ebb tide and the tidal flat was exposed to the sun directly. The other three mud samples (S3, S4, and S5 in Figure 2) were collected from under the sea water at different depths in different locations in the Ariake Sea. The sample collection was done in the last week of April 2006. The typical values of basic physicochemical properties of the mud samples collected from five study areas are tabulated in the Table 1.The mud samples were collected from the 0-0.2 m in the Ariake Sea.

Fig. 2. Map of the Ariake Sea & study area

3. Materials and methods

The vast tidal flat area of the Ariake Sea, which is 40 % of the total tidal area of Japan, is mainly muddy with high water content. The percentage of clay content is much higher than the sand or silt. To evaluate the temperature variation in different depths of the tidal mud, 5 numbers of thermocouples (Tokyo sokki kenkyojo Co., Ltd. Model no. N004853) were installed at 0.10 m, 0.20 m,0.50 m, 1.0 m and 2.0 m depth which were connected with data logger (TDS- 530) to store the continuous hourly data of the temperature at Higashiyoka tidal flat mud. The sensors were placed about 20 m away from the shore line. The data loggers were kept in a watertight box and put in a small ship which was tied with some anchor and moved upward and downward during the high tide and ebb tide, respectively. Every two days the automatically stored data was collected from the data logger in the ship. This field investigation was carried out from 1st April, 2006 to 8th April, 2006 at Higashiyoka tidal flat.

In order to measure the seasonal temperature variation, the data were collected from both Iida and Higashiyoka tidal flat, 20 m away from the shore line during the ebb tide once in every month. By inserting the thermocouple (3 m long and 0.96 cm diameter) vertically into the tidal flat upto 3.0 m depth and at each 0.10 m interval the data was measured. The thermocouple was connected with a battery and a digital display. The temperature data was displayed directly in degree celcius. The mud samples from tidal flat were collected during the ebb tide and about 20 m distance from the shore line. The sample was then sliced into specified layers in the laboratory to measure various properties in each layer. The sulfide content was measured following the standard method prescribed by the Japan fisheries resource conservation association. The instrument which was used to measure the sulfide content is the GASTEC 201L/H which was also used by Wu et al. (2003) to determine the sulfide content of the marine sediments.

In-situ samples were collected by inserting vertically a thin wall steel tube sampler with a diameter of 0.07 m and a length of 0.90 m at five sites. For sample collection from tidal flat region an amphibious ship was used. The mud samples from tidal flat were collected during the ebb tide and about 40 m distance from the shore line. For sample collection from inside the sea, a ship was used. The ship was stopped in the predetermined location which was fixed by the global positioning system (GPS). The diver dived into the sea and collected the mud samples by inserting the steel tube into the sea bed floor and capped the two openings of the tube. The sample was then sliced into 0.05 m layers in the laboratory to measure the thermal properties in each layer.

The thermal properties analyzer KD2 Decagon Devices, Inc. was used to measure the thermal properties. Thermal conductivity and thermal diffusivity were measured directly from the thermal properties analyzer.

Physicochemical Parameters	S1	S2	S3	S4	S5
Density ($\times 10^{-3}$ kg m^{-3})	2.71	2.69	2.68	2.69	2.64
Water content (%)	168	235	160	239	253
Liquid limit w_L(%)	130	150	107	149	142
Plasticity index I_p	73	87	67	89	88
Ignition loss (%)	11.9	13.3	14.4	12.6	13.7
pH	8.03	7.92	7.60	7.53	7.59
ORP (mV)	-40.7	-121.4	128	130	46.38
Acid volatile sulphide ($\times 10^{-3}$kg kg^{-1} dry-mud)	0.16	0.42	0.14	0.30	0.49
Salinity(kg m^{-3})	17	16	20	21	22
Grain size analysis(%)					
Sand	9	7	11	6	6
Silt	36	30	49	46	45
Clay	55	63	36	47	47

Table 1. Basic physicochemical properties of the samples

The volumetric heat capacity was calculated by the relation: Volumetric heat capacity = Thermal conductivity/thermal diffusivity.

4. Results and discussion

4.1 Daily variation of temperature

Figure 3 shows the variation of the tidal mud temperature at different depths from 1st April, 2006 to 8th April; 2006. It is seen that at 0.10 m and 0.20 m depth, the fluctuation of temperature was more prominent. However, from 0.50 m to 2.0 m depth, the diurnal variation was not so prominent. In the sub-surface region, the solar radiation affected the soil temperature more than the deeper part of the tidal mud.

This type of diurnal profile of temperature also agrees with the findings in Baeksu tidal flat in Korea (Yan-k et al. 2005). At 2.0 m depth, the temperature shows higher value than 1.0 m depth. This is probably due to the volumetric heat capacity of the tidal mud and the time lag for absorbing and releasing the heat during the summer and the winter. The peak temperature reached at different times at different depths. During the ebb tide, the time lag to reach the peak at different depths, is more than that at the high tide due to infiltration of sea water in the deeper depth.

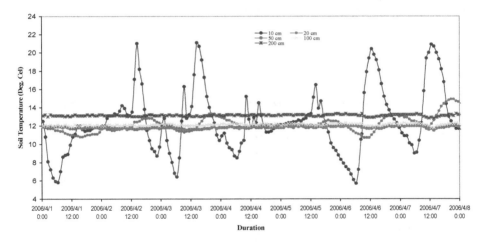

Fig. 3. Variation of diurnal temperature with depth in the Higashiyoka tidal flat

Figure 4 illustrates one day (24h) variation of tidal flat mud temperature influenced by the solar radiation. It is seen that at 0.1 m depth, the peak value was reached when the solar radiation was also at the peak. At night, the temperature did not show any variation both during the ebb tide and the high tide time. This proves that the tidal mud temperature is only influenced by the solar radiation in the subsurface region. The tidal mud temperature at subsequent depths reaches the peak at different times, , with the time lag increasing with depth. The peak temperature was reached about 2:00 PM and the value was about 17 ^{0} C at 0.10 m. The temperature at 0.50 m, 1.0 m and 2.0 m remained almost constant around 12-13 ^{0} C. It is concluded from this Figure that time lag increased with increasing depth but the rate of increasing decreased with the increasing depth. Thermal properties of the tidal flat mud govern this type of phenomenon.

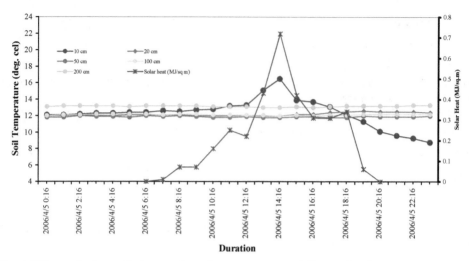

Fig. 4. Effects of solar radiation on the soil temperature in different depths

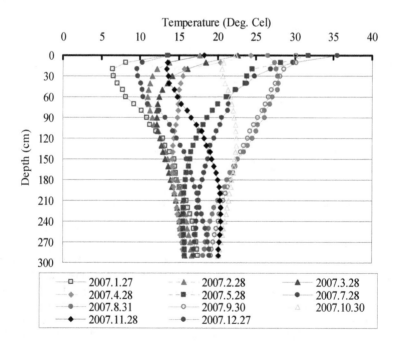

Fig. 5. Seasonal variation of temperature at Higashiyoka tidal flat

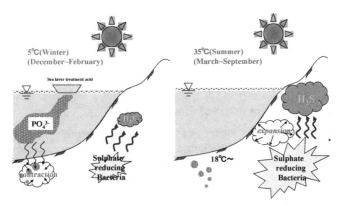

Fig. 6. Seasonal variation of temperature and consequently expansion and contraction tendency and acid treatment practice effects

4.2 Seasonal variation of temperature in tidal flat

Figure 5 shows that the seasonal variation of temperature at different depths at Higashiyoka tidal mud in 2007. During the spring and summer the surface temperature shows a higher value than the subsequent depths. During this time, heat was absorbed by the tidal mud and heat was transferred from the surface to the deeper part of the tidal mud. On the other hand, during winter and autumn the surface temperature was lower than the subsequent depths. During this time heat is released at the surface. During April, the variation was not so prominent. It showed almost straight line graph. Iida site also showed the same trend as with Higashiyoka site during the summer and the winter.

The acid treatment practice started during the winter season (December-February). In winter, the temperature drop down about 5^0 C. Due to the lowering of temperature the tidal flat mud showed a contraction. The acid treatment practice and the contraction tendency of the tidal mud occur during the same time. As a result, the chemicals used in the sea laver treatment agent entered into the tidal mud. On the other hand, during the summer the surface temperature reached about 38^0 C. The high temperature results in an expansion of the tidal mud. The tidal mud expansion causes easy movement of some biogenic gases generated inside the tidal mud.

4.3 Conceptual image of seasonal temperature effects and acid treatment practice on tidal flat

Figure 6 shows the conceptual image of the seasonal temperature variation and the acid treatment practice in the Ariake Sea. The various chemicals which are inside the sea laver treatment medicine enter into the tidal mud during the winter season due to the contraction effect of tidal mud. The chemicals and organic acid supply a lot of foods to the sulfate reducing bacteria. With this ample of foods and the convenient temperature during the spring and summer the sulfate reducing bacteria becomes very active and consequently produces hydrogen sulfide, sulfur di oxide, and as a result, the acid volatile sulfide (AVS) content increased. The AVS content at the Iida tidal flat area shows much higher than the safe limit for the living creatures in the tidal mud. The AVS represents a complex and dynamic biogeochemical system which is not defined simply by analysis of acid volatile sulfide materials (Richard and Morse, 2005). Actually there are many factors which are liable to produce AVS in some specific regions. However, the laboratory test showed that due to the acid treatment practice the AVS value increased in the tidal flat mud (Moqsud et al. 2007). So the conceptual image of acid treatment practice and the seasonal variation of temperature are thought to be rational.

4.4 Proposed mechanisms of pore water movement

Figure 7 illustrates the conceptual image of the pore water movement in the tidal mud due to the seasonal variation of temperature. During spring and summer, the temperature at the shallow depth of the Iida tidal mud of the Ariake sea was higher than that of deeper depth, whereas opposite phenomenon was found during autumn and winter. The temperature gradient in the mud causes pore water to move in the vapor phase from a higher temperature site to a lower temperature site. The vapor condenses at the lower temperature area and becomes water, which increases the total head and drives the water liquid phase from lower temperature site to the higher temperature site (Nassar et al. 2000). Aforementioned process is titled coupled heat-pore water vapor-pore water liquid flow, as shown in Fig. 7.

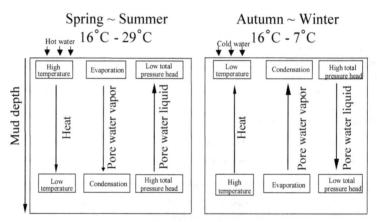

Fig. 7. Proposed concept of coupled heat-pore water vapor-pore water liquid flow in tidal flat

5. Thermal properties

5.1 Thermal conductivity variation with depth

Figure 8 shows the variation of thermal conductivity at different depths in the Ariake sea. In the samples of tidal flats (sample 1 and sample 2), the variation is more prominent than the other samples collected from deep sea. This is probably due to much turbulent in the tidal flat mud in the region and introduces various kinds of matter during the tidal water movement as well as the direct exposure to the sun light during the ebb tide. All the samples show great variations in the sub surface (0-0.20 m) region but less variation in deeper region. Thermal conductivity of mud varies with soil texture, water content and organic matter content (Hamdeh and Reeder, 2000). The water content of the Ariake mud is always over 130% in different depths, which indicates that the conductivity of the Ariake mud is not affected by the water content at different depths.

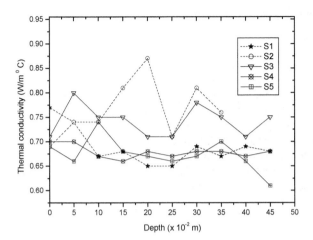

Fig. 8. Variation of thermal conductivity with depth in the Ariake sea

5.2 Thermal diffusivity variation with depth

Figure 9 shows the variation of thermal diffusivity with depth for all the Ariake mud. It is seen that in the tidal flats (sample 1 and sample 2); the thermal diffusivity varied much at the different depths. On the other hand in the case of deep sea mud sample (sample 3, sample 4 and sample 5) the thermal diffusivity was constant at different depths. This is due to a small chance in turbulence in the deep sea bed floor. However, in the tidal flat area, during the low tide, the tidal mud is exposed directly to the sunlight, and during the ebb tide, a lot of foreign matters come and disturb the homogeneity in the mud of the tidal mud layers. It is seen that in the deep sea mud, the value of thermal diffusivity is always in 0.12×10^{-6} m^2/s. In the tidal flat, the peak was reached at 0.13×10^{-6} m^2/s at different depths.

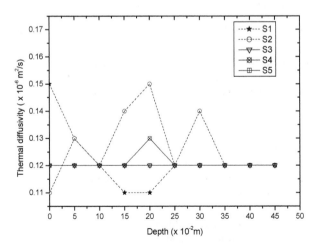

Fig. 9. Variation of thermal diffusivity with depth

5.3 Volumetric heat capacity variation with depth

The volumetric heat capacity of the tidal mud refers to the value which indicates the ability to store heat. If the volumetric heat capacity of a soil is high then the soil is more stable in terms of temperature change or the thermal environment. Figure 10 illustrates the variation of volumetric heat capacity with depth of the various samples. Sample 2 shows a great variation in volumetric heat capacity. The peak shows at 0.35 m depth and value is about 6.3 MJ/m³ °C. Clay soil generally has higher volumetric heat capacity than sandy soil for the same water content and soil density (Hamed, 2003). Volumetric heat capacity is very important for the acid infected tidal mud. Sulphate reducing bacteria (SRB) plays an important role in the geo-environmental condition of the Ariake Sea. These Bacteria like the layer where the volumetric heat capacity is higher (Moqsud et al.2006). Because in that layer it shows the more stable condition which is liked by the bacteria.

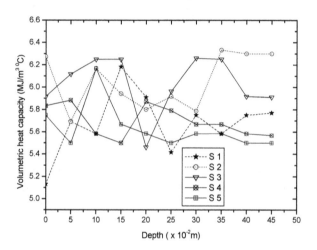

Fig. 10. Variation of volumetric heat capacity with depth

The temperature of underground soil is affected mainly by the soil thermal properties (Nassar et al., 2006) and these properties play a significant role in the geo-environmental condition in the global environment. The thermal properties of the mud are also induced by the mineralogical matter presence in the mud. The effects of this mineral matter on the thermal properties of the Ariake sea mud needs further study.

6. Conclusions

The temperature profiles, which result primarily from the molecular diffusion of heat through the sediment, resemble those typical of field soils and their form is similarly dependent on the thermal properties of the mud and the ambient meteorological conditions. The diurnal temperature variation is more visible near the surface (0.10 m and 0.20 m). The temperature increase gradually from morning, peak at noon and gradually decrease at afternoon. However, at 1.0 m and 2.0 m depth, the variation of temperature was not so prominent. This is due to the volumetric heat capacity and the thermal conductivity of the tidal mud. From the seasonal variation of temperature, it is seen that during late summer, the surface and subsurface temperature is always higher than the deeper depth of the mud while in the winter the opposite phenomenon occurs. The thermal properties of mud collected from tidal flat showed a different trend from the mud collected inside the sea due to the exposure to the sunlight and the tidal wave turbulation in the tidal flat areas. In this study an innovative idea has been adopted to explain the deterioration and the natural remediation of the tidal flat mud in the Ariake Sea. The proposed mechanisms for understanding the transient seepage of pore water liquid of the tidal mud which contributes to the transport of sea laver treatment acid in the tidal mud and also natural remediation of contaminated tidal mud of the Ariake Sea was described clearly. The seasonal variation of temperature and the volumetric heat capacity of the mud have played a significant role for the maintaining the deterioration and natural remediation in the Ariake sea mud.

7. References

Azad, K.A., Ohira, S.I., Oda, M. and Toda, K. (2005). On-site measurements of hydrogen sulphide and sulfur dioxide emissions from tidal flat sediments of the Ariake Sea, Japan. Atmospheric Environment, 39: 6077-6087.

Anandkumar K., Venkatesan R. and Prabha T. "Soil thermal properties at Kalpakkam in coastal south India", Earth Planet Science, Vol.110,No. 3,2001,pp. 239-245.

Ben-Dor,E and Banin A. "Determination of organic matter content in arid-zone soils using a simple loss-on-ignition method. Commun".Soil SciencePlant Analysis, Vol. 20,1989, pp.1675-1695

Blanchard G.F.,Guarini P. and Richard P. "Seasonal effect on the relationship between the photosynthetic capacity of intertidal microphytobenthos and temperature". Journal of Phycol.Vol.33, 1997, pp.723-728

De Vries D A. The thermal conductivity of soil. Meded. Landbouwhogesch, Wageningen, 1952.

De Vries D A. Thermal properties of soils. In: physics of plant Environment. Amsterdam, Holland, 1963.

Ekwue E., Stone R.,Maharaj V. and Bhagwat D.. "Thermal conductivity and diffusivity of four Trinidadian soils as affected by peat content." Transactions of the ASAE. Vol.48, No.5, 2005, pp.1803-1815.

Hamdeh H.N. "Thermal properties of soils as affected by density and water content." Biosystems engineering, Vol.86,No. 1,2003, pp. 97-102.

Hamdeh H.N and Reeder R C. "Soil thermal conductivity: effects of density, moisture, salt concentration, and organic matter". Soil science society of America Journal, Vol.64, 2000, pp. 1285-1290.

Hayashi S. And Du Y.J. (2005). Effect of acid treatment agent of sea laver on geo-environmental properties of tidal flat muds in the Ariake Sea. Journal of ASTM International, 3 (7), 52-59.

Holmer, M and P. Storkholm,. "Sulphate reduction and sulphur cycling in lake sediments: a review." Freshwater Biology, 46,2001,pp. 431-451.

Hosokawa Y, Kiebe E.,Miyoshi Y.,Kuwae T. and Furukawa K.. "Distribution of aerial filtration rate of short-necked clam in coastal tidal flat." The technical note of the port and harbour research institute, Vol. 844, 1996, pp.21 (in Japanese with abstract in English).

Kato, O. and Seguchi, M. (2001). The flow features and forming process of tidal flat in Ariake Sea. A report on the fifth international conference on the environmental management of enclosed coastal seas. EMECS 2001, 6-368.

Moqsud, M.A., Hayashi, S., Du, Y.J. and Suetsugu D. (2007). Impacts of acid treatment practice on geo-environmental conditions in the Ariake Sea, Japan. Journal of Southeast Asian geotechnical society, 38 (2): 79-86.

Moqsud M.A., Hayashi S., Du Y.J., Suetsugu D., Ushihara Y., Tanaka S. and Okuzono K. "Effects of acid treatment on geo-environmental conditions in the Ariake Sea, Japan." The 21st ICSW, March 26-29.,Philadelphia, USA. 2006, pp.436-445.

Moqsud M. A., Hayashi S., Du Y. J., Suetsugu D."Evaluation of thermal properties of the mud of the Ariake Sea." International symposium on lowland technology, September 14-16. (2006 a)Saga, Japan, pp. 263-268.

Nassar IN, Horton R,Flerchinger GN (2000) Simultaneous heat and mass transfer in soil columns exposed to freezing/thawing conditions. Soil Sci. 165 (3): 208-216.

Naidu A. and Singh D.. "A generalized procedure for determining thermal resistivity of soils." International Journal of thermal Sciences. Vol.43, 2004, pp. 43-51.

Ochsner T E., Horton R.,Ren T. "A new perspective on soil thermal properties." Soil science society of America Journal, Vol. 65, 2001, pp.1641-1647.

Rickard, D. and Morse, J.W. (2005). Acid volatile sulfide (AVS). Marine chemistry, 97:141-197.

Sorensen, J. and Jorgenson, B.B. (1987). Early diagnosis in sediments from Danish coastal water: microbial activity and Mn-Fe-S geochemistry. Geochimica et Cosmochimica Acta, 51: 1883-1890.

Wierenga P J., Nielsen D.R and Hagan R M. "Thermal properties of soil based upon field and laboratory measurements". Soil science society of America Journal, Vol.33, 1969, pp. 354-360.

Wu, S.S., Tsutsumi, H., Kita-Tsukamoto, K., Kogure, K., Ohwada, K. and Wada, M. (2003). Visualization of the respiring bacteria in sediments inhabited by *Capitella* sp.1. Fisheries Research, 69:170-175.

Yadav M R and Saxena G S.. "Effect of compaction and moisture content on specific heat and thermal capacity of soils". Journal of Indian society of soil science, Vol.21, 1973,pp. 129

Yadav M R and Saxena G S.. "Effect of compaction and moisture content on specific heat and thermal capacity of soils". Journal of Indian society of soil science, Vol.21, 1973,pp. 129

Yang-Ki C., Kim T., You,K.,Park,L., Moon, H., Lee,S.,Youn, Y., (2005). "Temporal and spatial variability in the sediment temperature on the Baeksu tidal flat,Korea." *Estuarine coastal and shelf science*, Vol. 65, pp. 302-308.

12

Solar Radiation Utilization by Tropical Forage Grasses: Light Interception and Use Efficiency

Roberto Oscar Pereyra Rossiello[1] and Mauro Antonio Homem Antunes[2]
[1]Soil Science Department, Federal Rural University of Rio de Janeiro
[2]Engineering Department, Federal Rural University of Rio de Janeiro
Brazil

1. Introduction

Neotropical grasslands or savannas cover almost half the surface of Africa and large areas of Australia, South America, India and Southeast Asia (FAO, 2010). In South America vegetation like savanna covers over 2.1 million km^2 mainly in central Brazil (also called Cerrado), Colombia and Venezuela. Tropical savannas are currently undergoing rapid and radical transformation due to human interventions on land-use patterns (Lehmann et al., 2009). In the last 30 years about 54 million hectares of native Cerrado vegetation have been replaced by cultivated pasture, mainly African grasses of the genus *Brachiaria* spp. (Boddey et al., 2004, Meirelles et al., 2011). At the same time, the majority of deforested land in the Amazon Basin has become cattle pasture, making forest-to-pasture conversion an important contributor to the carbon and climate dynamics of the region (Asner et al., 2004).The terminology here follows the International Forage and Grazing Terminology Committee (Allen et al., 2011):"the term grassland bridges pastureland and rangeland and may be either a natural or an imposed ecosystem. Grassland has evolved to imply broad interpretation for lands committed to a forage use".

At present it is unclear whether neotropical grasslands are governed by the same factors in Australia, Africa and South America, the three major continental regions of this biome (Lehmann et al., 2009). What is clear is that significant portions of native or cultivated grasslands on every continent have been degraded due to human activities (FAO, 2010). In the tropical regions of Brazil where it is estimated at least half of the cultivated pastures are degraded, the two main drivers of degradation processes are low soil fertility (especially low soil N) and excessively high animal stocking rates (Boddey et al., 2004).On a global scale, overgrazing alone can account for about 7.5 percent of grassland degradation (FAO, 2010).

There is consensus that the sustainability of pastoral ecosystems demands more appropriate livestock management practices. Besides, for the ecological management of these agro-systems, it is necessary to increase the level of understanding of the interactions between its biotic and abiotic components. Central to this understanding, the diverse facets of the interaction between solar radiation and grassland vegetation is of theoretical and practical importance. They are the subject of this chapter, and range from forage grass as monocultures to the complex interrelationships that exists among trees and grasses in silvipastoral agro-systems. In this chapter we will explore a few selected subjects within this broad chain of processes. Recently, focusing on the role of theory in plant science, Woodward (2011) noted that "the development of plant science is based on observations, the development of theories to explain these observations and the testing of these theories."Besides the need of theory to overcome empirical approaches, the theoretical basis is also functional for a better understanding of possibilities and limitations of the new available instrumentation from advances in remote sensing and other technologies for grassland monitoring and assessment. We give emphasis to the concept of sward canopy structure, discussing the central role of Leaf Area Index in the pasture trophic program via light interception. We also give some theoretical and practical emphasis on methodological aspects and procedures for measurements of canopy structure and radiation interception by vegetation. And finally we considered the efficiency use of radiation.

2. Leaf Area Index (LAI) and the G-function: Theoretical considerations

Before talking more specifically about leaf area index and radiation interception as key variables of pasture ecosystems, we need to review some basic concepts about electromagnetic nature of solar radiation underlying the discussion on theoretical and practical aspects of sward canopy structure.

2.1 Some solar radiation concepts

Energy is transferred by electromagnetic waves characterized by wavelength and frequency. The electromagnetic spectrum ranges from high frequency cosmic radiation to low frequency radio waves. For practical purposes the solar spectrum reaching the Earth comprises mostly the ultraviolet, visible and infrared radiation. The spectral regions of the solar spectrum are listed in Table 1.

Spectral region	Spectral range (nm)
Ultraviolet	10 to 400
Visible	400 to 700
Infrared	700 to 4000

Table 1. Spectral ranges of solar radiation with wavelengths in nanometers units $(1nm = 10^{-9}$ m).

The solar infrared radiation beyond 4000 nm that reaches the Earth is insignificant compared to that from 250 to 4000 nm. A few terms regarding solar radiation are considered here. The spectral region from 400 nm to 700 nm is also referred as the *photosynthetically active radiation* (PAR) as this is the region used by plants for photosynthesis. The region from 700 nm to 3000 nm is the reflected infrared, because the surfaces at the environment temperature do not emit in this part of the spectrum and from 1100 nm to 3000 nm is referred as the shortwave infrared.

The corpuscular theory of light is also of interest. It relates the amount of energy of the electromagnetic radiation in Joules (J) to a given wavelength. A quantum of energy is the amount of energy of a photon (a package of discrete energy of a single frequency) and is given by $E = h.v$, where h is the Planck's constant (6.626×10^{-34} J.s) and v is the frequency in Hertz. In the PAR region it is of interest to use the units of mol photon $m^{-2}s^{-1}$. If the sun is the radiation source of the PAR, 1 MJ m^{-2} is equivalent to 4.6 mol photon m^{-2} s^{-1} (Norman & Arkebauer, 1991).

The amount of energy that reaches a surface per unit area per time is called *irradiance* (E) and its unit is Watts per square meter (Wm^{-2}) being $1.0W = 1.0$ Js^{-1}.The *radiance* (L) is the flux density per area and per solid angle in steradian (sr) and has units of Wm^{-2} sr^{-1}. These two are related as $E=\pi L$ where π is the integral of a projected solid angle over the upper hemisphere, in units of sr. The *solar constant* is the amount of energy coming from the sun that reaches the Earth outside the atmosphere for a distance sun-Earth of 1 (one) Astronomical Unit (UA, the average distance between the sun and the Earth). The solar constant was determined by the World Meteorological Organization (WMO) as 1367 Wm^{-2}. From the total radiation at the top of the atmosphere 534.7 Wm^{-2} is in the PAR region.

In order to illustrate the distribution of this radiation we performed calculations for the day 23rd of September 2011 at PESAGRO weather station in Seropédica (22° 45' 28.37"S and 43°41' 5.47"W, Rio de Janeiro State, Brazil), at 13 UT (Universal Time). The sun-Earth distance was 1.0031UA and *solar zenith angle* (SZA, the angle from the sun direction to the vertical direction right above) was 34.4°. Considering the correction for the SZA the actual value of incoming radiation at the top of the atmosphere on a surface parallel to the Earth's surface was 1127.7 Wm^{-2}.Using the 6S model (Vermote et al., 1997) we found that the *total global* (from 250 to 4000 nm, *direct* plus *diffuse*) irradiance on a horizontal surface on the ground was 796.1 Wm^{-2}, which corresponds to 70.6% of the total radiation at the top of the atmosphere. From this amount of radiation reaching the horizontal surface, 356.2 $Wm^{-2}$was in the PAR region. The surface incoming PAR in that condition was 68.9% from direct beam and 31.1% from diffuse sky irradiance. If the surface is perpendicular to the solar beam the incoming direct irradiance in the PAR region was 298.8 Wm^{-2}, i.e., the transmission of PAR direct beam was 67.7%. These calculations using the 6S model considered a clear sky condition, tropical atmospheric model and continental aerosol model, with horizontal visibility of 15 km and target altitude of 34 m.

The incoming radiation on a plant canopy can be reflected, absorbed or transmitted to the soil background. From the conservation of energy it follows that the *incident energy* (Ei_λ) is equal to the sum of the *reflected* (Er_λ), *absorbed* (Ea_λ) and *transmitted* (Et_λ) fluxes:

$$Ei_\lambda = Er_\lambda + Ea_\lambda + Et_\lambda \tag{1}$$

Dividing these by the *incident energy*:

$$1 = \rho_\lambda + \alpha_\lambda + \tau_\lambda \tag{2}$$

where ρ_λ, α_λ and τ_λ are respectively the spectral *reflectance*, *absorbance* and *transmittance*. For remote sensing purposes it is considered the bidirectional reflectance, which is defined as reflectance acquired from a reflected directional flux divided by the directional incoming flux (Nicodemus et al., 1977). This is attained because the radiance that reaches the sensor is directional and under clear sky condition, the incoming flux can be considered a direct radiation flux.

2.2 Canopy structure definitions

According to Campbell & Norman (1989) "plant canopy structure is the spatial arrangement of above-ground organs of plants in a plant community" and as such includes the size, shape and orientation of all aboveground plant organs, making quantitative descriptions exceedingly difficult (Nobel et al., 1993). In practice, the leaf canopy structure of a plant community can be described in terms of *Leaf Area Index* (LAI), *leaf angle distribution* (LAD) and *leaf clumpiness*. The LAI concept was originally introduced by Watson in 1947 and is the one side leaf area per unit area of soil (m^2 leaf m^{-2} ground surface) and can be regarded as the number of leaf layers arranged above the ground. The LAD is the probability density of a leaf being in a certain angle in relation to the horizontal (or the leaf normal to the vertical). Clumpiness is related to how the leaves are distributed in the space. Area index, angle distribution and clumpiness can also be defined for stems (for stem area: SAI; for angle distribution: SAD) or any other aerial parts of the plant. Theoretical functions have been used in the literature to describe LAD for most situations present in nature, like those introduced by Wit (1965). It was expanded by Bunnik (1978) by adding the uniform and spherical types of LAD, which can be seen in Figure 1. Here a uniform distribution in azimuth is assumed, which means that the normal to the leaves are random regarding to the azimuth.

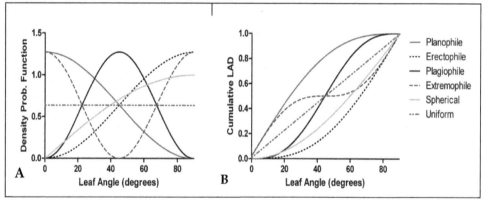

Fig. 1. Theoretical leaf angle distribution (LAD) of Bunnik (1978) for canopies. A) Density probability functions of leaf angle. B) Cumulative LAD.

The LAD represents the strategies that plants use to intercept radiation for plant processes. A planophile plant has most of the leaves near the horizontal and has a small variation of intercepted radiation for a range of solar zenith angles below 45 degrees. Erectophile canopies on the other hand, have large variations in interception and the gap fraction (fraction of openings in the canopy) is more frequent from close to nadir viewing. This way less radiation is intercepted when the sun is close to the zenith. Plagiophile canopies have most of the leaves around 45 degrees and conversely the extremophile canopies have most of the leaves around 0 and 90 degrees (respectively horizontal and vertical). The radiation interception by the extremophile canopy can lead to different results for the plant if the horizontal and vertical leaves are in different canopy layers.

The extinction of solar radiation by interception was initially described by Monsi & Saeki (1953, in English translation 2005) for horizontal leaf layers and generalized by Ross & Nilson in 1966 (Ross, 1981) to include leaf orientations other than horizontal. The rate of change of a downward direct flux of radiation (I_D) with the cumulative LAI at any height from the top of canopy is given by:

$$dI_D / dLAI = - \lambda_0 K_D I_D \tag{3}$$

where λ_0 is the leaf spatial distribution parameter, which is equal to 1 for leaf elements randomly distributed and leaf position is independent of other leaf positions, also referred as a Poisson canopy (Nilson, 1971, Baret et al., 1993). K_D is the *interception coefficient*, and is given by (Ross, 1981):

$$K_D = G(\theta_i, \theta_l, \varphi_i, \varphi_l) / \cos\theta_i \tag{4}$$

where $G(\theta_i, \theta_l, \varphi_i, \varphi_l)$ is the G-function, θ_i is the solar zenith angle, φ_i is the solar azimuth, φ_l is the leaf azimuth in relation to the solar azimuth, and θ_l is the leaf normal angle from the zenith direction, i.e., a vertical leaf has $\theta_l = \pi/2$ and a horizontal leaf has $\theta_l = 0$. The G-function represents the projected fraction of leaf area in the solar direction and is equal to the average value of the cosine of the angle between solar zenith angle and the leaf normal angle (θ_l) from the vertical.

Assuming that leaves have a random azimuth orientation (Ross, 1981), meaning that normal to the leaves are random regarding to the azimuth, the G-function is calculated as (Ross, 1981, Antunes, 1997):

$$G(\theta_i, \theta_l, \varphi_i, \varphi_l) = \int_0^{\frac{\pi}{2}} g'(\theta_l) \int_0^{2\pi} \frac{|\cos\delta|}{2\pi} d\theta_l d\varphi_l \tag{5}$$

where $g'(\theta_l)$ is the density probability of leaf angle between 0 and $\pi/2$ and δ is the angle between leaf normal and the sun. The cosine of δ is calculated as (Ross, 1981):

$$\cos\delta = \cos\theta_i\cos\theta_l - \sin\theta_i \sin\theta_l\cos\varphi_l \tag{6}$$

The spherical LAD is of special interest because the leaves are arranged in such a way in the canopy that the leaves from one square meter of soil can fit a surface of a sphere with the same area as the LAI. As a result, at any solar zenith angle, the fraction of leaf area projected towards the sun is always the same, which is 0.5. This means that the G-function for such a canopy is always 0.5 regardless of the sun's orientation. Maize canopy LAD has been measured in the field and found to be spherical (Antunes et al., 2001). Plants with a spherical LAD intercept the same amount of radiation regardless the direction of the solar beam and can be regarded as a good characteristic for a high productivity canopy.

Equation 3 can be solved by integrating for an entire layer of canopy yielding:

$$I_D = I_0\exp(-\lambda_0 G(\theta_i, \theta_l, \varphi_i, \varphi_l) LAI / \cos\theta_i) \tag{7}$$

where I_0 is the direct beam flux intensity at the top of the canopy, which can be set to be the fraction of direct beam above the canopy. In this equation the term $G(\theta_i,\theta_l,\varphi_i,\varphi_l) LAI/\cos\theta_i$ defines the mean number of contacts of the direct beam with the canopy elements (Nilson, 1971). Although the Equation 7 is similar to the Bouguer law (also referred as Beer's law) for radiation transmission in a turbid medium, in which K is the attenuation coefficient, the concept as applied here is different from Beer's law, since this equation defines only the amount (or fraction) of direct beam left after passing through a canopy layer with a defined leaf area, LAD and clumpiness.

2.3 Direct and indirect methods for determining canopy structure

Canopy structure can be determined by direct methods or estimated through indirect methods. The direct methods involve the measurement of leaf area, angle and position of leaves in the canopy or by destructive sampling. A group of methods involve the measurement of leaf area with optical area meter devices, scanners, hand-held planimeters or weighing of paper replicates (Nobel et al., 1993, Asner et al., 2003). Generally these methods allow the computation, separately, of the form, size and number of leaves (Bréda, 2003). Another way to obtain leaf area is by correlating either green or dry biomass to leaf area to find a conversion factor i.e. the *specific leaf area* (m^2 kg^{-1}). All these methods give a standard reference for the calibration or evaluation of indirect methods of determining LAI (Bréda, 2003) or vegetation cover (Schut & Ketelaars, 2003). On the other hand leaf angle distribution and leaf clumpiness determination are restricted to on site observations as they need to be carried out in undisturbed canopies.

Direct methods are cumbersome and time demanding. As a result, the indirect methods have been largely used for estimating canopy structure. These methods use the relationship with other more easily measurable parameters, such as canopy transmittance, green cover estimate or correlation with dry or green biomass. Gap-fraction inversion is amply used to estimate LAI (Welles & Norman, 1991). Gaps are obtained from devices specially designed for this (e.g., LAI-2000[*1]) or hemispherical photographs (e.g., CID 110[*1]). The LAI-2000 finds the gap fraction at five angles, making it possible to estimate also the average leaf angle. Ceptometers are also used to estimate LAI through gap fraction (López-Lozano et al., 2009). Inversion of light transmitted through the canopy, measured by a line quantum sensor, has also been used to estimate plant area index, a composition of leaves and stems of the plants (Cohen et al., 1997). Hemispherical photographs are usually taken looking upward (Rich, 1990) and the digital processing allows the estimation not only the total gap fraction but also to partition these gaps in different angles. However, the indirect methods based on light extinction through the canopy can be affected by canopy structure assumptions, exceptionally regarding the clumping of leaves (Larsen & Kershaw, 1996).

A sensitivity analysis of the LAI estimation using the LAI-2000, a ceptometer (AccuPAR[*1]) and hemispherical photographs was carried out by Garrigues et al. (2008), over 10 crops including pasture grasses. They found that the hemispherical photographs were the most robust technique, from many standpoints of evaluation, the least sensible to illumination conditions and thus can be applied for a large range of canopy structures.

2.4 Remote sensing of forage grasses

The use of remotely sensed data for monitoring forage grasses is of great interest due to its possibility of monitoring large areas and the broad range of sensors available, with varying

[*1] The mention of a trademark does not constitute an endorsement by the Federal Rural University of Rio de Janeiro of these products and does not imply approval for the exclusion of other suitable products.

spatial, spectral, temporal and radiometric resolutions. Remote sensing is also cost effective for monitoring large areas as it drastically reduces field work compared to *in situ* samplings. The rationale behind the use of remote sensing for vegetation monitoring relies on the fact that leaf reflectance varies throughout the solar spectrum (Knipling, 1970). The combination of leaf properties with the canopy structure and sun-viewing orientation lead to complex interactions that causes varying reflectance values at each spectral region, e.g., see Ollinger (2011) for a review.

Grass canopy biophysical properties and the illumination/viewing geometry affect reflectance observed at surface level (Walter-Shea et al, 1992). At the satellite level the reflectance values calculated from the radiance reaching the sensor undergoes the effects from the atmosphere, soil and litter background, canopy structure, bidirectional anisotropy, spatial heterogeneity, nonlinear mixing and topography (Myneni et al., 1995). These effects make it much more complex to establish fixed relationships between canopy parameters (like LAI) and canopy reflectance. Despite this limitation, remote sensing has been largely used for monitoring and canopy parameter estimation. Two approaches that have been largely used to estimate canopy parameters from remotely sensed data, involve the use of vegetation indices and inversion of canopy radiative transfer models.

2.4.1 Vegetation indices

Vegetation indices are band combinations of remotely sensed data that have some relation with canopy parameters. The first introduced was the *simple ratio vegetation index* (SR or RVI) between the reflectance of near infrared band (ρ_{NIR}) and the reflectance of the red band (ρ_{red}). Many other indices have been introduced after the SR. The idea behind the indices was that they could remove soil background effects from remotely sensed data while keeping the high sensitivity of NIR reflectance to canopy LAI. The commonly used vegetation index is the *normalized difference vegetation index* (NDVI), which is given by:

$$NDVI = \frac{\rho_{NIR} - \rho_{red}}{\rho_{NIR} + \rho_{red}} \tag{8}$$

One index that has been used is the *soil adjusted vegetation index* (SAVI), which is calculated through the following equation (Huete, 1988):

$$SAVI = \frac{(1+L)*(\rho_{NIR} - \rho_{red})}{\rho_{NIR} + \rho_{red} + L} \tag{9}$$

where L is the canopy background adjustment factor. If L is disregarded SAVI reduces to NDVI. Vegetation indices are affected by the atmosphere (Myneni & Asrar, 1994). Thus Huete et al. (2002) introduced the *enhanced vegetation index* (EVI) which is calculated as:

$$EVI = G \frac{\rho_{NIR} - \rho_{red}}{\rho_{NIR} + C_1 * \rho_{red} - C_2 * \rho_{blue} + L} \tag{10}$$

where G is a gain factor, C_1 and C_2 are the coefficients of the aerosol resistance term and ρ_{blue} is the reflectance in the blue band. The G, C_1 and C_2 have been set to be 2.5, 6 and 7.5, respectively (Huete et. al. 2002). The authors also used a value of L equal to 1. The C_1 and C_2 terms use the blue band to correct for aerosol influences in the red band. The EVI was developed to optimize the vegetation signal by improving the sensitivity in high biomass conditions, taking out the background effects (leaf litter or soil) and by reducing the effects of the atmosphere. EVI has been mostly used for images of the Moderate-Resolution Imaging Spectroradiometer (MODIS) sensor for monitoring vegetation.

The rationale behind the vegetation indices is that increasing vegetation LAI increases reflectance in the NIR due to a low leaf absorption in this spectral region, while in the red region a decrease occurs. It has been shown that most of the indices saturate for LAIs above around 3 (Tucker, 1979). However, the shape of the curve of fAPAR (the fraction of absorbed PAR) is the same as NDVI versus LAI. Sellers (1987) showed that the SRVI has a nearly linear relationship to fAPAR. Myneni et al. (1992) showed a nearly linear relationship between fAPAR and NDVI and a simple linear model relating fAPAR to top of canopy NDVI has been proposed (Myneni & Williams, 1994). Global datasets of time series of NDVI and EVI are available (e.g, Lhermitte et al., 2011), which make them highly attractive for monitoring grasslands over large areas and for estimating canopy conditions.

2.4.2 Canopy radiative transfer model inversion

Canopy radiative transfer models have been developed to simulate bidirectional reflectance factor of vegetation canopies (BRF). One simulated BRF ($BRF_j(S)$) can be represented by a function or algorithm, f, of subsystem characteristics (aj, bj, cj, dj, ej) (Goel & Strebel, 1983, Goel, 1988):

$$BRF_j(S) = f(a_j, b_j, c_j, d_j, e_j) \tag{11}$$

where aj, bj, cj, dj and ej define the source, the atmosphere, the vegetation, the soil and the sensor subsystems, respectively. The source is characterized by the solar zenith and azimuth angles and the total flux intensity (normalized to one), the atmosphere is characterized by the direct plus diffuse radiation, the canopy is characterized by the leaf reflectance (ρ_P) and transmittance (τ_P), LAI, LAD and the leaf spatial distribution parameter (λ_0), the soil is characterized by the soil reflectance (ρ_s) and the sensor is characterized by the view zenith (VZA) and view azimuth angles.

The inversion process consists of deriving a function or algorithm, g, that will yield the set of canopy parameters {cj}, as a function of the observed canopy BRF (BRF(O)) and the other subsystem characteristics (Goel & Strebel, 1983):

$$\{c_j\} = g(BRF_j(O), a_j, b_j, d_j, e_j) \tag{12}$$

Soil reflectance {dj} may also be derived through the inversion process along with the vegetation canopy parameters. The numerical inversion of a canopy radiative transfer model involves the minimization of differences between a set of simulated and observed BRF values acquired under different illumination/viewing geometries. Canopy parameter values that give the lowest difference between BRF(S) and BRF(O) are the estimated canopy values.

One limitation of the inversion process is that the number of observed values must be at least equal to the number of canopy parameters to be retrieved. This makes the process difficult for satellite image applications as most of the sensors collect single illumination/viewing geometry. This also limits the number of estimated parameters, which in most applications is the LAI. The validity of the inversion process has been validated for estimating canopy parameters of grasses using field radiometry observations (Privette et al., 1996). They have demonstrated the feasibility of estimating LAI, LAD, leaf reflectance and transmittance, total canopy *albedo* (the reflected radiation integrated over the hemisphere and the solar spectrum) and the fraction of absorbed photosynthetically active radiation. Inversion has also been used to estimate LAI and leaf chlorophyll for grassland (Darvishzadeh et al., 2008) and for sugar beet canopies (Jacquemoud et al., 1995) using spectroradiometer data and the PROSAIL model developed by Jacquemoud & Baret (1990).

3. Canopy structure and radiation flux interactions

In this section we initially present a brief description of the morphological characteristics of some forage grasses that are used for dairy and meat production in pasture-based systems, address some features relative to LAI and radiation interception measurements, and finally, we use a stoloniferous (*Cynodon* spp.) and a tufted (*Pennisetum purpureum*) grass for discussing experimental work about the extinction coefficient, leaf and stem area development and canopy angular distributions.

3.1 Typical growth habits of tropical forage grasses

In terms of tropical perennial forage grasses, two main morphogenetic groups with typical growth habits are usually recognized: *i*) tussock grasses, which are grasses that produce tillers, and have an erect and clumped growth form; and *ii*) creeping grasses, spreading by stolons, rhizomes or both(Skerman & Riveros 1989, Cruz & Boval, 2000,Van de Wouw et al., 2009).The first group includes tufted species as *Pennisetum purpureum, Panicum maximum, Andropogon gayanus, Hyparrhenia rufa* and *Brachiaria brizantha* while the examples of the second group are: *Brachiaria humidicola, Brachiaria mutica, Digitaria decumbens* and *Cynodon nlemfuensis* (Figure 2). Of course there are also species which can be defined as morphological intermediates, examples are: *Dichanthium aristatum, Dichanthium annulatum, Bothriochloa pertusa* and *Digitaria decumbens* (Cruz & Boval, 2000), and less common are: *Brachiaria* (Syn. *Urochloa*) *decumbens*. As pointed out by Cruz & Boval (2000) these morphological intermediates have the capacity to develop into stolons when growing as isolated plants, but in dense stands, most of the stems do not reach the ground but grow laterally at the top of the canopy.

3.2 Measurement of solar radiation interception

The instantaneous or daily integrated PAR absorbed at each level by a grass canopy (APAR) is the main factor determining the rate of carbon assimilation of individual leaves (Nobel et al., 1993). Therefore it is an important input for canopy photosynthesis models, once at the ecosystem level, leaf canopy is the unit of photosynthesis (Nouvellon et al., 2000, Hikosaka, 2005). Conceptually APAR, as an expression of the energy flux available, is the result of the difference between the net radiation above the canopy and the net radiation below the canopy (Norman & Arkebauer, 1991). It is calculated as:

$$APAR = (Ei_\lambda - Erc_\lambda) - (Et_\lambda - Ers_\lambda) \qquad (13)$$

where *Erc* and *Ers* are the symbols for radiation reflected by the canopy and the soil surfaces, respectively, and λ = 400-700 nm. For most purposes APAR is approximated by a more easily estimable quantity, the intercepted PAR (IPAR) which expresses a difference between incoming PAR (*Ei*) and the radiation transmitted to the bottom of the canopy (*Et*).

$$IPAR = E_i - E_t \qquad (14)$$

One measure of light interception efficiency is given by the relationship between IPAR and the total incident PAR at the top of the canopy, named fractional IPAR (fIPAR).

Fig. 2. Some tropical perennial forage grasses. A) Porto Rico Stargrass (*Cynodon nlemfuensis*), showing the stoloniferous growth habit at the edge of a experimental plot (September 2011, end of dry season). B) Dairy heifers grazing on the same pasture, during the rainy season (January 2011). C) Regrowth of napier grass (*Pennisetum purpureum*) sixty days after cutting; D) with subsequent growth, napier grass shows a like-cane growth habit, with erect culms. E) A marandu palisadegrass (*Brachiaria brizantha* cv. Marandu) sward in a dairy farm in Southern Minas Gerais State. (Photographs: courtesy by Dr. Sérgio T. Camargo Filho, Researcher of PESAGRO, Rio de Janeiro).

Soil nitrogen (N) has a strong effect on plant growth. Many studies, worldwide have shown that crop N uptake is co-regulated by both soil N supply and biomass accumulation processes (Fernandes & Rossiello, 1995; Hikosaka, 2005; Lemaire et al., 2007).When water supply is non-limiting, both carbon and nitrogen capture and use processes are closely linked with one another by the development of leaf area and the pattern of intercepted radiation (Lemaire et al., 2007; Giunta et al., 2009), since about half of leaf nitrogen is invested in photosynthetic proteins (Ghannoum et al., 2005; Hikosaka, 2005). Because of these interrelationships, we also consider the roles of N nutrition in the processes of interception and use of solar radiation by forage grasses.

In Figure 3, are presented the data related to fractional PAR intercepted by swards of Tifton 85 bermudagrass (*Cynodon* spp.), during field measurements to evaluate the effects of nitrogen fertilization on several physiological and morphological traits of the grass. Data were selected due to its simplicity of expression, as they show clearly that an increase in availability of N causes a temporal acceleration of the fractional IPAR by the sward, according to a logistic pattern. In practical terms we can say that N accelerates the canopy closure. Thus, at the high level of application of N, a fractional IPAR value of 0.9 was rapidly obtained, about three weeks after the beginning of the rest period. In contrast, in the same period, the control treatment did not surpass the level of 0.5 fIPAR.

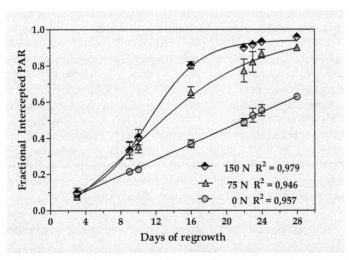

Fig. 3. Temporal variations of instantaneous fractional PAR interception values by Tifton 85 bermudagrass canopies under three rates of nitrogen fertilization. Experimental plots (4m x 4 m) were established in July 2008 on a Typic Fragiudult soil in Seropédica, RJ, Brazil. Several measurements were performed between November 30th and December 28th (rainy season) using optical instrumentation. After twenty-eight days, when the 150-N plots reached a level of 0.95 fIPAR (mean of four replicates) the regrowth cycle was finished by cutting. Environmental conditions during the period were as follows: mean solar radiation: 15.9 MJ m^{-2} day^{-1}, mean air temperature: 25.4° C and total rainfall: 222.5 mm (E. Barbieri Junior & R. Rossiello, unpublished data).

These results can be attributed to the effects of N on the morphogenetic traits responsible for the structural features in this type of pasture (Cruz & Boval, 2000), where axillary meristems develop as horizontal stolons (Figure 2A) under high levels of sunlight and good water supply. Particularly, in this case, nitrogen stimulated significantly (p≤0.05) canopy height growth rates, tiller population density and leaf area development (data not shown).

Some issues related to measurement procedures are pertinent. Photosynthetically active radiation was measured at the top of the sward canopy using a single quantum sensor (LI-191SA) while at the bottom the transmitted PAR was recorded with a LI-191 SA line quantum sensor connected to a LI-250A light meter (LI-COR Inc., Nebraska, USA). The sensor was inserted at the soil surface level regardless how much of dead material was present. This was possible due to young age of this hybrid bermudagrass with a small amount of dead material accumulated at the base. However, older perennial pastures may have sizeable amounts of dead phytomass accumulated on the bottom of the canopy (Le Roux et al., 1997, Guenni et al., 2005, Sbrizzia & Silva, 2008).In stoloniferous species, after 4 or 5 weeks of growth under non-N-limiting conditions, the loss of leaf biomass as a consequence of changes in allocation patterns can account for half of the leaf tissues produced (Cruz & Boval, 2000). In these situations an appropriate evaluation of IPAR may be a substantial problem. As an example, let us consider data on vertical light distribution in the pasture of Porto Rico Stargrass showed in Figure 2A. Measurements were taken under clear sky using optical equipment described above (Figure 4). Results showed that canopy light interception at 12.5 cm and at sensor level heights were 0.873 and 0.986 respectively, i.e. dead phytomass layer was responsible for about 11% of fIPAR (Figure 4A). On the other hand, in a plot adjacent, vegetated by Swazi grass (*Digitaria swazilandensis* Stent), the same variables have values of 0.930 and 0.988 respectively (profile not shown) reflecting morphological differences among the structural components of the two pastures. As noted previously by many researchers, when measuring grass light interception with optical sensors, it is nearly impossible to position the sensor under the grass canopy without disturbing it (Russell et al., 1989). One possible way of avoiding this problem, when disturbance is very apparent (Figure 4B), is to use a single sensor screwed to a transparent ruler graduated (Figure 4C), it is a solution more functional at plot than at field scale. Under field conditions, it may be more interesting to consider the bottom of the sward canopy a given "cut level" above the horizon of standing dead material, knowing however that the amounts of dead material accumulated are seasonally determined.

3.3 Leaf area index, extinction coefficient and angular distribution of canopy elements

Interception of PAR is modified by canopy architecture as represented by the extinction coefficient, k (Bréda, 2003, Zhou et al., 2003). For simulation purposes in canopy photosynthesis and radiation interception models, a fixed value for k is sometimes assigned (Thornley, 2002). However, research with real canopies has shown that this coefficient varies seasonally, in line with changes in traits such as leaf angle, canopy height or LAI (Bréda, 2003, Polley et al., 2011). A fixed value of k may be appropriate for estimating values of ceiling LAIs, i.e. when fIPAR is around 0.90 and the crop growth rate is near its maximum. However, during the previous vegetative growth, in several instances, it has been shown that k changes as sward architecture changes. Nouvellon et al. (2000), working with shortgrass ecosystems in northwestern Mexico found that the k value for diffuse and global radiation decreased as LAI sward increased. We found the same trend in canopies of Tifton 85 bermudagrass modified by nitrogen fertilization. In our study, decreases in k_{PAR} with sward height seemed to fit a linear pattern (Figure 5). In this case, sward height is a direct surrogate for herbage biomass or foliage density, structural properties with which it is highly correlated (Oliveira et al., 2010).

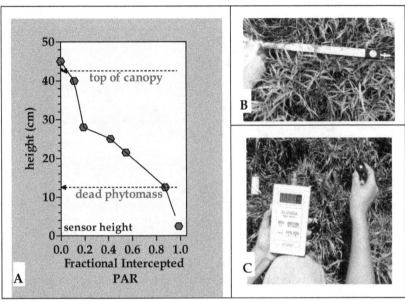

Fig. 4. Vertical light distribution in canopies of two stoloniferous species. A) Profile of a Porto Rico Stargrass sward (same as Figure 2A). At the site, mean heights of canopy and dead phytomass were 43 cm and 12.5 cm respectively. Sensor height is 2.5 cm and therefore, when facing upward, this is the location of its sensitive surface nearest ground surface. B) Plot of Swazi grass (*Digitaria swazilandensis* Stent) near the anterior. The proper deployment of the line quantum sensor is impeded by a dense layer of dead material so that its sensitive surface lies suspended at 7.5 cm from the soil surface. C) A better option may be to move a simple sensor through a vertical length of the canopy (Mean height: 29 cm). Measurements were taken on day 21st of September 2011 at PESAGRO Experimental Dairy Farm, Seropédica, RJ, Brazil, at the time corresponding to SZA between 24.8° and 28.0°.

Preliminarily we must recognize that k_{PAR} data as shown in Figure 5 are, to some degree, oversized. This fact is a result of using an optical approach for the measurement of intercepted PAR. The instrumentation used in this method does not discriminate between leaves and stems, or among green, senescent or dead leaf blades (Asner et al., 2003, Bréda, 2003). Overestimation arises because k values are calculated with the transmission values of the whole standing foliage, but with LAI values that includes only green leaf blades, which is the so-called "true LAI"(He et al., 2007). However the magnitude of this overestimation is a matter of debate because it is species- specific (Bréda, 2003, Guenni et al., 2005, He et al.,

2007), and in our case it is assumed that it is distributed equally among the treatments since the canopy leaf to stem ratio was almost invariant (Figure 5). Given this, the data indicate that under nitrogen influence, there is a structural change in the canopy towards a more erectophile condition.

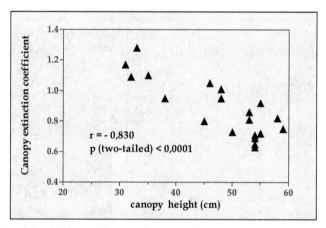

Fig. 5. Canopy heights of Tifton 85 bermudagrass related to corresponding extinction coefficients (k_{PAR}) values, after a regrowth period of 35 days (From January 25th to March 1st 2007). Variations in these traits were induced by nitrogen fertilization (0, 75,150,227 and 300 kg N-urea/ha). Concurrent values of Green Leaf Area Index ranged from 0.78 to 4.19.Mean leafiness (leaf blade: stem ratio) was 1.09 ± 0.05 and did not vary significantly ($p > 0.05$) among applied N doses (A.P. Oliveira & R. Rossiello, unpublished data).

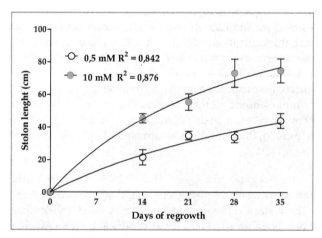

Fig. 6.Maximum stolon length of clonal propagules of Tifton 85 bermudagrass grown in Hoagland solution culture, in response to N levels (0.5 or 10 mM) and days of regrowth in a controlled growth environment . Photosynthetic photon flux density: 450 μmol photon m^{-2} s^{-1}, air temperature (day/night): 30/24° C, photoperiod: 12 h. (R. Rossiello, unpublished data).

As the theory predicts we can suppose that this result is due to changes in LAD, but how the changes take place is unclear. A possible interpretation is that by influencing canopy development, nitrogen modifies the light spectrum within the canopy with consequences for the stolon differentiation (Cruz & Boval, 2000). In *Cynodon* species or cultivars, large variation in length and number of stolons might be due to the very plastic response of stolons to light intensity and nutrient availability (Van de Wouw et al., 2009). Willemoes et al.(1987) observed that bermudagrass stolons irradiated with red light showed an upward curvature and an increase in leaf and internode lengths in comparison with those grown in darkness or under red plus far red radiant flux. Thus even in the low levels of photosynthetic irradiance existing in the middle of Tifton 85 canopies, high N availability in the growth medium could increase stolon elongation process, as can be inferred from results obtained in controlled environmental conditions (Figure 6).

In fact, light fluxes fluctuating deeper in the canopy, with variables red/far red ratios, in the presence of a growth substrate rich in N, could form the basis of the canopy response showed in Figure 5. Of course, we can also suppose that this type of response could be a consequence of absence of grazing pressure on shoot morphology. However, in a very different context, Pinto et al. (1999) working with Tifton-85 swards being continuously grazed by sheep found that taller swards (more lenient grazing) had the lowest senescence rates and suggested that changes in sward structure with increasing sward height could be promoting changes in the canopy light environment. Clearly this is an area that deserves more ecophysiological research.

Season has a strong influence on canopy structural properties due to the seasonal course of solar elevation and the associated changes in ratios of diffuse to direct solar beam. Kubota et al. (1994) observed a large structural change of napier grass canopy with growth. Young shoots of the cultivar Merkeron were transplanted in a field and grown for 102 days. During this growth period LAI increased from 0.7 to 15.4 while k decreased gradually from 1.1 to 0.38 due to elongation and erection of stems (large increase in the frequency of stalks with angles of 80-90° relative to the soil surface, see Figure 1B). These results indicate that in this grass, changes from a planophile to an erectophile growth pattern (see Figures 2C-D) are accomplished by correlative variations in SAD. This type of modification protects lower leaves from heavy shading, allowing the canopy to approximate an optimum LAI throughout the growth period (Kubota et al., 1994).Besides this, a long duration of vegetative growth are regarded as the main causes of high productivity of this species, with aboveground dry mass yields of 60 tons/ha/year (Morais et al., 2009). Zhou et al. (2003) working with sugarcane (*Saccharum* spp.) cultivars in Zimbabwe, Africa, obtained different results. They found that the k_{PAR} values of four cultivars (calculated by solving for k in the light extinction expression, as in Figure 5) increased (although not significantly, p>0.05) with increasing crop age, with a mean from 0.47 at 87 days to 0.64 at 116 days after planting. Additionally, it was observed that high stalk population cultivars intercepted more PAR than low stalk population cultivars because they had more intercepting leaf surfaces, but leaf size seemed less important than tiller population to explain differential patterns of PAR interception among cultivars. However, no information about possible differences in stalk angular distribution was given. In still other situations, there may be compensations between LAD and SAD during the growing season, so that the net effect of shifts in canopy angular distribution on light interception is decreased. This was the case in the above work of Nouvellon et al. (2000) who observed that early in the season, LAD was highly erectophile and shifted towards a less erectophile condition during the seasonal growth. However, this trend was compensated by a higher contribution of the highly erectophile stems (LAD↓ and SAD↑) to the total plant area index in the later stages of plant development. These findings suggest that, although in most cases the leaf angular distribution is the predominant factor in solar radiation interception; in some situations the role of steam angular distribution cannot be ignored.

There are few works dealing with values of LAI and light interception of tropical and subtropical grasses under conditions of cutting or grazing. In grazed pastures, leaf tissues are subjected to discrete defoliation events, the frequency and intensity of which greatly affect the physiology of plants and therefore the rate at which new leaf tissues are produced (Lemaire & Agnusdei, 2000). As a general rule, recommendations for grazing management are made in order to preserve a residual LAI suitable for the plants to continue growing thus maintaining the persistence of the herbage resource. In this context, the height of post-grazing residue is one of the determining factors of the regrowth rates in tropical pasture grasses, especially for tussock grasses as *Pennisetum purpureum* (Zewdu et al., 2003). In this species the dynamics of tillering in terms of tiller classes also influences growth rates and herbage accumulation (Skerman & Riveros, 1989). Carvalho et al. (2007) studied the effect of these two variables on the seasonal patterns of leaf area development and light interception, in an experiment performed at an experimental field of Embrapa in Coronel Pacheco, MG, Brazil (21° 33′ S, 43°06′ W, 410 m).Two post-grazing residues (50 and100 cm) and two tiller classes (basal and aerial) were combined in a split-plot arrangement, from October 2002 to April 2003. Selected results of this work are shown in Figure 7.

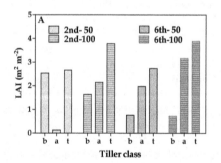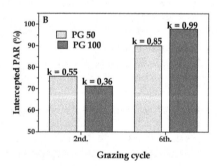

Fig. 7. Effects of post-grazing residue (50 and 100 cm) and tiller class on leaf area index (LAI) and PAR interception of napier grass swards during two grazing cycles. A) Leaf area index of basal (b), aerial (a) and total (t: basal + aerial) tillers affected by sward height post-grazing residue and grazing cycle. B) Canopy PAR interception values, evaluated in pre-grazing conditions, in response to the same variables. Number at the top indicates the value of extinction coefficient (*k*). LAI was determined destructively, according to: LAI = leaf area.tiller^{-1} x tiller number.m^{-2}. IPAR was measured using LI-COR optical sensors. Grazing cycles: second, from November 3th to December 6th 2002; sixth, from March 15th to April 17th 2003. (Adapted from Carvalho et al., 2007).

During the spring there was a larger appearance of basal tillers in swards managed at 50 cm post-grazing residue. Conversely, population densities of aerial tillers were predominant in the summer months. Interestingly, within each residue height, LAIs values were practically the same in the second and sixth grazing cycles, with different contributions of both tiller types (Figure 7A). However, LAI variations and PAR interception were not strongly related throughout the experimental period. The progressive increase in k values from 2nd to 6th cycle indicates that over the grazing cycles the foliage acquired a more planophile arrangement linked to a seasonal change in plant architecture. This shift was mediated by the afore-mentioned proliferation of aerial tillers which have a lower insertion angle than basal tillers and make up a flatter canopy (Carvalho et al., 2007). The marked dominance of aerial tillers in the last grazing cycle was apparently responsible for a greater PAR interception in the pre-grazing condition especially in pastures managed with100cm of residue (Figure 7B). However the authors do not exclude the contribution of dead material in this response. Qualitatively similar results were obtained by Giacomini et al.(2009) working with marandu palisadegrass (*Brachiaria brizantha* cv. Marandu, Figure 4E) subjected to intermittent stocking.

4. Radiation use efficiency

Monteith (1972) showed that phytomass production under tropical climate conditions is correlated with the amount of photosynthetically active radiation (PAR) absorbed by plants. This finding provides the basis for deriving the concept of ecosystem gross primary productivity (GPP). Ecosystem GPP may by calculated using algorithms that employ the light-use efficiency (LUE) concept (Polley et al., 2011). LUE (ε) is a conversion factor or the ratio of GPP to APAR (Equation 13). Following this concept, we have:

$$GPP = APAR \times \varepsilon PAR = PAR \times fAPAR \times \varepsilon PAR \qquad (15)$$

where fAPAR is the fraction of PAR that is absorbed by the grass canopy. From this identity, we can infer that green or dry biomass could be increased when radiation absorption or use efficiency, or both, are maximized. However, Norman & Arkebauer (1991) considered two meanings for the term "use efficiency": *i*) mass of CO_2 fixed per unit of absorbed photosynthetically active radiation, or *photosynthetic light-use efficiency* and *ii*) mass of dry matter (DM) produced per unit of absorbed photosynthetically active radiation or *dry matter light efficiency* which is the same as Equation 15. As noted by these authors, the second definition is more problematic as it involves both maintenance and growth respiration terms, which may not depend on light directly. Despite this objection, this agronomic definition is the most frequently cited in radiation use research, where IPAR can replace APAR (Norman & Arkebauer, 1991). According to Albrizio & Steduto (2005) for a given species and environment, RUE is approximately a constant value during the growth season, provided that: *i*) respiration is proportional to photosynthesis; *ii*) photosynthesis response to irradiance is essentially linear at the canopy scale and *iii*) no substantial change in the chemical composition of biomass occurs during the growth cycle considered. Under non-limiting water and nutrient conditions, all of these conditions can be met for tropical forage grasses, however, to date there are few available data. Guenni et al. (2005) reported RUE values for five *Brachiaria* species that ranged (not significantly) from 1.3 to 1.7 g DM (MJ IPAR)$^{-1}$ for *B. brizanta* and *B. humidicola*, respectively.

5. Conclusions

Multi-functionality approach recognizes that grasslands have to be considered not only as a means for providing animal products for increasing human population. Also, other additional ecosystem functions as enhancement of carbon sequestration or mitigation of greenhouse gas emissions should be considered by farmers as a way of capitalizing new opportunities to diversify the forage-livestock system (Lemaire, 2007, Sanderson et al., 2007). However returning to the first page of this chapter we remember that degradation processes in tropical grassland are advancing over wide areas through overgrazing on poor soils. So it seems that recuperation and multi-functionality concepts in pasture ecosystems will transit a long way together. We state the central role that the leaf canopy structure, expressed as leaf area index, plays in terms of intercepting solar radiation. Despite this, there are very few studies comparing different alternatives of estimating LAI in forage plants, particularly those of tropical climate. There are several possibilities to apply technologies already available in the generation of new methods. Some of these include: *i*) use of remotely sensed data for monitoring canopy parameters (vegetation indices as NDVI, SAVI and others), *ii*) measurements of foliage cover through digital color photographs taken vertically above the plant canopy using a stationary camera stand (Rotz et al., 2008), iii) a more intensive and creative use of the gap fraction methods including examination of hemispherical photographs for estimates of foliage angular distribution and canopy leaf area. Regarding this last information, the orientation of foliage elements (stems and leaves) is an important piece of information for describing light penetration in canopies especially for tussock grasses.

6. Acknowledgments

The authors acknowledge the National Research Council for Scientific and Technological Development of Brazil (CNPq) and Carlos Chagas Foundation of Rio de Janeiro (FAPERJ) for granting graduate scholarships and providing funds for the research projects that supported this paper.

7. References

Albrizio, R. & Steduto, P. (2005). Resource use efficiency of field-grown sunflower, sorghum, wheat and chickpea I. Radiation use efficiency. *Agricultural and Forest Meteorology*, Vol.130, No.3-4 (June 2005), pp. 254–268, ISSN 0168-1923

Allen V.G., Batello, C., Berretta, E.J., Hodgson, J., Kothmann, M., Li, X., McIvor, J., Milne, J., Morris, C., Peeters, A. & Sanderson, M. (2011). An international terminology for grazing lands and grazing animals. *Grass and Forage Science*, Vol.66, No.1 (March 2011), pp. 2-28, online ISSN 1365-2494

Antunes, M.A.H. (1997). *A vegetation canopy radiative transfer model and its use to estimate canopy leaf area index and absorbed fraction of photosynthetically active radiation*. Ph.D. Dissertation, University of Nebraska - Lincoln. 181 p. Paper AAI9812346

Antunes, M.A.H., Walter-Shea, E.A. & Mesarch, M.A. (2001). Test of an extended mathematical approach to calculate maize leaf area index and leaf angle distribution, *Agricultural and Forest Meteorology*, Vol.108, No.1 (May 2001), pp. 45–53, ISSN 0168-1923

Asner, G.P., Scurlock, J.M.O.& Hicke, J.A.(2003). Global synthesis of leaf area observations: implications for ecological and remote sensing studies. *Global Ecology & Biogeography*, Vol.12, No.3(May 2003), pp. 191-205, OnlineISSN: 1466 - 8238

Asner, G.P., Townsend, A.L., Bustamante, M.M.C., Nardoto, G.B. & Olander, L.P.(2004). Pasture degradation in the central Amazon: linking changes in carbon and nutrient cycling with remote sensing. *Global Change Biology*, Vol.10, No.5 (May 2004), pp. 844-862, Online ISSN 1365-2486

Baret, F., Andrieu, B. & Steven, M.D. (1993). Gap frequency and canopy architecture of sugar beet and wheat crops. *Agricultural and Forest Meteorology*, Vol.65, No.3-4 (August 1993), pp. 261-279, ISSN 0168-1923

Boddey, R.M., Macedo, R., Tarré, R.M., Ferreira, E., Oliveira, O.C., Rezende, C. de P., Cantarutti, R.B., Pereira, J.M., Alves, B.J.R. & Urquiaga, S. (2004). Nitrogen cycling in *Brachiaria* pastures: the key to understanding the process of pasture decline. *Agriculture, Ecosystems &Environment*, Vol.103, No.2 (July 2004) pp. 389–403, ISSN 0167-8809

Bréda, N.J.J. (2003). Ground-based measurements of leaf area index: a review of methods, instruments and current controversies. *Journal of Experimental Botany*, Vol.54, No.392 (November 2003), pp. 2403-2417, ISSN 0022-0957

Bunnik, N.J.J. (1978). *The multispectral reflectance of shortwave radiation by agricultural crops in relation with their morphological and optical properties.* Doctoral Dissertation. University of Wageningen, 1978. 175 p.

Campbell, G.S. & Norman, J.M.(1989). The description and measurement of plant canopy structure In: *Plant canopies: their growth, form and function*, G. Russell, B. Marshall & P.G. Jarvis (eds.), pp. 1-19, Cambridge University Press, ISBN 0 521 39563 1, Cambridge, UK

Carvalho, C.A.B., Rossiello, R.O.P., Paciullo, D.S.C., Sbrissia, A.F. & Deresz, F. (2007). Tiller classes on leaf area index composition in elephant grass swards. *Pesquisa Agropecuária Brasileira*, Vol.42, No.4 (April 2007), pp. 557-563, ISSN 1678-3921 (In Portuguese with abstract in English)

Cohen, S., Rao, R.S. & Cohen, Y. (1997). Canopy transmittance inversion using a line quantum probe for a row crop. *Agricultural and Forest Meteorology*, Vol.86, No.3-4 (September 1997), pp. 225-234, ISSN 0168-1923

Cruz, P. & Boval, M.(2000). Effect of nitrogen on some morphogenetic traits of temperate and tropical perennial forage grasses, In: *Grassland ecophysiology and grazing ecology*, G. Lemaire, J. Hodgson, A. de Moraes, C. Nabinger & P.C. de F. Carvalho (Eds.), pp. 151-168,CAB International, ISBN 0851994520, Wallingford, UK.

Darvishzadeh, R., Skidmore, A., Schlerf, A. &Atzberger, C. (2008). Inversion of a radiative transfer model for estimating vegetation LAI and chlorophyll in a heterogeneous grassland. *Remote Sensing of Environment*, Vol.112, No.5 (15 May 2008), pp. 2592-2604, ISSN 0034-4257

FAO (2010). Challenges and opportunities for carbon sequestration in grassland systems. A technical report on grassland management and climate change mitigation.Integrated Crop Management Vol.9-2010, 59 p., ISBN 978-92-5-106494-8, Rome, Italy.

Fernandes, M.S. & Rossiello, R.O.P.(1995). Mineral nitrogen in plant physiology and plant nutrition. *Critical Reviews in Plant Sciences*, Vol.14, No.2 (March 1995) pp.111-148, ISSN 0735-2689.

Garrigues, S., Shabanov, N.V. Swanson, K., Morisette, J.T., Baret, F. & Myneni, R.B. (2008). Intercomparison and sensitivity analysis of Leaf Area Index retrievals from LAI-2000, AccuPAR, and digital hemispherical photography over croplands. *Agricultural and Forest Meteorology*, Vol.148, No.8-9 (4 July 2008), pp. 1193–1209, ISSN 0168-1923

Giacomini, A.A., Silva, S.C. da, Sarmento, D.O.L., Zeferino, C.V., Trindade, J.K. da, Souza Júnior,S.J., Guarda, V.A., Sbrizzia, A.F., Nascimento Júnior,D. do. (2009).Components of the leaf area index of marandu palisadegrass swards subjected to strategies of intermittent stocking. *Scientia Agricola*, Vol.66. No.6 (November /December 2009), pp. 721-732, ISSN 0103-9016

Ghannoum, O., Evans, J.R., Chow, W.S., Andrews, J.T., Conroy, J.P.& Von Caemmerer, S. (2005).Faster rubisco is the key to superior nitrogen-use efficiency in NADP-malic enzyme relative to NAD-malic enzyme C4 grasses. *Plant Physiology*, v.137,No.2 (February 2005), pp. 638–650, ISSN 0032-0889

Giunta, F., Pruneddu, G & Motzo, R. (2009). Radiation interception and biomass and nitrogen accumulation in different cereal and grain legume species. *Field Crops Research*, Vol.110, No.1 (January 2009), pp. 76-84, ISSN 0378-4290

Goel, N.S. (1988). Models of vegetation canopy reflectance and their use in estimation of biophysical parameters from reflectance data. *Remote Sensing Reviews*, Vol.4, No.1, pp. 1-212, ISSN 0275-7257

Goel, N.S. & Strebel, D.E. (1983). Inversion of vegetation canopy reflectance models for estimating agronomic variables. I. Problem definition and initial results using the Suits model. *Remote Sensing of Environment*, Vol.13, No.6 (December 1983), pp. 487-507, ISSN 0034-4257

Guenni, O., Gil, J.L. & Guedez, Y. (2005). Growth, forage yield and light interception and use by stands of five *Brachiaria* species in a tropical environment. *Tropical Grasslands*, Vol.39, No.1 (March 2005), pp. 42-53, ISSN 0049-4763

He, Y., Guo, X., & Wilmshurst, J. F. (2007). Comparison of different methods for measuring leaf area index in a mixed grassland. *Canadian Journal of Plant Science*, Vol.87, No.4 (October 2007) pp. 803-813, ISSN 0008-4220

Hikosaka, K. (2005). Leaf canopy as a dynamic system: ecophysiology and optimality in leaf turnover. *Annals of Botany*, Vol.95, No.3 (February 2005), pp. 521–533, ISSN 0305-7364

Huete, A., Didan, K., Miura, T., Rodriguez, E.P., Gao, X. & Ferreira, L.G. (2002). Overview of the radiometric and biophysical performance of the MODIS vegetation indices. *Remote Sensing of Environment*, Vol.83, No.1-2 (November2002), pp. 195-213, ISSN 0034-4257

Huete, A.R. (1988). A soil adjusted vegetation index (SAVI). *Remote Sensing of Environment*, Vol.25, No.3 (August 1988), pp. 295 309, ISSN 0034-4257

Jacquemoud, S. & Baret, F. (1990). PROSPECT: A model of leaf optical properties spectra. *Remote Sensing of Environment*, Vol.34, No.2 (November 1990), pp. 75-91, ISSN 0034-4257

Jacquemoud, S., Baret, F., Andrieu, B., Danson, F.M. & Jaggard, K. (1995). Extraction of vegetation biophysical parameters by inversion of the PROSPECT + SAIL models on sugar beet canopy reflectance data. Application to TM and AVIRIS sensors.*Remote Sensing of Environment*,Vol.52, No.3 (June 1995), pp. 163-172, ISSN 0034-4257

Knipling, E.B. (1970). Physical and physiological basis for the reflectance of visible and near-infrared radiation from vegetation. *Remote Sensing of Environment*, Vol.1, No.3 (Summer 1970), pp. 155-159, ISSN 0034-4257

Kubota, F., Matsuda, M., Agataand, W. & Nada, K.(1994). The relationship between canopy structure and high productivity in napier grass, *Pennisetum purpureum* Schumach. *Field Crops Research*,Vol.38, No.2 (August 1994), pp.105-110, ISSN 0378-4290

Larsen, D.R. & Kershaw, J.R. Jr. (1996). Influence of canopy structure assumptions on predictions from Beer's law. A comparison of deterministic and stochastic simulations. *Agricultural and Forest Meteorology*, Vol.81, No.1-2 (September 1996), pp. 61-77, ISSN 0168-1923

Lehmann, C.E., Ratnam, J. & Hutley, L.B. (2009).Which of these continents is not like the other? Comparisons of tropical savanna systems: key questions and challenges.*NewPhytologist*, Vol.181, No.3 (February 2009), pp. 508-511, ISSN 1469-8137

Lemaire, G.(2007). Research priorities for grassland science: the need of long term integrated experiments networks. *Revista Brasileira de Zootecnia*, Vol.36 Special supplement (July 2007), pp. 93-100, ISSN 1516-3598

Lemaire, G. & Agnusdei, M.(2000). Leaf tissue turnover and efficiency of herbage utilization. In: In: *Grassland ecophysiology and grazing ecology*, G. Lemaire, J. Hodgson, A. de Moraes, C. Nabinger & P.C. de F. Carvalho (eds.), pp. 265-287,CAB International, ISBN 0851994520, Wallingford, UK.

Lemaire, G., Van Oosterom, E., Sheehy, J., Jeuffroy, M.H., Massignam, A. &Rossato, L.(2007). Is crop nitrogen demand more closely related to dry matter accumulation or leaf area expansion during vegetative growth? *Field Crops Research*, Vol.100, No.1 (January 2007) pp. 91–106,ISSN 0378-4290

Le Roux, X., Gauthier, H., Bégué, A. & Sinoquet, H. (1997).Radiation absorption and use by humid savanna grassland: assessment using remote sensing and modelling.*Agricultural and Forest Meteorology*, Vol.85, No.1-2 (June 1997), pp. 117-132, ISSN 0168-192385 (1997)

Lhermitte, S., Verbesselt, J., Verstraeten, W.W. &Coppin, P. (2011). A comparison of time series similarity measures for classification and change detection of ecosystem dynamics. *Remote Sensing of Environment*, in press, ISSN 0034-4257

López-Lozano, R., Baret, F., Cortázar-Atauri, I.G., Bertrand, N. & Casterad, M.A. (2009). Optimal geometric configuration and algorithms for LAI indirect estimates under row canopies: The case of vineyards. *Agricultural and Forest Meteorology*, Vol.149, No.8 (3 August 2009), pp. 1307–1316, ISSN 0168-1923

Meirelles, M.L., Franco, A.C., Farias, S.E.M. & Bracho, R. (2011). Evapotranspiration and plant-atmospheric coupling in a *Brachiaria brizantha* pasture in the Brazilian savannah region. *Grass and Forage Science*, Vol.66, No.2 (June 2011), pp. 206-213, Online ISSN 1365-2494.

Monsi, M. &Saeki, T. (2005). On the factor light in plant communities and its importance for matter production. *Annals of Botany*, Vol.95, No.3 (February 2005), pp. 549–567. ISSN 1095-8290

Monteith, J.L.(1972). Solar radiation and productivity in tropical ecosystems. *Journal of Applied Ecology*, Vol.9, No.3 (December 1972), pp. 747-766, Online ISSN 1365-2664

Morais, R.F., Souza, B.J., Leite, J.M., Soares, L.H.B., Alves, B.J.R., Boddey, R.M., Urquiaga, S.(2009). Elephant grass genotypes for bioenergy production by direct biomass combustion. *Pesquisa Agropecuária Brasileira*,Vol.44, No.2 (February 2009), pp. 133-140, ISSN 0100-204X

Myneni, R.B. & Williams, D.L. (1994). On the relationship between FAPAR and NDVI. *Remote Sensing of Environment*, Vol.49, No.3 (September 1994), pp. 200-211, ISSN 0034-4257

Myneni, R.B. &Asrar, G. (1994). Atmospheric effects and spectral vegetation indices. *Remote Sensing of Environment*, Vol.47, No.3 (March 1994), pp. 390-402, ISSN 0034-4257

Myneni, R.B., Asrar, G., Tanre, D. &Choudhury, B.J. (1992). Remote sensing of solar radiation absorbed and reflected by vegetated land surfaces. *IEEE Transactions on Geoscience and Remote Sensing*, Vol.30, No.2 (March 1992), pp. 302-314, ISSN 0196-2892

Myneni, R.B., Maggion, S., Iaquinta, J., Privette, J.L., Gobron, N., Pinty, B., Kimes, D.S., Verstraete, M.M. & Williams, D.L. (1995). Optical remote sensing of vegetation: modeling, caveats, and algorithms. *Remote Sensing of Environment*, Vol.51, No.1 (January 1995), pp. 169-188, ISSN 0034-4257

Nicodemus, F.E., Richmond, J.C., Hsia, J.J., Ginsberg, I.W.& Limpers, T. (1977).*Geometrical considerations and nomenclature for reflectance*. Washington, DC, U.S. Department of Commerce. 52 p. (NBS Monograph 160)

Nobel, P.S., Forseth, I.N. & Long, S.P. (1993). Canopy structure and light interception. In: *Photosynthesis and production in a changing environment: a field and laboratory manual*, D.O.Hall, J.M.O. Scurlock, H.R. Bolhàr-Nordenkampf, R.C. Leegood & S.P. Long (eds.), pp. 79-90, Chapman & Hall, ISBN 0 412 42900 4, London, UK.

Norman, J.M. & Arkebauer, T.J.(1991). Predicting canopy light-use efficiency from leaf characteristics. In: *Modelingplant and soil systems*. J.Hanks & J.T. Ritchie (eds.), pp. 125-143, American Society of Agronomy, Inc., Crop Science Society of America, Inc., Soil Science Society of America, Inc., ISBN 0-89118-106-7, Madison, USA.

Nouvellon, Y, Begué, A.,Moran, M.S., Seen, D.L., Rambal, S., Luquet, D., Chehbouni &Inoue, Y. (2000). PAR extinction in shortgrass ecosystems: effects of clumping, sky conditions and soil albedo. *Agricultural and Forest Meteorology*, Vol.105, No.1-3 (November 2000), pp. 21–41, ISSN 0168-1923

Oliveira, A.P.P., Rossiello, R.O.P., Galzerano, L., Costa Júnior, J.B.G., Silva, R.P. & Morenz, M.J.F. (2010).Responses of the grass Tifton 85 to the application of nitrogen: soil cover, leaf area index, and solar radiation interception. *Arquivo Brasileiro de Medicina Veterinária e Zootecnia*, Vol.62, No.2 (April 2010), pp.429-438, ISSN 0102-0935(In Portuguese with abstract inEnglish)

Ollinger, S.V. (2011). Sources of variability in canopy reflectance and the convergent properties of plants. *New Phytologist*, Vol.189, No.2 (January 2011), pp. 375–394, ISSN 1469-8137

Pinto, L.F.M., Da Silva, S.C., Barioni, L.G., Pedreira, C.G.S., Carnevalli, R.A., Sbrissia, A.F., Carvalho, C. A. B., Fagundes, J.L., Watanabe, L.H. & Machado, F.C. (1999). Herbage accumulation dynamics in grazed swards of Cynodon spp. In: de Moraes, A., Nabinger, C., Carvalho, P.C.de F., Alves, S.J. & Lustosa,S.B.C. (eds) *Proceedings of International Symposium on Grassland Ecophysiology and Grazing Ecology*, pp. 353-356, Universidade Federal do Paraná, Curitiba, Brazil.

Polley, H.W., Phillips, R.L., Frank, A.B., Bradford, J.A., Sims, P.L., Morgan, J.A. & Kiniry, J.R. (2011). Variability in light-use efficiency for gross primary productivity on Great Plains grasslands. *Ecosystems*, Vol.14, No.1 (January 2011), pp. 15–27, ISSN 1432-9840

Privette, J.L., Myneni, R.B.& Hall, F.G. (1996). Optimal sampling conditions for estimating grassland parameters via reflectance model inversions, *IEEE Transactions On Geoscience and Remote Sensing*, Vol.34, No.1 (January 1996),pp. 272-283, ISSN 0196-2892

Rich, P.M. (1990). Characterizing plant canopies with hemispherical photographs. *Remote Sensing Reviews*,Vol.5, No.1 (January 1990), pp. 13-29, ISSN 0275-7257

Ross, J. (1981).*The Radiation Regime and Architecture of Plant Stands*. Dr. W. Junk. Publishers, ISBN 906193 897x, The Hague, Netherlands

Rotz, J.D., Abaye, A.O., Wynne, R.H., RAyburn, E.B., Scaglia, G., Phillips, R.D. (2008). Classification of digital photography for measuring productive ground cover. *Rangeland Ecology and Management*, Vol.61, No.2 (March 2008), pp. 245-248, ISSN 1550-7424

Russell, G., Jarvis, P.G. & Monteith, J.L.(1989). Absorption of radiation by canopies and stand growth. In: Plant canopies: their growth, form and function, G. Russell, B. Marshall & P.G. Jarvis (eds.),pp. 21-39, Cambridge University Press, ISBN 0 521 39563 1, Cambridge, UK.

Sanderson,M.A.,Goslee, S.C.,Soder, K.J., Skinner, R.H.,Tracy, B.F. & Deak, A.(2007).Plant species diversity, ecosystem function, and pasture management—A perspective. *Canadian Journal of Plant Science*, Vol.87, No.3 (July2007) pp. 479-487, ISSN 0008-4220

Sbrissia, A.F. & Silva, S.C. (2008).Comparison of three methods for estimating leaf area index of marandu palisadegrass swards under continuous stocking. *Revista Brasileira de Zootecnia*, Vol.37, No.2 (February 2008), pp. 212-220, ISSN 1516-3598 (In Portuguese with abstract in English)

Schut, A.G.T. & Ketelaars, J.J.M.H. (2003). Monitoring grass swards using imaging spectroscopy.*Grass and Forage Science*, Vol.58, No.3 (September 2003), pp. 276–286, ISSN: 1365-2494

Sellers, P.J. (1987). Canopy reflectance, photosynthesis, and transpiration, II.The role of biophysics in the linearity of their interdependence. *Remote Sensing of Environment*, Vol.21, No.2 (March 1987), pp. 143-183, ISSN 0034-4257

Skerman, P.J. & Riveros, F. (1989). *Tropical grasses*. FAO Plant Production and Protection Series No.23, ISBN 92-5-101128-1, Rome.

Thornley, J.H.M. (2002). Instantaneous canopy photosynthesis: analytical expressions for sun and shade leaves based on exponential light decay down the canopy and an acclimated nonrectangular hyperbola for leaf photosynthesis. *Annals of Botany*, Vol.89, No.4 (April 2002), pp. 451-458, ISSN 0305-7364

Tucker, C.J. (1979). Red and photographic infrared linear combinations for monitoring vegetation. *Remote Sensing of Environment*, Vol.8, No.2 (May 1979), pp. 127-150, ISSN 0034-4257

Van de Wouw, M., Mohammed, J., Jorge, M.A. & Hanson, J. (2009). Agro-morphological characterisation of a collection of Cynodon. *Tropical Grasslands*, Vol.43, No.3 (November 2009), pp. 151-161, ISSN 0049-4763

Van Esbroeck,G.A., Hussey, M.A. & Sanderson, M.A. (1997). Leaf appearance rate and final leaf number of switchgrass cultivars. *Crop Science*, Vol.37, No.3 (May-June1997), pp. 864-870, ISSN 0011-183X

Vermote, E.F., Tanre, D., Deuze, J.L., Herman, M. & Morcrette, J.J. (1997). Second simulation of the satellite signal in the solar spectrum, 6S: An overview. *IEEE Transactions on Geoscience and Remote Sensing*, Vol.35, No.3 (May 1997), pp. 675-686, ISSN 0196-2892

Walter-Shea, E.A., Blad, B.L., Hays, C.J., Mesarch, M.A., Deering, D.W. & Middleton, E.M. (1992). Biophysical properties affecting vegetative canopy reflectance and absorbed photosynthetically active radiation at the FIFE site.*Journal of Geophysical Research: Atmospheres*,Vol.97, No.D17, pp. 18925-18934, 0148–0227

Welles, J.M. & Norman, J.M. (1991). Instrument for indirect measurement of canopy architecture. *Agronomy Journal*, Vol.83, No.5 (September-October 1991), pp. 818–825, ISSN 0002-1962

Willemoes, J.G., Beltrano, J. & Montaldo, E.R.(1987). Stolon differentiation in *Cynodon dactylon* (L.) pers. mediated by phytochrome. *Environmental and Experimental Botany*,Vol.27, No.1 (January 1987), pp. 15-20, ISSN 0098-8472

Wit, C.T. de (1965). *Photosynthesis of Leaf Canopies*. Agricultural Research Reports 663, 57p., Centre for Agricultural Publications and Documentation, Wageningen,

Woodward, I. F. (2011). Theory in plant science. *New Phytologist*, Vol.192, No.2 (October 2011), pp. 303-304,online ISSN 1469-8137

Zewdu, T, Baars, R.T.M., Yami,A. (2003).Effect of plant height at cutting and fertilizer on growth of Napier grass (*Pennisetum purpureum*). *Tropical Science*, Vol.43, No.1 (March 2003), pp. 57-61, Online ISSN 1556-9179

Zhou, M.M., Singels, A. & Savage, M.J. (2003).Physiological parameters for modelling differences in canopy development between sugarcane cultivars. *Proceedings of the South African Sugar Technologist´s Association*, Vol.77, pp. 610- 620.

13

Effects of Solar Radiation on Animal Thermoregulation

Amy L. Norris and Thomas H. Kunz
Boston University
USA

1. Introduction

Solar radiation affects all aspects of the Earth's environment both directly and indirectly. Radiation from the sun produces a wide range of wavelengths that reach Earth and are absorbed or reflected from animate and inanimate surfaces. Visual, ultraviolet and infrared wavelengths, and solar intensity are the primary variables measured to evaluate their effects on the behavioral, thermoregulatory, and cellular responses of organisms. This review will focus on the effects of solar radiation on animal thermoregulation and the various methods used by scientists to assess the effects of infrared radiation on skin temperatures.

1.1 Ectotherms vs. endotherms

All organisms regulate their internal body temperature (T_B) to maintain a relatively constant temperature within a small range; in effect the rate of heat gained or produced must be balanced with heat lost to the environment. Animals are generally divided into ectotherms or endotherms based on how they maintain internal T_B. Ectotherms, previously referred to as "cold-blooded", are animals such as invertebrates, amphibians, reptiles, and fish who regulate their T_B externally primarily through behavioral mechanisms that alter heat exchange between their bodies and the environment. Endotherms such as birds and mammals, once described as "warm-blooded", are able to generate heat internally. Comparatively, leaky cell membranes, a greater amount of enzymes, and insulation (hair, feathers, and subcutaneous fat) contribute to endotherms having a higher basal metabolic rate than ectotherms, allowing them to maintain a relatively constant T_B separate from their ambient surroundings — known as homeothermy or euthermy (Nagy, 2004). Notwithstanding, some larger invertebrates and fish employ muscle activity to perform some degree of homeothermy for short periods and some small mammals and birds relax their thermogenic abilities to enter states of torpor or may even hibernate at certain times of year — a thermoregulatory strategy known as daily or seasonal heterothermy.

1.2 Heat transfer and thermoregulation

Each species has an optimal temperature where cellular processes are ideally maintained to optimize energy expenditure. Heat transfer across the body surface must be balanced lest an individual succumb to extreme heat gain or loss, a possibility especially in extreme environments such as polar or desert regions. Most organisms attempt to remain within a favorable range of temperatures. For homeotherms, this is known as the thermal neutral zone (TNZ). This optimal range of ambient temperatures typically lies between 30-42°C for homeothermic or euthermic organisms (Speakman, 2004). Within this range, metabolic rate is minimal. Outside the TNZ, metabolic energy is required to maintain T_B within the optimal range. Metabolic activity involving temperature-dependent enzyme activity will not function properly if the animal becomes hypo- or hyperthermic. Thus, thermoregulation is an integral part of an organism's energy balance.

Visual and ultraviolet wavelengths of solar radiation may be reflected or absorbed by animal surfaces, producing distinctive coloration and/or synthesis of vitamin D in terrestrial vertebrates, but infrared radiation is absorbed directly. Infrared radiation can increase T_B near to or excelling the upper critical temperature (UCT). The UCT is the greater temperature limit where behavioral modifications are not enough to inhibit heat absorption and therefore energy must be expended in the attempt to dump excess heat. The temperature range from the UCT to the upper lethal temperature (ULT), the T_B where an organism can no longer thermoregulate and dies of overheating, is known as the zone of evaporatory cooling where evaporation of metabolic water is the most efficient, and sometimes only, method available for transferring excess heat. On the other hand, the lower critical temperature (LCT) is the turning point at which more heat is lost to the environment than is normally metabolically produced. The lower lethal temperature (LLT) is the extreme cold temperature where an animal can no longer produce enough heat and dies of hypothermia. The range of temperatures between the LCT and LLT is known as the zone of metabolic regulation, where heat production through metabolic processes such as shivering or non-shivering thermogenesis is necessary to increase T_B. Figure 1 depicts a general TNZ graph, LCT, UCT, and zones of energy use (Randall, 2002).

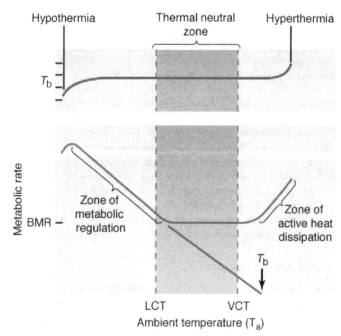

Fig. 1. Thermal neutral zone graph showing lower and upper critical temperature, zones of metabolic regulation and evaporation. Source: Randall/Eckert Animal Physiology 5ed.

Heat is transferred by four mechanisms: radiation, conduction, convection, and evaporation. Many species combine methods to absorb or transfer heat as efficiently as possible. Solar radiation plays a large part in determining not only ambient and body temperatures but animal behavior as well. A wide variety of species have developed methods to reduce the cost of thermoregulation by behaving certain ways: seeking shade, burrowing, panting, gular fluttering, wing flapping when exposed to temperatures above the UCT, entering short bouts of torpor or longer bouts of hibernation, increasing insulation, or migrating.

1.2.1 Radiation

The sun is Earth's principal source of radiative energy. With a surface temperature of approximately 6000K, the sun's electromagnetic radiation transmits some of this heat in the form of infrared radiation to our atmosphere and to varied surfaces on Earth (Speakman, 2004). Infrared radiation, undetectable to the human eye, is the peak wavelength emitted by objects with a surface temperature of -20° to 40°C. Special cameras sensitive to infrared wavelengths allow humans to record thermal images and videos of infrared radiation emitted by both inanimate objects and organisms, making it possible to effectively measure the surface temperature of whatever is within the camera's field of view (McCafferty, 2007). The amount of radiated heat detected depends upon factors such as the emissivity and absorptivity of an inanimate object or organism. Emissivity is defined relative to what is known as a black body, a perfect emitter with an emissivity of 1.0. Most animals have an emissivity value within 0.90-0.98, often dependent on surface properties such as fur or skin color. Surfaces either reflect or absorb light to varying degrees contingent on pigment levels and texture that in turn affect emissivity. Dark colors absorb energy within the infrared spectrum, increasing the absorptivity of inanimate objects or organisms compared to light colors which reflect visual wavelengths of solar energy. Water content, often high in living organisms, additionally contributes to an animal's elevated emissivity, since water is itself an excellent emitter.

1.2.2 Conduction

Heat transfer between two solid objects in contact with one another is known as conduction. Heat energy travels down a thermal gradient and thus is conducted from a warmer object to a cooler one. This attribute also allows for conduction within a single body, if the core of an object or organism is warmer than the surface layer for example, heat will be conducted along the gradient. The rate at which this transmission occurs depends on several factors such as the material properties of both items, distance heat must travel, and the actual surface area that is physically in contact. Like radiation, thermal conductivity relies on certain properties of the materials in question. Insulation is the opposite of conductivity, meaning objects or organisms having little to no insulation may have a high thermal conductivity. This increases the rate and likelihood of heat transfer through conduction. Ectotherms such as insects, amphibians, reptiles, and fish have little insulation, making them more likely to gain or lose heat to their surroundings. In a given environment, a sun warmed rock in an otherwise cool environment can be critical to an ectothermic animal trying to remain active. In fact, insulation would be detrimental to ectotherms relying on external heat sources, because any such barrier would slow the rate of heat transfer into the body (Speakman, 2004). Endotherms such as birds and mammals have varying degrees of insulation made from feathers, fur, and/or fat deposits allowing them to retain their internal heat and thus rely less upon external heat sources.

The ratio of surface area to the volume of an organism is also an important component of heat transfer through conduction. Animals not only gain or lose heat via conduction with surfaces and objects in the environment they come in contact with, but also from the core of the body outwards toward the skin surface. Larger animals create thermal inertia, requiring less energy to balance heat loss and so have a low thermal conductance. Smaller animals lose heat rapidly and need more energy per gram of tissue to maintain heat balance with their higher thermal conductivity. Behavioral adjustments that change the amount of exposed surface area as well as physiological responses can increase or decrease an animals' thermal conductivity in either short periods of time or for seasonal modifications.

1.2.3 Convection

Rather than two solid objects adjacent with one another, convection requires one fluid coming in contact with another fluid (water or air) of a different temperature. Rather than heat transferring at the molecular level or through electromagnetic waves, heat is dispersed via the bulk motion of fluids (Cengel, 2003). Once again the temperature gradient between the two surfaces influences the rate of heat transfer. Conduction and convection along the human body is illustrated in Figure 2. Water conducts heat 23-25 times greater than air, making water an efficient medium for heat loss exhibited by both animals and humans moving towards sources of water during periods of intense solar radiation and high ambient temperatures (T_A). Wind is also an efficient mode of convection. When cool air flows over the warmer skin surface of an animal, this may allow that organism to remain active longer on warm, sunny days. Convection even occurs within the body. Blood transfers heat throughout the body, bringing warmth from the core to the extremities, or bringing cooled blood from the surface back to overheated organs or muscle tissue.

1.2.4 Evaporation

Evaporation uses the energy required to convert liquid water to gas, allowing organisms to transfer heat even if the T_A is greater than T_B. This is often the only mechanism available to organisms during prolonged exposure to solar radiation that efficiently dumps excess heat in an effort to maintain a safe T_B. Both environmental as well as metabolic water (Figure 2) can be used to transfer heat from an organism's body to the surrounding air (Gupta, 2011). Though highly effective, if the air is saturated with moisture already, such as during periods of high relative humidity or in constantly humid environments, evaporation as a thermoregulatory mechanism is rendered almost useless because net evaporation ceases when the air can no longer absorb additional moisture (Speakman, 2004).

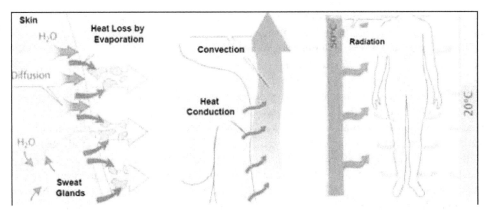

Fig. 2. Heat loss via evaporation, convection, conduction, and radiation using the human body as an example. Source: Gupta, 2011.

1.3 Ultraviolet radiation

The emitted solar spectrum contains ultraviolet (UV) radiation that is separated into different wavelength bands: UVA (315-400 nm), UVB (280-315 nm), and UVC (200-280 nm). UVC and 70-90% of UVB wavelengths are typically blocked by the ozone layer (Rafanelli et al., 2010; Schaumburg et al., 2010). However, due to ozone deterioration there is an increase in UVB levels, especially over Antarctica. Certain environments, such as polar areas, are more susceptible to varying levels of UV radiation (Rafanelli et al., 2010). Organisms across all taxa are directly or indirectly affected by UV radiation. Such responses are discussed in more detail later in this chapter.

1.4 Infrared technology

Not too long ago, infrared technology was out of reach of the scientific community. Cameras with infrared (IR) imaging were large, poorly maneuverable as field equipment, and expensive. However, it is currently possible to obtain hand held thermal cameras for under $2,000. The introduction of smaller, manageable, and relatively inexpensive equipment has opened the gates for biologists to begin using infrared imaging in their research. Already a number of experiments have validated their use and lauded their capability to be invaluable to research. Thermal images can be obtained to document the variation on an organism's body surface owing to different types and properties of insulation, or other thermal or physiological properties of an organism's body. This provides researchers with added insight on both the physiological and behavioral effects that solar radiative heat gain may have on thermoregulatory responses of animals.

2. Ectotherms

2.1 Invertebrates

Insects and other invertebrates live within their own microclimes and just a small change in position can affect their T_B. Some species of ants, for example, move between nest areas during seasonal changes or use objects such as leaves or rocks to alter their insulative properties. Fire ants are able to raise their nest temperature by modifying their nest shape relative to the sun's angle, while wood ants use solar radiation in combination with behavioral modifications to maintain ideal nest temperatures (Vogt et al., 2008; Galle, 1973). Non-communal insects such as grasshoppers bask to increase their T_B (O'Neill & Rolston, 2007). Butterflies in general often bask or use ground contact to gain heat by conduction and convection then seek shade or minimize their exposed wing surface area to either facilitate heat loss or decrease heat absorption (Clench, 1966). Basking monarch butterflies take to the air during periodic cloud cover before returning to their clusters (Calvert et al. 1992). The lack of direct sunlight is enough for them to expend energy, creating heat through movement of their body. The difference in temperature from direct solar radiation to cloud cover likely stimulates this adaptive response in butterflies, causing them to return to the safety of the colony in the chance that any sustained cloud cover could lead to lower thoracic temperatures during migrational flight (Calvert et al., 1992). Several species of Arctic butterflies select specific basking substrates and orient their wings perpendicular the sun, allowing them to absorb heat in order to continue locomotion even at very low T_A (Kevan & Shorthouse, 1970). Without the ability to maintain their threshold temperature during flight, stranded butterflies could be subjected to increased predation and risk exposure to cold stress.

Honeybees use solar radiation at T_As under 30°C, the minimum thermal threshold, as an alternative to generating energy to raise thoracic temperatures. Relying on solar radiation to increase T_B enhances muscle efficiency during flight and allows the bees' suction pump to function at low T_A. If overheating occurs, bees actively seek water to ingest, using the liquid to cool thoracic temperatures (Kovac et al., 2010). Other insects such as wasps also use solar radiation to increase thoracic temperature so their own active production of heat is reduced (Kovac et al., 2009).

Insects are not the only invertebrates to be affected by solar radiation; those living within the intertidal zone of oceanic coastlines are subjected to solar rays as well. Although many organisms can avoid direct solar radiation by moving to safe hiding places between rocks, in tide pools, or in the sand via burrowing, others are limited in their ability to use locomotion to do so. For example, marine snails will flee from areas exposed to direct solar radiation and move to cooler areas close by often in shade or under the waterline (Chapperon & Seuront, 2001). Periwinkles are unable to flee the declining water level and are thus equipped to withstand relatively short periods of desiccation and thermal stress. In fact, they often orient themselves frontally or dorsally facing the sun to limit the amount of sunlight hitting their lateral surfaces. By reducing the surface area perpendicular to the radiation, periwinkles can maintain a lower T_B when subjected to the desiccating heat and solar intensity during low tides (Muñoz et al., 2005).

2.2 Amphibians and reptiles

Small amphibians and reptiles must be careful to maintain a certain T_B and are often affected by microclimate changes. It is well known that ectotherms use basking to supplement heat gain, making solar radiation an important thermoregulatory tool (Nagy,

2004). Turtles will often choose habitats with access to direct solar radiation and bask on sunny days (Dubols et al., 2009). Ectotherms will frequently expose a maximum amount of surface area toward the sun, regulating the interception of solar radiation through body orientation. Some reptiles have dark patches to aid in heat absorption. Certain species of lizards orient their bodies either parallel or perpendicular to the sun's rays based on T_A. Lizards living in high altitudes increase basking frequency, are more likely to orient perpendicular to the sun, and restrict durations of activity to minimize the range of temperatures to which they are exposed (Adolph, 1990; Hertz & Huey, 1981). However, this can raise their chances of being captured by a predator. Kestrels in Norway return to their nests with an increasing amount of lizards, which typically peaks in midday. Solar intensity, as well as T_A, often influences the abundance of lizards available to kestrels due to possible patterns in lizard behavior (Steen et al., 2011). The giant tortoise from South Africa is another species known to orient itself based on solar radiation. When facing away from the sun, the carapace casts a shadow on the head and neck area allowing the tortoise to inhibit the rate at which its T_B increases. This maintenance of a larger thermal gradient between the skin surface and T_A allows the tortoise to forage for longer periods under direct solar radiation (Coe, 2004). Some species of toads living at high altitudes are more active on sunny days, raising their T_B by 20° or more in some cases (Lambrinos & Kleier, 2003). Heat gain using solar radiation also aids in digestion and increases growth rates of toads and other anurans at both extreme and moderate elevations. When food is withheld, toads end basking and allow their T_B to decline (Lillywhite et al., 1973). Dependence on solar radiation makes it possible for most species of reptiles and amphibians to live at high elevations in tropical regions.

3. Endotherms

Evaporative cooling, relying on environmental or body water, is an important mechanism used by many endotherms to avoid overheating. Numerous species have evolved behaviors that facilitate heat loss or minimize heat gain from solar radiation. Evaporation can occur passively through the skin of mammals and birds. Animals lacking sweat glands rely on saliva spread onto the surface of their body or may pant. By changing their posture or orientation to the sun, some animals expose a larger percentage of their surface area to cooler substrates allowing heat to be conducted away from their bodies. Solar radiation correlates with changing postures or orientations to reveal more of their surface area and is employed by some species such as sea lions to increase or decrease exposure to solar radiation. Anteaters, whose prey is not very energy rich, often use solar radiation to offset metabolic costs of thermoregulation by avoiding sunlit areas on hot days. They also switch their foraging behavior to nocturnal periods when days are exceptionally hot (de Sampaio et al., 2006).

3.1 Ratio of surface area to volume

Organisms with a large ratio of surface area to volume such as small passerines and mammals must be careful to maintain heat balance, as they are prone to rapid heat exchange. They require a large amount of energy intake to aid in heat balance; birds and bats even more so due to the high metabolic cost of flight. Treecreepers, a type of bird found in forests of Spain, selectively forage on certain areas of trees depending on the T_A. In warm temperatures the birds tend to forage on shaded areas whereas they prefer areas exposed to sunlight in cooler temperatures (Carrascal et al., 2001). By maintaining T_B using behavioral thermoregulation the birds are able to save energy that can be allotted to other expenses such as predator avoidance, especially since foraging in sunlit patches increases their predation risk. In fact, the overall abundance and species richness of birds in a montane forest was calculated to be mainly a function of solar radiation (Huertas & Diaz, 2001). By choosing areas higher in available sunlight, small birds can lower their metabolic cost of thermoregulation and improve survival in cold temperatures. In warm environments small birds are able to balance thermoregulation by avoiding sunlit areas. Verdins are able to reduce their metabolic rates by shifting to wind and solar-shielded microsites, allowing for a highly reduced rate of evaporative water loss. In order to maintain efficient use of energy and water, verdins must actively select thermoregulatory beneficial areas within their habitat (Wolf & Walsburg, 1996).

3.1.1 Torpor

Small mammals often use torpor as a thermoregulatory strategy. By decreasing T_B and metabolic rate, small animals can substantially reduce energy expenditure. Though it requires energy to arouse from torpor, some mammals can choose sites exposed to solar radiation during the day and are able to bask in sunlight, using the passive absorption of radiant heat to rewarm their body and profit from the energy saved while torpid (Warnecke et al., 2008).

3.2 Birds

Birds lack sweat glands so they must rely on other mechanisms of heat reduction. Even something as simple as changing orientation in relation to the sun can have a large effect in reducing heat gain. Flying is an energetically expensive mode of locomotion that can generate high heat loads. Herring gulls, for example, are able to lessen heat gained by reducing exposed surface area and orienting their bodies in a way that presents only the white-feathered surface, allowing for a greater degree of reflection rather than absorption of solar energy (Lustwick et al., 1978). The Great Knot, a shorebird from Australia, raises its back feathers to initiate a possible increase in convective or cutaneous cooling (Battley et al., 2003). Cormorants often spread their wings on land; most likely a wing-drying strategy but also a mechanism used to dissipate heat. During periods of low wind these cormorants rely on the sun to dry their plumage. This behavior saves cormorants the large energetic cost it would take to dry their feathers using metabolic heat to evaporate the water (Sellers, 1995).

Depending on the environment, many incubating birds that are restricted to their nest site must tolerate direct solar radiation. For example, the Heermann gull resorts to ptilomotor responses, posture changes, and increased respiration through a gaping mouth, employing several mechanisms of heat dissipation to offset heat gain from direct radiation as well as conduction from the nest substrate (Figure 3) (Bartholomew & Dawson, 1979).

Fig. 3. Thermoregulatory posture changes in the Heerman gull. Source: Bartholomew & Dawson, 1979.

Kentish plovers abandon their nests when exposed to high T_A and solar intensity. This species rushes to a nearby water source to soak their ventral surface, quickly reducing their T_B by convection. Although this raises the possibility that egg temperature will increase to dangerous levels in the absence of the individuals that are incubating the egg, overall this behavior allows the plovers to stay at their nests for longer periods (Amat & Masero, 2009). Adult birds and their chicks nesting in extreme environments such as the arctic tundra are exposed to long periods of sunlight along with low T_A, low sun angles and unobstructed wind. By exposing more surface area (i.e. facing away from the sun), Greater Snow goose chicks, which have yellow down in comparison to the grey down and white feathers of older chicks and adults, are able to increase the amount of radiative heat gain. The plumage of younger goslings may even absorb solar radiation at a higher rate than adult plumage (Fortin et al., 2000). Grackle chicks exhibit shade seeking behavior, moving around the edges of the nest in attempts to keep their highly vascularized heads shaded. These chicks also orient their bodies toward the sun to diminish absorptive surface area which is a behavior also seen in chicks of Ferruginous hawks, gulls, and adult titmice and chickadees (Glassey & Amos, 2009; Tomback & Murphy, 1981 Bartholomew & Dawson, 1954 Wood & Lustick, 1989 in Glassey & Amos 2009). Moderating heat load is extremely important, especially by nestlings and small passerines, for without such behavior or access to shade in high T_{AS} small birds can succumb to heat stress within 20 min (Glassey & Amos, 2009).

3.3 Mammals

Most species of mammals, particularly larger ones, are well insulated with fur, fat, or a combination of both. Often this creates conflicting thermoregulatory demands depending on the gradient between body and ambient temperature as well as environmental variables such as solar radiation, wind speed, humidity etc. A variety of behavioral responses to solar radiation exposure have evolved in mammals.

3.3.1 Behavioral

Many mammals use simple methods such as body orientation or posture changes to balance radiant heat gain from solar radiation. The black wildebeest, which inhabits the savannah, a habitat that has little natural shade, orients itself either parallel or perpendicular to the sun depending largely upon skin temperature. As the intensity of solar radiation increases, wildebeests are more likely to change their position when standing so their bodies are parallel to the sun's rays. By minimizing the surface area exposed to solar radiation in a parallel orientation, wildebeests absorb 30% less radiant heat than if they stood in a perpendicular stance (Maloney et al., 2005). The angle of the sun, both daily and seasonal, also affects wildebeest body orientation. During the cooler season, when solar angle and intensity is reduced, the preference of body orientation decreases. By reducing heat gain these large mammals are able to lessen the energy needed for evaporative cooling. However, this orientation behavior decreases when there is a reliable source of water available (Maloney et al., 2005). Other African species, such as the eland, blue wildebeest, and impala also use preferential body orientations, positioning their bodies parallel to the sun in summer and perpendicular in winter (Figure 4) (Hetem et al., 2011). Once again, energetic demands drive these behaviors relative to the amount of direct solar radiation.

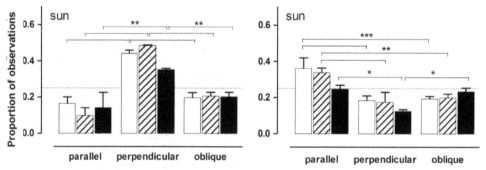

Fig. 4 Proportion of observations (mean7SD) in which eland (open bars), blue wildebeest (hatched bars) and impala (black bars) orientated parallel, perpendicular and oblique to incident solar radiation in winter (left panels) and summer (right panels). The dotted line represents the proportion expected if orientation was random (0.25); significant differences between orientations indicated by *P<0.05, **P<0.01, ***P<0.001. Source: Hetem et al., 2011

Marine mammals that haul out on land, such as the pinnipeds, face a thermoregulatory challenge when it comes to reproductive or molting periods. Developed for efficient locomotion and thermoregulation in an aquatic environment they often have thick fur, blubber, or a combination of both to protect them from frigid temperatures beneath the ocean surface. On land, however, these insulative properties, along with a low ratio of surface area to volume, become a hindrance to heat transfer across the body surface. The flippers of sea lions are long, hairless, and often thought to have a thermoregulatory function. South Australian fur seals, New Zealand and California sea lions respond to solar radiation by regulating the amount of flipper surface area that is exposed to solar radiation (Beentjes, 2006; Gentry, 1973). As temperature and solar intensity increases, sea lions unfold from a prone position with flippers tucked beneath their body to a dorsal up position with all four flippers laid out on the sand and exposed to the air for conductive and convective heat transfer (Figure 5) (Beentjes, 2006; Gentry, 1973).

Of course, the best method to inhibit solar radiation is to block or avoid it. Cape ground squirrels use their own tails to shade their bodies as well as orient themselves so that their backs are oriented toward the sun when the T_A exceeds 40°C. T_B is actually reduced over 5°C, allowing these squirrels to continue foraging for longer periods when exposed to direct solar radiation (Bennett et al., 1984). Male South Australian fur seals exhibit shade seeking behavior, not only blocking solar radiation but also transferring heat via conduction to a cool rocky substrate (Gentry, 1973). Marine mammals as large as the northern elephant seal, which periodically haul out on land, are able to use sand to block some of the solar radiation they would otherwise be subjected to during breeding or molting periods. This behavior known as sand-flipping, observed in some species of pinnipeds that use their fore-flippers to scoop sand up onto their backs, increases in New Zealand sea lions as solar radiation increases (Beentjes, 2006). The layer of moist and/or cooler sand facilitates heat transfer through conduction as well as creates a barrier against direct solar radiation. Southern sea lions have been observed to dig their foreflippers into the cool substrate to shield themselves from the sun's rays (Campagna & Le Boeuf, 1998). Flipper waving is another behavior often seen in several species of pinnipeds, exhibited by seals lying on their side, raising a flipper and sometimes moving it back and forth, using convection to diffuse heat from the body before letting it rest again. As solar radiation warms the substrate, heat gain increases through conduction and reflection, making behaviors such as posture changes, sand-flipping, and flipper waving by hauled-out pinnipeds increasingly beneficial to their overall energy balance. For some pinnipeds, solar radiation is often an indirect or combined stressor when associated with T_A and/or wind speed. Intense solar radiation, such as the levels recorded near the equator and other tropical regions, may only be tolerated if evaporative cooling is used, which may partially explain why tropical pinnipeds are often found near upwellings of cold water rather than in warm ocean currents.

Daily and seasonal migrations are often a result, at least in part, of solar radiation intensity. The marked ibex, which lives in arctic environments, has a low tolerance for heat gain. During the summer, males change their behavior, feeding mainly in the early morning rather than midday or evening. As solar radiation increases throughout the day, the marked ibex migrate to higher elevations, thus using the cooler air to reduce heat gain. Throughout the afternoon and evening, this species feeds in the cooler hours of the day before moving to higher altitude as T_A increases (Aublet et al, 2009).

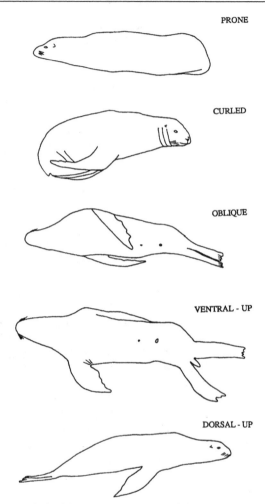

Fig. 5. Postures (prone, curled, oblique, ventral-up, and dorsal-up) used by New Zealand sea lions as solar radiation and ambient temperature increase. Source: Beentjes, 2006.

Seasonal migration can also be affected by solar radiation. Models using data collected over long periods or seasons are created to calculate the range of intensities of solar radiation over a large area such as a nature reserve or park. Collared pandas within Foping Nature Reserve, China have been radio tracked and their distributions in the park were overlaid with a map of 12 months of solar radiation data. During the warm months pandas moved to areas with lower solar radiation, whereas during the colder months they were recorded in areas of higher radiation. The model suggested that solar radiation does affect the distribution of giant pandas (Liu et al., 2011).

Smaller mammals are prone to losing heat rapidly due to their high ratio of surface area to volume but they are also susceptible to losing water through evaporation at a rapid rate. Degus, a species of rodent found in arid and semiarid environments, minimize the distances they travel out in the open, away from the shelter of scrub brush. One hypothesis for this behavior is predator avoidance, but another possibility is to avoid heat gain in areas subjected to direct solar radiation. Degus also exhibit seasonal changes in behavior, reducing activity near midday in summer while remaining steadily active during winter. Temperatures ≥30°C have not been measured in the microhabitat beneath the shrubs, preventing degus from reaching hyperthermic body temperatures while they remain sheltered from solar radiation. In order to both avoid predation and maintain efficient heat balance degus use the microhabitats underneath shrubs, especially during periods of high ambient temperature (Lagos et al., 1995).

Various species of bats roost in trees, exposed to sunlight and other environmental variables during day, unlike cave dwelling bats who are contained in a relatively stable microclimate and shielded from solar radiation. The wing membranes of bats, like sea lion flippers, are naked and incorporate a large amount of overall body surface area. This makes bat wings a likely tool for thermoregulation. Flying foxes, found mainly in the tropics, likely lack sweat glands and are exposed to the high temperatures and humidity of the forests they inhabit. These bats exhibit wing fanning and body licking, incorporating evaporation and convection to facilitate heat loss (Ochoa-Acuña & Kunz, 1999). Through exposure of greater amounts of wing surface, flying foxes can increase the area available for heat transfer as required during periods of increasing body and ambient temperature. Other species have special adaptations to aid them in releasing excess heat. The Brazilian free-tailed bat is known to fly during periods of daylight in the warmer environments it inhabits as well as taking part in long migrations. These bats have a unique vascular radiator lacking any insulative fur along their flanks (Reichard et al., 2010a). By flushing the area with warmed blood Brazilian free-tailed bats are able to efficiently dump heat when necessary while conserving body heat in the high altitudes they forage in by shunting blood away from the radiators (Reichard et al., 2010; Reichard et al., 2010b).

3.3.2 Physiological

Solar radiation is not only important when researching wild mammals but domesticated animals as well, especially livestock left out in large pastures with little to no shade. When cattle, goats, and sheep are exposed to the sun, they experience greater heat loads than those present in enclosed shelters or in shaded areas (Al-Tamimi, 2007; Brosh et al., 1998; Sevi et al., 2001). For some species of cattle, a high heat load during the summer results in reduced growth and reproductive rates, causing a decrease in overall productivity (Brosh et al., 1998). Dairy cows experience a decrease in fertility when under severe heat stress, more so in the summer than the winter (De Rensis & Scaramuzzi, 2003; Schütz et al., 2009). In addition to behaviors like shade-seeking, physiological responses such as increased respiration rate, heart rate, skin temperature, and of course high T_B are often the first signs of heat stress resulting from direct solar radiation. Above the UCT, evaporation through respiratory and cutaneous water loss can aid in reducing T_B. Blood vessels dilate near the skin surface allowing increased blood flow to areas that facilitate heat loss through multiple modes of heat transfer. This response is only efficient as long as blood temperature is less than T_A; because it is the thermal gradient that drives heat loss across the skin barrier. Heat loss can be supplemented by behavioral changes such as feeding in the late afternoon or even at night so that the heat increment of feeding is produced during cooler periods, allowing heat loss through both conduction and radiation. Increased rates and total intake of water is often observed in livestock, compensating for the water lost through evaporation. Some species of goats and cattle are able to reduce their T_Bs in the early morning as a preparatory strategy for the increased amount of solar radiation they will be exposed to during the warmer parts of the day (Al-Tamimi, 2007; Brosh et al., 1998).

3.3.2.1 Evaporation

Many animals are able to use their own body parts, fluids, or environment to reduce heat absorption or expedite heat loss. Since water transfers heat 25X faster than air, metabolic as well as environmental water is used in evaporative cooling. It is beneficial to have an external water source nearby as the rate of body water turnover increases with solar radiation in most species, though at different levels depending on body mass and other characteristics (Figure 6) (King et al., 1975). Environmental water is an excellent source if available and many animals, including humans, use rivers, lakes, ponds, and the ocean to transfer excess heat. South Australian, Northern fur seals and Stellar sea lion females are able to tolerate solar radiation by keeping themselves wet, via mass movements to the shoreline while New Zealand sea lions immerse their hind flippers in tide pools or stay within the splash zone (Beentjes, 2006). Humans and other organisms that have sweat glands secrete the plasma portion of their blood onto the surface of their skin where it evaporates, the heat energy being used to transform liquid to gas. Organisms lacking sweat glands must devise other means of cooling their T_B effectively. Dogs, cats, even sea lions, pant when they become overheated, using the evaporation of their saliva and moist tissue to dump heat. Kangaroos lick their own wrists where the skin is thinner and blood vessels are closer to the surface, aiding heat balance in the high solar and arid regions of Australia (Dawson et al., 2000). South Australian fur seal males unable to gain access to water due to territorial defense use their own urine for evaporatory cooling. After urinating on the rocks, males wet their ventral side and rear flippers, then lie on their side and raise a hindflipper into the air to enable convective heat loss. Female Stellar sea lions are often seen to huddle around small tidepools with increasing temperatures (Gentry, 1973).

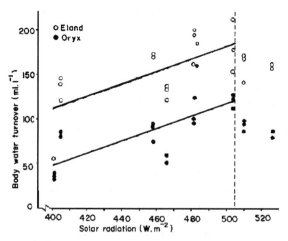

Fig. 6. Relation between daily total body water turnover and solar radiation in domestic Eland and oryx under African ranching conditions. Source: King et al., 1975.

4. Reproduction

Breeding demands a large amount of energy intake by both sexes, even more so when males must physically compete with one another to gain access to one or several females. Pinnipeds all have blubber layers of varying thickness to insulate them against cold water. Elephant seals are no exception, with males weighing over 2 tons. Male northern elephant seals must compete for reproductive access to females and physical combat among competing males can last upwards to 45 minutes (Norris et al., 2010). Blubber acts as insulation both in and out of water but is much heavier on land. Combative males produce a great amount of extra heat that must be expelled during and immediately after physical interactions with other males. Weaker males may retreat to the ocean for multiple cooling avenues that aid in thermoregulation but alpha males, if they wish to maintain their access to females and increase mating success, cannot leave their harem and thus are subjected to environmental variables such as T_A, solar radiation, wind, and humidity. The only heat loss mechanisms readily available to alpha males on land are conduction and radiation to a cooler substrate and convection from prevailing winds (Norris et al., 2010).

Solar radiation has a significant effect on the circulatory physiology of male northern elephant seals. Infrared thermal images show that certain areas of skin function as thermal windows, vasodilating the blood vessels and shunting warm blood directly to the skin surface, facilitating heat transfer to the air by convection or substrate by conduction. By increasing skin temperature in specific areas, males are able to increase the temperature gradient more so than if blood was perfused along the entire body surface. On warm days, conduction and radiation work against the males, thus on days with high solar radiation and low wind the males are inactive, allowing skin temperature to rise 42°C (Figure 7) (Norris et al., 2010). In this manner they can lower their metabolic rate and conserve energy. Even pups of the California sea lion have been observed to sleep and stay still during periods of intense solar radiation for the same reason. On days when wind speed is high and/or clouds block direct solar radiation, male elephant seals are more likely to be active or engage in combat behavior, enabling them to rely on convection or even evaporation via precipitation to increase the rate of heat transfer (Norris et al., 2010).

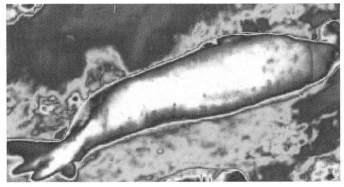

Fig. 7. False color thermal image of a male northern elephant seal; max skin temperature of 42°C. Source: Norris unpublished data

Males of some pinniped species claim a territory rather than a harem and must rely on the habitat within their territory to entice or gain access to reproductive females. For example, South Australian fur seals exhibit several behavioral mechanisms to dissipate excess heat before finally abandoning their territory for access to water as T_B approaches a certain thermal threshold often correlated with substrate (rock) temperature. Reproductive success of male New Zealand fur seals may depend on how much water is present within their territory. Males with areas including tide pools or along the shoreline have more access to females than males with no water in their territory. Male Northern fur seals that abandon their territories suffer a 50% reduction in mating success, whereas males of Southern Australian fur seal with no access to water average only 1.7 copulations per male compared to 3.6 copulations per male near a water source (Gentry, 1973). Southern sea lion males with territories lacking water access experience half the copulation frequencies of males with available water. During their forays to the water line, female Southern Australian fur seals expose themselves more frequently to males versus females that remain stationary and thus may contribute to an increased pregnancy rate (Campagna & Le Boeuf, 1998).

5. Ultraviolet radiation

Ultraviolet light has a wavelength shorter than visible light and is named as such due to the emitted electromagnetic waves with frequencies above that of visible violet light. UVA and UVB wavelengths emitted by the sun affect animal behavior and physiology. UVA radiation is exploited by some pollinating insects to facilitate the detection of flower condition for nectar production in pollination. UVB radiation may affect reproduction and development as well as synthesis of vitamin D (Schaumburg et al., 2010). Overexposure to UVB is known to increase mutation rates in individual cells of whole organisms. The targets of UVB damage within living cells include nucleic acids, proteins, and lipids. Damage to these can inhibit DNA and cellular processes as well as impair cell membranes.

5.1 Aquatic organisms

Wavelength and intensity of solar radiation are modified as they travel through water but at shallow depths UV radiation is only able to penetrate a certain distance. Organisms found in epipelagic and littoral zones include phyto- and zooplankton, invertebrates, fish eggs and larvae, as well as entire ecosystems such as coral reefs. Some species of zooplankton change their vertical migration patterns within aquatic habitats to avoid exposure to UV radiation. Others respond by increasing pigmentation, an energetically costly process that can increase the organism's visibility to predators (Häder et al., 2007). The photochemical efficiency of some algae, such as those responsible for red tide, decreases when subjected to high solar radiation for short periods of time. However, if the algae are exposed to UV radiation for prolonged periods, they may be able to acclimate to this extended time by increasing protein repair and synthesizing UV-absorbing compounds (Guan et al., 2011). UV radiation induces decreased production of biomass such as phytoplankton, which can lead to a reduced capacity to sequester carbon dioxide, a gas that causes increased insolation of the Earth's atmosphere and is postulated as a major contributor to climate change (Hoffert & Caldeira, 2004).

Although many ectotherms are able to use solar radiation as a heat source, ultraviolet radiation can have detrimental affects, especially on certain early development stages. In larval stages of some fish, UV radiation can affect development, increase mutation rates, or cause skin and ocular damage. When exposed to full solar UV radiation, yellow perch eggs actually perish before they hatch (Häder et al., 2007). Invertebrates such as sea urchins are also inhibited by UVB radiation at different life stages. In some species apoptosis or abnormal development occurs in the embryos, while adults of other species exhibit a covering behavior, using pieces of debris to block direct contact with solar rays penetrating through shallow water (Häder et al., 2007; Nahon et al., 2009). UVB radiation is known to decrease survivability of amphibian embryos or larvae depending on species (Häkkinen et al., 2001).

The mortality of coral reefs throughout the world has been well documented and is partly attributed to rising ocean temperatures, pollution, and UV radiation. Bleaching occurs at a thermal threshold and the coral dies soon after the photosynthetic energy producing zooxanthellae are expelled. Corals exposed to high levels of UV radiation receive damage to both the symbiotic zooxanthellae and the coral tissue (Lesser & Farrell, 2004). Pathways involving carbon fixation and photochemistry in the zooxanthellae along with DNA damage and necrosis of the host coral tissues are the results of thermal stress due to the high irradiance of solar radiation. The presence of high UV radiation can lower the bleaching threshold, decreasing the time it takes to bleach in an environment that may otherwise not be as stressed. However, some coral reef species are able to sequester substances acquired through their diet into UV-absorbing elements (Dunne & Brown, 2001).

5.2 Terrestrial organisms

UVB radiation is known to both negatively and positively affect species. While growth of leaves and stems are inhibited by UVB along with reduced daytime seedling emergence and biomass in early stages of growth, some species of plants exposed to the radiation are less likely to be attacked by leaf beetles (Ballere' et al., 1996). Other plants such as soybeans also receive reduced herbivorous damage when exposed to UVB (Zavala et al., 2000). Thrips, insects that feeds on plant leaves, actually avoid exposure to UVB solar radiation, suggesting that insects can behaviorally respond to that particular wavelength presence (Mazza et al., 1999). UVB can also have an indirect effect upon some species of insect larvae, as those who eat UVB radiated plant material are found to have decreased growth rates and suffer more mortalities versus larvae that feed on non-radiated plant matter (McCloud & Berenbaum, 1994).

Humans are sensitive to solar radiation and may experience sunburn, heat strokes, eye diseases, and skin cancer when overexposed. Low doses of UV radiation are required for vitamin D synthesis and can be used to treat some illnesses. By regulating dosages of UVB radiation to humans with allergic chronic dermatitis, nickel sensitivity, and psoriasis, physicians are able to successfully treat and suppress hypersensitivity. Photoimmunology is a relatively new field and human experiments are often rare due to ethical guidelines and research protocols. Immune system responses by humans to prolonged exposure to UV radiation are usually detrimental. UV damage often includes changes in intracellular signaling, T-cell numbers in exposed skin and inhibition of natural killer cell activity

(Duthie et al., 1999). Even humans with varying levels of pigmentation in their skin, such as those of diverse racial descent, are affected differently when exposed to UVB (Coelho et al., 2009; Duthie et al., 1999). Langerhans cells, antigen-presenting cells within the epidermis, die due to membrane disruption and organelle damage in Celtic descendants while cells in the darker skin of Aboriginal or Asian Australians are depleted by apoptosis (Duthie et al., 1999). The melanin produced in the skin when humans are exposed to UV radiation is actually able to absorb UV, shielding nuclear DNA (Coelho et al., 2009). Long term exposure can lead to degeneration of skin cells, underlying fibrous tissue and/or blood vessels leading to premature skin aging and skin cancer. Other possible effects include ocular inflammation and cataracts that can cause blindness. Most tissue damage is due to high levels of UVB wavelengths while more indirect damage via reactive oxygen intermediates affecting DNA, proteins, and lipids is caused by UVA radiation (Figure 8) (Rafanelli et al., 2010).

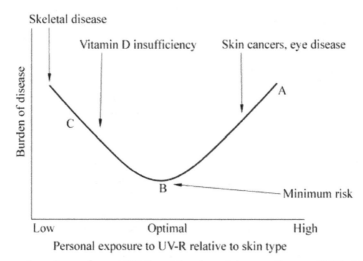

Fig. 8. Relationship of exposure to UV-R and burden of disease. Source: WHO, Ultraviolet radiation and the INTERSUN Programme in Rafanelli et al., 2010

Vitamin D is necessary for healthy growth and function of most terrestrial vertebrates. Exposure to direct sunlight allows UVB photons to enter the skin and begin a chain reaction resulting in the formation of vitamin D. This nutrient is essential for efficient intestinal absorption of calcium, especially in humans (Holick, 2008). However, an organism's diet is often a supplemental source of vitamin D in addition to providing the building blocks of vitamin D synthesis. In bats, diet strongly affects the levels of circulating vitamin D metabolite. Certain cave dwelling species, receiving little to no direct sunlight, actually have the highest recorded concentrations in vertebrate taxa. Unlike the plant visiting bats, which have levels not even sufficient for humans, sanguivorous and piscivorous species have access to dietary vitamin D, increasing their serum concentrations of the vitamin D metabolite (Southworth et al., 2009). Ectotherms, who often bask in sunlight, are able to use this behavior to regulate their own vitamin D levels when dietary intake is insufficient

(Karsten et al., 2009). Deficiency of vitamin D in humans can lead to increased risks of bone fractures from osteoporosis, common cancers, cardiovascular disease, autoimmune and infectious diseases (Holick, 2008).

6. Infrared thermal cameras

A specialized sensor in units of Watts/m^2 normally measures solar radiation while a UV meter is needed to determine UV radiation either by wavelength or a UV index. The effects of solar and UV radiation on organisms are harder to quantify. Observed behaviors can be statistically associated with monitored solar intensity and/or UV radiation to calculate whether there is a significant effect. Physiological responses however, must be measured with other equipment or using procedures involving immobilization, radioactive tracers etc. Such methods can cause stress to the animal in question or establish a bias in some way as is often the case with attaching equipment to wild animals or bringing them into a laboratory setting. Often the variable being monitored is some form of temperature, be it core/body, skin, or intramuscular. Temperature is often the first physical measurement affected by solar radiation. Infrared radiation (IR), simply put, is the thermal emission of an object or organism. Thermal infrared cameras are able to detect various wavelengths in the infrared spectrum that can be accurately measured based on the temperature of a surface using the physical laws of radiative transfer.

The use of thermal cameras is relatively new to scientific field research. In the past, thermal cameras were expensive, large, bulky, and hindered by the need for a large external power source. As the technology has progressed, the IR detector itself has become smaller and thermal cameras have been used for military applications, on airplanes, helicopters, and today the individual soldier. Currently a handheld IR camera can be purchased for under $2,000 and is often used by electricians, engineers, and construction companies to detect problem areas in work zones and other facilities. Scientists too, are now able to purchase high definition IR cameras that can be easily carried into the field and wirelessly linked with data loggers and tablets.

Thermal cameras allow for a non-invasive scrutiny of an organism's skin surface temperature, giving the ability to see temperature variability across the body rather than just one to a few individual areas recorded by a thermocouple or other device (Figure 9) (McCafferty, 2007; Norris et al., 2010). Clear links between heat loss and areas of the body such as the head and appendages have been established due to thermal imaging (McCafferty, 2007; McCafferty et al., 2011). Behavioral responses to solar radiation and other environmental variables can be paired with changes in skin surface blood circulation. The surrounding environment is imaged at the same moment as the organism being studied, allowing for temperature measurements of the substrate to be collected for future analysis. Thermograms of insects such as bees make measuring individual or hive temperatures quick and simple without having to use smoke or gaseous chemicals that take time and may affect the bees' thermoregulatory properties (Kovac et al., 2010). Thermal video of emerging bats can be recorded and their flight trajectories tracked using specialized computer analysis programs (Hristov et al., 2008). Thermal imaging can even be used for plant and crop research, measuring temperature in relation to solar radiation absorption and water treatment (Jones et al., 2009).

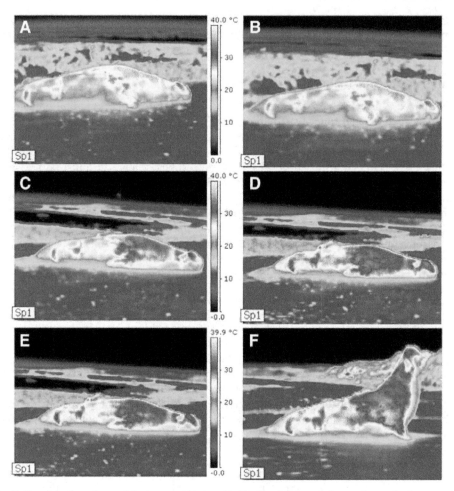

Fig. 9. Post combat thermal images sequence (A-F) of an alpha male northern elephant seal taken at intervals of ~1min immediately following a combat lasting 11 min. Skin temperature variability is clearly displayed over time. Source: Norris et al., 2010.

The use of thermal cameras for biological study has been well validated in the past for both individual studies and those involving groups of animals (McCaffery, 2007). Using either the IR camera or computer software paired with it, maximum, minimum, and mean temperatures can be measured and calculated rapidly and efficiently. However, distance from the organism, emissivity, relative humidity, and T_A must be known to reduce errors in the camera's detection of surface temperature. Solar and UV radiation measurements can now be statistically paired with the temperature data, skin variability in temperature, thermal gradient between skin temperature and ambient/substrate temperature, and other environmental or physiological measurements. Use of IR cameras eliminates the need to capture or handle animals in physiological, behavioral, and ecological studies involving thermoregulation.

7. Conclusions

Solar radiation is essential for life, transferring energy to plants that form the basis of food webs. Assorted wavelengths emitted by the sun initiate a variety of responses across most species. Organisms have evolved thermoregulatory behaviors and responses to concentrated or extended periods of solar radiation as well as to changes in ambient temperature, which is also affected by solar intensity. Ultraviolet radiation can have both beneficial and detrimental effects on organisms depending on dosage and wavelength. Infrared radiation can be used as a measure of surface temperature using thermal infrared cameras, now relatively affordable, and have been validated for scientific research.

8. Acknowledgments

We wish to thank the organizers of this volume for inviting us to prepare this chapter. Preparation of this paper was supported by a grant from the Office of Naval Research to John Baillieul, Margrit Betke, Calin Belta, Yannis, Paschalidis, and Thomas Kunz.

9. References

Adolph, S.C. (1990). Influence of behavioral thermoregulation on microhabitat use by two sceloporus lizards. *Ecology*, Vol.71, No.1, (February 1990), pp. 315-327, ISSN 0012-9658

Al-Tamimi, H.J. (2007). Thermoregulatory response of goat kids subjected to heat stress. *Small Ruminant Research*. Vol.71, No. 1-3, (August 2007), pp. 280-285, ISSN 0921-4488

Amat, J.A., Masero, J.A. (2009). Belly-soaking: a behavioural solution to reduce excess body heat in the Kentish plover *Charadrius alexandrines*. *Journal of Ethology*. Vol.27, No.3, (September 2009), pp. 507-510, ISSN 0289-0771

Aublet, J.F., Festa-Bianchet, M., Bergero, D., Bassano, B. (2009). Temperature constraints on foraging behavior of male alpine ibex (*Capra ibex*) in summer. *Oecologia*. Vol.159, No.1, (February 2009), pp. 237-247, ISSN 0029-8549

Ballere', C.L., Scopel, A.L., Stapleton, A.E., Yanovsky, M.J. (1996). Ultraviolet-B radiation affects seedling emergence, DNA integrity, plant morphology, growth rate, and attractiveness to herbivore insects in *Datura ferox*. *Plant Physiology*. Vol.112, No.1, (September 1996), pp. 161-170, ISSN 0032-0889

Bartholomew, G.A., Dawson, W.R. (1979). Thermoregulatory behavior during incubation in Heermann's gulls. *Physiological Zoology*. Vol. 52, No.4, (October 1979), pp. 422-437, ISSN 0031-935X

Battley, P. F., Rogers, D. I., Piersma, T., Koolhaas, A. (2003). Behavioral evidence for heat-load problems in great knots in tropical Australia fuelling for long distance flight. *Emu*. Vol.103, No.2, (June 2003), pp. 97-103, ISSN 0158-4197

Beentjes, M. P. (2006). Behavioral thermoregulation of the New Zealand sea lion (Phocarctos hookeri). *Marine Mammal Science*. Vol.22, No.2, (April 2006), pp. 311-325, ISSN 0824-0469

Bennett, A.F., Huey, R.B., John-Alder, H., Nagy, K.A. (1984). The parasol tail and thermoregulatory behavior of the cape ground squirrel *Xerus inauris*. *Physiological Zoology*. Vol.57, No.1, (January-February 1984), pp. 57-62, ISSN 0031 935X

Brosh, A., Aharoni, Y., Degen, A. A., Wright, D., Young, B.A. (1998). Effects of solar radiation, dietary energy, and time of feeding on thermoregulatory responses and energy balance in cattle in a hot environment. *Journal of Animal Science*. Vol.76, No.10, (October 1998), pp. 2671-2677, ISSN 0021-8812

Calvert, W. H., Brower, L., P., Lawton, R. O. (1992). Mass flight response of overwintering monarch butterflies (Nymphalidae) to cloud-induced changes in solar radiation intensity in Mexico. *Journal of the Lepidopterists' Society*. Vol.46, No.2, (August 1992), pp. 97-105, ISSN 0024-0966

Campagna, C. and Le Boeuf, B. J. (1988). Thermoregulatory behaviour of southern sea lions and its effect on mating strategies. *Behaviour*. Vol.107, No.1/2, (November 1988), pp. 72-90, ISSN 0005-7959

Carrascal, J., Diaz, A. Huertas, D., Mozetich, I. (2001). Thermoregulation by treecreepers: trade-off between saving energy and reducing crypsis. *Ecology*. Vol.82, No.6, (June 2001), pp. 1642-1654, ISSN 0012-9658

Cengel, Y.A. (2nd ed). (2002). *Heat Transfer-A Practical Approach*. McGraw-Hill, ISBN 0072458933, New York, U.S.A.

Chapperon, C., Seuront, L. (2011). Behavioral thermoregulation in a tropical gastropod: links to climate change scenarios. *Global Change Biology*. Vol.17, No.4, (April 2011), pp. 1740-1749, ISSN 1354-1013

Clench, H.K. (1966). Behavioral thermoregulation in butterflies. *Ecology*. Vol.47, No.6, (November 1966), pp. 1021-1034, ISSN 0012-9658

Coe, M. (2004). Orientation, movement and thermoregulation in the giant tortoises (Testudo (Geochelone) of Aldabra atoll, Seychelles: animals, *Transactions of the Royal Society of South Africa: Proceedings of a Colloquium on Adaptations in Desert Fauna and Flora*, Vol.59, No.2, (March 2010), pp.73-77, ISSN 0035-919X

Coelho, S.G., Choi, W., Brenner, M., Miyamura, Y., Yamaguchi, Y., Wolber, R., Smuda, C., Batzer, J., Kolbe, L., Ito, S., Wakamatsu, K., Zmudzka, B.Z., Beer, J. Z., Miller, S.A., Hearing, V.J. (2009). Short- and long-term effects of UV radiation on the pigmentation of human skin. *Journal of Investigative Dermatology Symposium Proceedings*. Vol.14, No.1, (August 2009), pp. 32-35, ISSN 1087-0024

Dawson, T.J., Blaney, C.E., Munn, A.J., Krockenberger, A., Maloney, S.K. (2000). Thermoregulation by kangaroos from mesic and arid habitats: influence of temperature on routes of heat loss in eastern grey kangaroos (Macropus giganteus) and red kangaroos (Macropus rufus). *Physiological and Biochemical Zoology* Vol.73, No.3, (June 2000), pp. 374-381, ISSN 1522-2152

De Rensis, F., Scaramuzzi, R.J. (2003). Heat stress and seasonal effects on reproduction in the dairy cow-a review. *Theriogenology*. Vol.60, No.6, (October 2003), pp. 1139-1151, ISSN 0093-691X

de Sampaio, C., Camilo-Alves, P., de Miranda Mourao, G. (2006). Responses of a specialized insectivorous mammal (*Myrmecophaga tridactyla*) to variation in ambient temperature. *Biotropica*. Vol.38, No.1, (January 2006), pp. 52-56. ISSN 0006-3606

Dubols, Y., Blouln-Demers, G., Shipley, B., Thomas, D. (2009). Thermoregulation and habitat selection in wood turtles *Glyptemys insculpta*: chasing the sun slowly. *Journal of Animal Ecology*. Vol.78, No.5, (September 2009), pp. 1023-1032, ISSN 0021-8790

Dunne, R.P., Brown, B.E. (2001). The influence of solar radiation on bleaching of shallow water reef corals in the Andaman sea, 1993-1998. *Coral Reefs*. Vol.20, No.3, (November 2001), pp. 201.210, ISSN 0722-4028

Duthie, M.S., Kimber, I., Norval, M. (1999). The effects of ultraviolet radiation on the human immune system. *British Journal of Dermatology*. Vol.140, No.6, (June 1999), pp. 9950-1009, ISSN 1365-2133

Fortin, D., Larochelle, J., Gauthier, G. (2000). The effect of wind, radiation and body orientation on the thermal environment of Greater Snow goose goslings. *Journal of Thermal Biology*. Vol.25, No.3, (June 2000), pp. 227-238, ISSN 03064565

Galle, L. (1973). Thermoregulation in the nests of *Formica pratensis* Retz. (Hymenoptera: Formicidae). *Acta Biology*. Vol.19, No.47, (3), pp. 139-142, ISSN 0020-1812

Gentry, R. (1973). Thermoregulatory behavior of eared seals. *Behaviour*. Vol.46, No.1, (January 1973), pp. 73-93, ISSN 0005 -7959

Glassey, B., Amos, M. (2009). Shade seeking by the common grackle (*Quiscalus quiscula*) nestlings at the scale of the nanoclimate. *Journal of Thermal Biology*. Vol.34, No.2, (February 2009), pp. 76-80, ISSN 03064565

Guan, W.C., Li, P., Jian, J.B., Wang, J.Y. (2011). Effects of solar ultraviolet radiation on photochemical efficiency of *Chaetoceros curvisetus*. (Bacillariophyceae). *Act Physiologiae Plantarum*. Vol.33, No.3, (May 2011), pp. 979-986, ISSN 0137-5881

Gupta, S. (2011). All weather clothing, In: *World of Garment Textile Fashion*. 25.09.2011, Available from: http://www.fibre2fashion.com/industry-article/11/1040/all-weather-clothing1.asp

Häder, D.P., Kumar, H.D., Smith, R.C., Worrest, R.C. (2007). Effects of solar UV radiation on aquatic ecosystems and interactions with climate change. *Photochemical Photobiology Science*. Vol.6, No.3, (March 2007), pp. 267-285, ISSN 0031-8655

Häkkinen, J., Pasanen, S., Kukkonen, J.V.K. (2001). The effects of solar UV-B radiation on embryonic mortality and development in three boreal anurans (*Rana temporaria, Rana arvalis* and *Bufo bufo*). *Chemosphere*. Vol.44, No.3, (July 2001), pp. 441-446, ISSN 0045-6535

Hertz, P.E., Huey, R.B. (1981). Compensation for altitudinal changes in the thermal environment by some Anolis lizards on Hispaniola. *Ecology*. Vol.62, No.3, (June 1981), pp. 515-521. ISSN 0012-9658

Hetem, R.S., Strauss, M., Heusinkveld, B.G., de Bie, S., Prins, H.H.T., van Wieren, S.E. (2011). Energy advantages of orientation to solar radiation in three African ruminants. *Journal of Thermal Biology.*, Vol.36, No.7., (October 2011), pp. 452-460, ISSN 0306-4565

Hoffert, M.I., Caldeira, K. (2004). Climate Change and Energy, In: *Encyclopedia of Energy*, C. J. Cleveland, (Ed.), 359-380, Elsevier Academic Press. ISBN 012176480X, Boston, U.S.A.

Holick, M.F. (2008) Sunlight, UV Radiation, Vitamin D, and Skin Cancer: How much sunlight do we need? In: *Sunlight, Vitamin D and Skin Cancer*. J. Reichrath pp.1-15 Springer ISBN 978-0-387-77574-6 New York, U.S.A.

Hristov, N.I., Betke, M., Kunz, T.H. (2008). Applications of thermal infrared imaging for research in aeroecology. *Integrative and Comparative Biology*, Vol.48, No.1, (July 2008), pp. 50-59 ISSN 1540-7063

Huertas, D.L., Diaz, J.A. (2001). Winter habitat selection by a montane forest bird assemblage: the effects of solar radiation. *Canadian Journal of Zoology*. Vol.79, No.2, (February 2001), pp. 279-284, ISSN 0008-4301

Jones, H.G., Serraj, R., Loveys, B.R., Xiong, L., Wheaton, A., Price, A.H. (2009). Thermal infrared imaging of crop canopies for the remote diagnosis and quantification of plant responses to water stress in the field. *Functional Plant Biology*. Vol.36, No.11, (November 2009), pp. 978-989, ISSN 1445-4408

Karsten, K.B., Ferguson, G.W., Chen, T.C., Holick, M.F. (2009). Panther chameleons, Furcifer pardalis, behaviorally regulate optimal exposure to UV depending on dietary vitamin D3 status. *Physiological and Biochemical Zoology* Vol.82, No.3, (June 2009), pp. 218-225, ISSN 1522-2152

Kevan, P.G., Shorthouse, J.D. (1970). Behavioural thermoregulation by high arctic butterflies. *Arctic*. Vol.23, No.4, (January 1970), pp. 268-279 ISSN 1923-1245

King, J.M., Kingaby, G.P., Colvin, J.G., Heath, B.R. (1975). Seasonal variation in water turnover by oryx and eland on the Galana Game Ranch research project. *African Journal of Ecology*. Vol.12, No.3-4, (December 1975), pp. 287-296, ISSN 1365-2028

Kovac, H., Stabentheiner, A., Schmaranzer, S. (2009). Thermoregulation of water foraging wasps (*Vespula vulgaris* and *Polistes dominulus*). *Journal of Insect Physiology*. Vol.55, No.10, (October 2009), pp. 959-966, ISSN 0022-1910

Kovac, H., Stabentheiner, A., Schmaranzer, S. (2010). Thermoregulation of water foraging honeybees-balancing of endothermic activity with radiative heat gain and functional requirements. *Journal of Insect Physiology*. Vol.56, No.12, (December 2010), pp. 1834-1845, ISSN 0022-1910

Lagos, V.O., Bozinovic, F., Contreras, L.C. (1995). Microhabitat use by a small diurnal rodent (*Octodon degus*) in a semiarid environment: thermoregulatory constraints or predation risk? *Journal of Mammalogy*. Vol.76, No.3, (August 1995), pp.900-905, ISSN 0022-2372

Lambrinos, J.G., Kleier, C.C. (2003). Thermoregulation of juvenile Andean toads (*Bufo spinulosus*) at 4300m. *Journal of Thermal Biology*. Vol.28, No.1, (January 2003), pp. 15-19, ISSN 0306-4565

Lesser, M.P., Farrell, J.H. (2004). Exposure to solar radiation increases damage to both host tissues and algal symbionts of corals during thermal stress. *Coral Reefs*. Vol.23, No.3, (September 2004), pp. 367-377, ISSN 0722-4028

Liu, X., Cheng, X., Skidmore, A. K. (2011). Potential solar radiation pattern in relation to the monthly distribution of giant pandas in Foping Nature Reserve, China. *Ecological Modeling*. Vol.222, No.3, (February 2011), pp. 645-652, ISSN 0304-3800

Lillywhite, H.B., Licht, P., Chelgren, P. (1973). The role of behavioral thermoregulation in the growth energetics of the toad *Bufo boreas*. *Ecology*. Vol.54, No.2, (March 1973), pp. 375-383, ISSN 0012-9658

Lustwick, S., Battersby, B., Kelty, M. (1978). Behavioral thermoregulation: orientation toward the sun in herring gulls. *Science*. Vol.200, No.4337, (April 1978), pp. 81-83, ISSN 0036-8075

Maloney, S.K., Moss, G., Mitchell, D. (2005). Orientation to solar radiation in black wildebeest (*Connochaetes gnou*). *Journal of Comparitive Physiology*. Vol.191, No.11, (November 2005), pp. 1065-1077, ISSN 0340-7594

Mazza, C. A., Zavala, J., Scopel, A.L., Ballare', C.L. (1999). Perception of solar UVB radiation by phytophagous insects: behavioral responses and ecosystem implications. *Proceedings of the National Academy of Science*. Vol.96, No.3, (February 1999), pp. 980-985, ISSN 0027-8424

McCafferty, D.J. (2007). The value of infrared thermography for research on mammals: previous applications and future directions. *Mammal Review*. Vol.37, No.3, (July 2007), pp. 207-223, ISSN 0305-1838

McCafferty, D.J., Gilbert, C., Paterson, W., Pomeroy, P.P., Thompson, D., Currie, J.I., Ancel, A. (2011) Estimating metabolic heat loss in birds and mammals by combining infrared thermography with biophysical modeling. *Comparative Biochemistry and Physiology, Part A*. Vol.158, No.3, (March 2011), pp. 337-345, ISSN 1095-6433

McCloud, E.S., Berenbaum, M.R. (1994). Stratospheric ozone depletion and plant-insect interactions: effects of UVB radiation on foliage quality of *Citrus jambhiri* for *Trichoplusia ni*. *Journal of Chemical Ecology* Vol.20, No. 3, (March 1994), pp. 525-539, ISSN 0098-0331

Muñoz, L.P., Finke, G.R., Camus, P.A., Bozinovic, F. (2005). Thermoregulatory behavior, heat gain and thermal tolerance in the periwinkle *Echinolittorina peruviana* in central Chile. *Comparative Biochemical Physiology*. Vol.142, No.1, (September 2005), pp. 92-98, ISSN 1095-6433

Nagy, K. (2004). Heterotrophic Energy Flows, In: *Encyclopedia of Energy*, C. J. Cleveland, (Ed.), 159-172, Elsevier Academic Press. ISBN 012176480X, Boston, U.S.A.

Nahon, S., Castro Porras, V.A., Pruski, A.M., Charles, F. (2009). Sensitivity to UV radiation in early life stages of the Mediterranian sea urchin *Sphaerechinus granularis* (Lamarck). *Sci Tot Environment*. Vol.407, No.6, (March 2009), pp. 1892-1900, ISSN 0048-9697

Norris, A. L., Houser, D. S., Crocker, D. E. (2010). Environment and activity affect skin temperature in breeding adult male northern elephant seals (*Mirounga angustirostris*). *Journal of Experimental Biology*. Vol.213, No.24, (December 2010), pp. 4205-4212, ISSN 0022-0949

Ochoa-Acuña, H., Kunz, T.H. (1999). Thermoregulatory behavior in the small island flying fox, *Pteropus hypomelanus* (Chiroptera: Pteropodidae). *Journal of Thermal Biology*. Vol.24, No.1, (February 1999), pp. 15-20, ISSN 0306-4565

O'Neill, K.M., Rolston, M.G. (2007). Short-term dynamics of behavioral thermoregulation by adults of the grasshopper *Melanoplus sanguinipes*. *Journal of Insect Science*. Vol.7, No.27, (May 2007), pp. 1-14, ISSN 1536-2442

Rafanelli, C., Damiani, A., Benedetti, E., De Simone, S., Anav, A., Ciataglia, L., Di Menno, I. (2010). UV Solar Radiation in Polar Regions: Consequences for the Environment and Human Health, In: *UV Radiation in Global Climate Change*. J.R. Slusser & D.L. Schomldt, (Ed.), 73-105, Springer, ISBN 978-3-642-03313-1 Berlin, Germany

Randall, D.J., Burggren, W.W., French, K., Eckert, R. (2002). *Animal Physiology Mechanisms and Adaptations* (5 ed.), W.H. W.H. Freeman and Co. ISBN 0716738635, New York, U.S.A.

Reichard, J.D., Prajapati, S.I., Austad, S.N., Keller, C., Kunz T.H. (2010a). Thermal windows on Brazilian free-tailed bats facilitate thermoregulation during prolonged flight. *Integrative and Comparative Biology*, Vol.50, No.3, (September 2010), pp. 358-370 ISSN 1540-7063

Reichard, J.D., Fellows, S., Frank, A.J., Kunz, T.H. (2010b). Thermoregulation during flight: body temperature and sensible heat loss from free-ranging Brazilian free-tailed bats (*Tadarida brasiliensis*). *Physiological and Biochemical Zoology*, Vol.50, No.6, (November 2010), pp. 358-370, ISSN 1522-2152

Schaumburg, L.G., Poletta, G.L., Imhof, A., Siroski, P.A. (2010). Ultraviolet radiation induced genotoxic effects in the broad snouted caiman, *Caiman latisrostris*. *Mut Research*. Vol.700, No.1-2, (July 2010), pp. 67-70, ISSN 0027-5107

Schütz, K.E., Rogers, A.R., Cox, N.R., Tucker, C.B. (2009). Dairy cows prefer shade that offers greater protection against solar radiation in summer: shade use, behavior, and body temperature. *Applied Animal Behavioral Science*. Vol.116, No.1, (January 2009), pp. 28-34, ISSN 01681-591

Sellers, R.M. (1995). Wing spreading behaviour of the cormorant *Phalacrocorax carbo*. *Ardea*. Vol.83, No.1, (April 1995), pp. 27 36. ISSN 0373-2266

Sevi, A., Annicchiarico, G., Albenzio, M., Taibi, L., Muscio, A., Dell'Aquila, S. (2001). Effects of solar radiation and feeding time on behavior, immune response, and production of lactating ewes under high ambient temperature. *Journal of Dairy Science*. Vol.84, No.3, (March 2001), pp. 249-640, ISSN 0022-0302

Southworth, L.O., Matthieu, J., Chen, T.C., Holick, M.F., Kunz, T.F. (2009). Natural variation of 25-hydroxyvitamin D in free-ranging New World tropical bats (Chiroptera). *Acta Chiropterologica*, Vol.11, No.2, (December 2009), pp. 451-456, ISSN

Speakman, J. (2004) Thermoregulation, In: *Encyclopedia of Energy*, C. J. Cleveland, (Ed.), 125-137, Elsevier Academic Press. ISBN 012176480X, Boston, U.S.A.

Steen, R., Løw, L.M., Sonerud, G.A. (2011). Delivery of common lizards (*Zootoca* (*Lacerta*) *vivipara*) to nests of Eurasian kestrels (*Falco tinnunculus*) determined by solar height and ambient temperature. *Canadian Journal of Zoology*. Vol.89, No.3, (March 2011), pp. 199-205, ISSN 0008-4301

Vogt, J.T., Wallet, B., Freeland, T.B. (2008). Imported fire ant (Hymenoptera: Formicidae) mound shape characteristics along a north-south gradient. *Environmental Entomology*. Vol.37, No.1, (February 2008), pp. 198-205, ISSN 0046-225X

Warnecke, L., Turner, J., Geiser, F. (2008). Torpor and basking in a small arid zone marsupial. *Naturwissenschaften*. Vol.95, No.1, (January 2008), pp. 73-78, ISSN 0028-1042

Wolf, B.O. & Walsburg, G.E. 1996. Effects of radiation and wind on a small bird and implications for microsite selection. *Ecology*. Vol.77, No.7, (October 1996), pp. 2228-2236, ISSN 0012-9658

Zavala, J.A., Scopel, A.L., Ballare', C.L. (2000). Effects of ambient UV-B radiation on soybean crops: impact on leaf herbivory by *Anticarsia gemmatalis*. *Plant Ecology*. Vol.156, No.2, (October 2001), pp. 1-10, ISSN 1385-0237

14

Effects of Solar Radiation on Fertility and the Flower Opening Time in Rice Under Heat Stress Conditions

Kazuhiro Kobayasi
Shimane University
Japan

1. Introduction

This chapter focuses upon the effects of solar radiation on the flower opening time and panicle temperature, both of which significantly affect heat-stress-induced sterility in rice, although solar radiation affects crop growth rates and production through photosynthesis, the fundamental physiological function of green plants. Three possible traits that may mitigate the damage caused by high temperature during the flowering period in rice have been proposed (Zhao et al., 2010). The first trait is the formation of a long dehiscence at the base of the anther, but this trait is under genetic control so that it is not affected by solar radiation. The second trait is early morning flower opening to avoid high temperatures during anthesis. The flower opening time can be affected by solar radiation (Kobayasi et al., 2010). The third trait is panicle temperature, which can also be affected by solar radiation as well as microclimates and cultivar-related factors such as transpirational conductance and panicle shape (Yoshimoto et al., 2005). The latter two traits are discussed in this chapter.

Two experiments were conducted in Shimane Prefecture, Japan to reveal the roles of solar radiation in determining the flower opening time and panicle temperature. In section 2, the effects of solar radiation on the flower opening time have been examined. Early morning opening of rice flowers is a beneficial response to avoid heat-stress-induced sterility because the sensitivity of rice flowers to high temperatures decreases during the 1-hr period after flower opening (Satake & Yoshida, 1978). The effects of solar radiation on the flower opening time were evaluated using generalized linear models. Cultivar differences in the contribution of solar radiation to the flower opening time were estimated. Furthermore, the roles of the diurnal change in solar radiation on determining the flower opening time were examined. In section 3, the effects of solar radiation on panicle temperature have been examined. Atmospheric temperature is typically used to estimate temperature-dependent stress. However, solar radiation also affects plant tissue temperature. An empirical model to estimate the contribution of solar radiation to panicle temperature was developed using generalized linear models.

2. Experiment 1: Effects of solar radiation on the flower opening time in rice

Global climate change poses a serious challenge to crop production around the world. In rice (*Oryza sativa* L.), temperatures higher than 34 °C at the time of flowering may induce flower sterility and decrease yields, even in temperate regions such as southern Japan, if the cropping season is not changed to avoid such temperatures (Horie et al., 1996; Kim et al., 1996). Crop simulation models (Horie et al., 1996) have suggested that yields of currently grown rice varieties in southern Japan would be reduced by up to 40% under future climate scenarios. Serious yield losses in rice due to flower sterility occurred in the Yangtze Valley of China in 2003, when the temperatures during the hottest summer in the region's history affected the reproductive stage of rice cultivated in this region (Wang et al., 2004).

Early morning opening of rice flowers is a beneficial response to avoid sterility caused by heat stress during anthesis because the sensitivity of rice flowers to high temperatures decreases during the 1-hr period after flower opening (Satake & Yoshida, 1978). Thus, a flower opening time, 1 hr earlier than normal may reduce the risk of sterility because it may lead to anthesis before the air temperature reaches the critical level; air temperature can rise at a rate of higher than 3 °C hr^{-1} starting at approximately 1000 (Nishiyama & Blanco, 1980). A controlled environment experiment revealed that flowers of 'Milyang 23' that opened earlier in the morning and at a lower temperature had higher seed sets than those that opened later (near midday) and at higher temperatures (Imaki et al., 1987).

Thus, the selection of cultivars with early flower opening times is an important method for reducing heat-induced sterility (Ishimaru et al., 2010; Jagadish et al., 2008; Nishiyama & Blanco, 1980). For example, the flowers of O. glaberrima Steud. open earlier than those of O. sativa L. (Jagadish et al., 2008; Nishiyama & Blanco, 1980), and the flowers of interspecific hybrids between O. glaberrima and O. sativa open earlier than those of O. sativa (Nishiyama & Satake, 1981). Reduced sterility in the early morning flowering line subjected to rising temperatures during anthesis in the greenhouse was attributed to an earlier flowering time compared with 'Koshihikari' (Ishimaru et al., 2010).

Although the flower opening time is under genetic control, it is affected by aspects of the weather such as solar radiation and air temperature (Hoshikawa, 1989; Imaki et al., 1983; Jagadish et al., 2007, 2008; Kobayasi et al., 2010; Nakagawa & Nagata, 2007; Nishiyama & Satake, 1981). The relationship between the flower opening time and solar radiation has been researched under field conditions (Kobayasi et al., 2010). Using correlation analysis, most japonica cultivars showed a negative correlation between solar radiation and flower opening times; this correlation showed that higher solar radiation resulted in earlier flower opening times. However, the indica cultivar 'IR72' did not show a negative correlation between the solar radiation and flower opening time. This result suggests that the contribution of solar radiation to the flower opening time is different among ecotypes, and compared with japonica cultivars, indica cultivars have lower sensitivity in determining the flower opening time by solar radiation. Moreover, the response of flower opening to high temperature differs among rice cultivars. The flower opening time occurs earlier at high temperatures in 'Milyang 23', whereas it occurs later in 'Nipponbare' (Imaki et al., 1983). In a study of indica cultivars, the flower opening time occurred approximately 45 min earlier at higher temperatures (Jagadish et al., 2007). Furthermore, the relationship between solar radiation and air temperature should be incorporated into the analysis of the flower opening time because the amount of solar radiation is one of the most important factors in determining air temperature. It has been suggested that synergistic effects on the flower opening time may exist between temperature and light (Kobayasi et al., 2010). In addition to solar radiation and air temperature, other weather factors such as vapor-pressure deficit and wind speed (Tsuboi, 1961) affect the flower opening time. However, the combined effects of air temperature, solar radiation, vapor-pressure deficit, and wind speed on the flower opening time of various rice genotypes remain unclear, particularly under field conditions; this limits our ability to predict the flower opening time.

The cycle of solar radiation may also affect the flower opening time as well as the amount of solar radiation. Most studies on the effects of weather factors on the flower opening time have been conducted under controlled environments with artificial light conditions (Imaki et al., 1983; Jagadish et al., 2007, 2008; Nishiyama & Blanco, 1981) and the flowers of rice plants grown in a glasshouse or a growth chamber have been reported to open 1–2 hr later than those grown outdoors (Imaki et al., 1982). This suggests that solar radiation substantially affects the flower opening time; not only the strength of solar radiation but also light conditions can influence the flower opening time because the duration of anthesis increases under continuous light or dark conditions (Hoshikawa, 1989). The light intensity and cycle of light and dark may affect the flower opening time. The effects of a diurnal cycle of light on the flower opening time in dicotyledonous *Pharbitis nil* flowers have been experimentally studied under artificially controlled conditions (Kaihara & Takimoto, 1979).

In this section, first, the role of solar radiation in determining the flower opening time was evaluated using correlation analysis between solar radiation and the flower opening time. The correlations between the flower opening time and solar radiation averaged hourly, from 0500 to 1000 were analyzed. Second, we used general linear models to separately evaluate the effects of the type of cultivars, air temperature, solar radiation, vapor-pressure deficit, and wind speed on the flower opening time. In the second analysis, we used two types of 1-hr time spans: a 1-hr time span based on Japan Standard Time, and a 1-hr time span based on the time of mean flower opening times in each cultivar; this eliminated the effects of the circadian rhythm. Finally, we attempted to detect the roles of the diurnal change in solar radiation in determining the flower opening time.

2.1 Materials and methods

Experiment 1 was conducted on an experimental field of Shimane University in Matsue, Shimane Prefecture, Japan (35°29′N, 133°04′E, 4 m above sea level) in 2010. Three indica cultivars and four japonica cultivars with wide ranges of flower opening times (Kobayasi et al., 2010) were used (Table 1). 'Xiaomazhan' had the earliest flower opening time in Japan (Kobayasi et al., 2010) and in Jiangsu Province, China (Zhao et al., 2010) among indica and japonica cultivars. Among japonica cultivars, 'Milky Queen' had the earliest flower opening time in Japan (Kobayasi et al., 2010) and in Jiangsu Province, China (Zhao et al., 2010). Although indica cultivars are not commonly planted in Japan, we used 'IR72' and 'Takanari' because the flowers of indica cultivars open earlier than those of japonica cultivars (Imaki et al., 1987).

The soil type at the study site was alluvial sandy clay. On three occasions, 30-day-old seedlings grown in nursery boxes were manually transplanted to the experimental field of Shimane University to obtain flowering plants under different weather conditions. Table 1 shows the seeding dates, ecotype, origin, and measurement periods for the flower opening time and weather factors. The planting density was 22.2 hills m^{-2} (one seedling per hill, 15-cm hill spacing, and 30-cm row spacing). The area of each plot was 12.8 m^2. A basal dressing of 4.0 g m^{-2} of N, 4.9 g m^{-2} of P$_2$O$_5$, and 4.3 g m^{-2} of K$_2$O was applied. Top dressing was not used. Culture methods such as irrigation and pesticide used followed the standard local practices for rice production in Shimane Prefecture.

Cultivar	Ecotype	Origin	Seeding dates	Measurement period
'Fujihikari'	japonica	Japan	April 15, May 7	July 13-20, 23-27
'Koshihikari'	japonica	Japan	April 15, May 7, 28	July 26-31, August 5-11, 17-21
'Xiaomazhan'	indica	China	April 15, May 7, 28	July 25-29, August 3-7, 20-25
'Milky Queen'	japonica	Japan	May 7, 28	August 5-11, 17-21
'IR72'	indica	Philippines	May 7, 28	August 24-29, September 5-11
'Takanari'	indica	Japan	May 28	August 28- September 2
'Asahi'	japonica	Japan	May 7, 28	September 2-9

Table 1. Ecotype, origin, seeding dates and measurement periods for the seven *O. sativa* cultivars used in Experiment 1.

2.1.1 Measurements of the flower opening time and weather factors

Physical stimuli such as touch may promote flower opening in rice (Tsuboi, 1961). To avoid this phenomenon, we used digital photographs of the panicles instead of physical inspections. The panicles were photographed at 10-min intervals with waterproof digital cameras (Optio W30, Pentax, Tokyo, Japan) to determine the flower opening time of the seven cultivars. The photographs were automatically taken using cameras. We put the camera on a tripod and used a built-in electronic timer to control the measurement intervals. We recorded the time of anther extrusion of all observable flowers (more than 30% of flowers in a panicle) on the obverse side of approximately 10 panicles per day per cultivar. The medians of anther extrusion time among all observed flowers in each panicle were calculated. The flower opening time was defined as the mean of the medians per day. Recording the anther extrusion time of the flowers behind panicles and leaves was occasionally difficult.

We measured air temperature, solar radiation, relative humidity, and wind speed every 5 min using a wireless weather station (Wireless Vantage Pro, Davis Instruments, Hayward, CA, USA), which was located at the experimental field of Shimane University (http://www.ipc.shimane-u.ac.jp/weather/station/i/home.html). The weather station was installed at a distance of approximately 5 m from observation plots. The ground surface below the station was covered with grass. We installed a thermometer, hygrometer, solarimeter, barometer, and an anemometer at heights of 150, 150, 180, 150, and 300 cm, respectively. At our study site, the sun rose between 0502 (July 13) and 0547 (September 11); thus, we did not measure solar radiation before twilight. Vapor-pressure deficits were calculated from air temperature and relative humidity using the method based on microclimate (Buck, 1981).

2.1.2 Statistical analysis

Pearson's correlation analysis was used to identify the relationship between solar radiation and the flower opening time. In this analysis, hourly average of solar radiation values between 0500 and 1000 were used. Correlation analysis was restricted to the period between 0500 and 1000 because sunrise hours during the flower opening time observation periods were around 0500, and flowers of 'Xiaomazhan' usually started to open before 1000. It is possible that the span in which flowers respond to solar radiation before flower opening are different among cultivars, i.e. cultivars with early morning flowering would respond earlier in the morning to solar radiation. Hourly average of solar radiation values were used over seven 1-hr periods based on the mean flower opening time in each cultivar (successive 1-hr periods from 7 hr before the mean flower opening time until 1 hr before the mean flower opening time).

The collected data were analyzed by means of generalized linear models and multiple regression procedures using SPSS (Version 14J for Windows, SPSS Japan Inc., Tokyo, Japan). Because inherent relationships exist among weather factors, relatively high correlations may exist among solar radiation, air temperature, vapor-pressure deficit, and wind speed. Therefore, to evaluate their individual effects as well as cultivar effects on the flower opening time, generalized linear models were used. In this analysis, solar radiation, air temperature, vapor-pressure deficit, and wind speed values were averaged over five 1-hr periods (0500–0600, 0600–0700, 0700–0800, 0800–0900, and 0900–1000). The relative contribution of each weather component to the flower opening time was determined using their standardized partial regression coefficients, and the overall strength of the relationships was quantified using the multiple correlation coefficients. The contribution of solar radiation, air temperature, vapor pressure deficit, and wind speed to the flower opening time in each cultivar was estimated by substituting the obtained weather data in the generalized linear models, and the relative contributions of solar radiation, air temperature, vapor pressure deficit, and wind speed to the flower opening time among the cultivars were evaluated. To examine the role of diurnal changes in solar radiation in determining the flower opening time, multiple correlation coefficients from 1800 on the day before flowering to 0900 on the flowering day were calculated for each cultivar.

2.2 Results

In 2010, record heat and sunshine occurred in Japan because a strong North Pacific High covered the country during summer. The total solar radiation between 0000 and 1200 during the flower opening time observation periods ranged from 1.0 to 13.3 MJ m^{-2}. Sunshine hours in the former and latter halves of August 2010 were 6.58 and 9.75, respectively. The normal sunshine hour in August is 6.47. This suggests that average solar radiation during the flower opening time observation periods was higher than usual in that year. The vapor-pressure deficit between 0000 and 1200 during the flower opening time observation period ranged from 1.9 to 18.3 hPa.

The maximum and minimum air temperature between 0000 and 1200 during the flower opening time observation period ranged from 24.6 to 36.7°C and from 19.4 to 27.3°C, respectively (Fig. 2). In particular, the mean air temperatures between the latter part of July and August in 2010 reached record-high levels. The maximum and minimum air temperature increased rapidly from mid-July through early August. The maximum air temperature was higher than 34°C (temperatures >34°C at the time of flowering may induce flower sterility and decrease yield) for 23 days, however, in normal years, the maximum air temperature was rarely higher than 34°C.

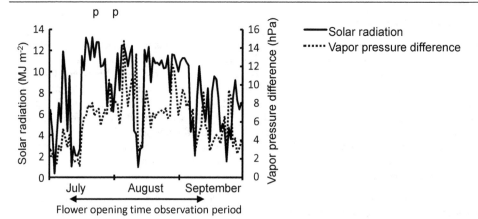

Fig. 1. Total solar radiation and average vapor-pressure deficit between 0000 and 1200 during the observation period.

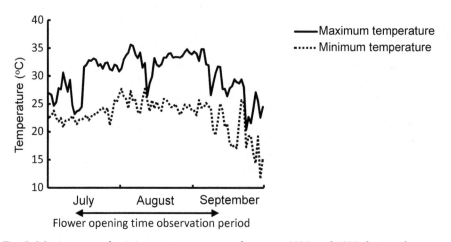

Fig. 2. Maximum and minimum temperatures between 0000 and 1200 during the observation period.

Average wind speed and atmospheric pressure ranged from 0.0 to 1.4 m s^{-1} and from 1001 to 1017 hPa, respectively, between 0000 and 1200 during the flower opening time observation period. The wind speed and atmospheric pressure were stable because a strong North Pacific High covered Japan during the summer of 2010.

Significant correlations existed between the meteorological variables. In particular, the correlation between air temperature and vapor-pressure deficit was high ($r = 0.715$, $p < 0.001$). The correlation between solar radiation and air temperature and that between solar radiation and vapor-pressure deficit were also relatively high ($r = 0.532$, $p < 0.001$; $r = 0.664$, $p < 0.001$, respectively). Wind speed and atmospheric pressure had relatively weak correlations ($r < 0.5$) with other meteorological variables.

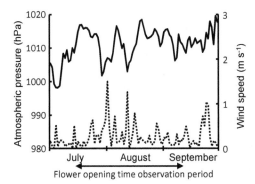

Fig. 3. Average vapor-pressure deficit and average wind speed between 0000 and 1200 during the observation period.

2.2.1 Variations in the flower opening time during the observation periods

Although flower opening time is under genetic control, wide ranges in flower opening time were recorded in all seven cultivars. Cultivar 'Xiaomazhan' flower opening time recorded 1-2 hr earlier than other cultivars. The flowers of the indica cultivars ('Xiaomazhan', 'IR72', and 'Takanari') opened earlier than those of the japonica cultivars ('Fujihikari', 'Koshihikari', 'Milky Queen' and 'Asahi'). The flowers of 'Milky Queen' did not open earlier than those of the other japonica cultivars although its flowers opened early in Japan (Kobayasi et al., 2010) and China (Zhao et al., 2010). The range in the flower opening time was higher than 2 hr in 'Fujihikari', 'Xiaomazhan', and 'IR72'.

2.2.2 Correlations between solar radiation and the flower opening time

The six cultivars other than 'Takanari' showed negative correlations between the flower opening time and mean hourly solar radiation for every hour between 0500 and 1000 (Fig. 4). The cultivars can be classified into two groups. One group showed negative, high correlations and comprised the indica cultivar 'IR72' and japonica cultivars 'Fujihikari', 'Asahi', and 'Milky Queen'. The other group showed relatively weak correlations and comprised the indica cultivars 'Takanari' and 'Xiaomazhan' and the japonica cultivar 'Koshihikari'.

Cultivar	Flower opening time	Mean flower opening time	Range in the flower opening time
'Fujihikari'	0958–1250	1124	172
'Koshihikari'	1049–1244	1136	115
'Xiaomazhan'	0726–0944	0858	138
'Milky Queen'	1053–1226	1132	93
'IR72'	0946–1152	1030	126
'Takanari'	1011–1109	1038	58
'Asahi'	1053–1205	1122	72

Table 2. Variations in the flower opening time, mean flower opening time, and range in flower opening time expressed as Japan Standard Time for seven *O. sativa* cultivars in 2010.

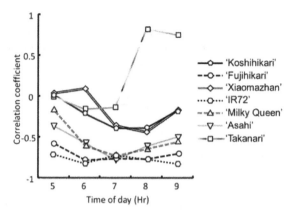

Fig. 4. Correlation coefficients between the flower opening time and hourly solar radiation for each hour from 0500 to 1000 in seven *O. sativa* cultivars.

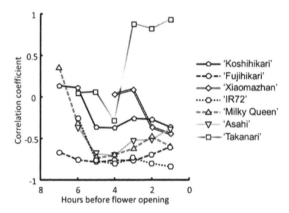

Fig. 5. Correlation coefficients between the flower opening time and hourly solar radiation for each hour over seven 1-hr periods based on the mean flower opening time in each cultivar (successive 1-hr periods from 7 hr before mean flower opening time until 1 hr before flower opening time)

To examine the difference in response hours to solar radiation before flower opening among cultivars, correlation coefficients between the flower opening time and hourly solar radiation were calculated for each hour over seven 1-hr periods based on the mean flower opening time in each cultivar (successive 1-hr periods from 7 hr before mean flower opening time until 1 hr before flower opening time; Fig. 5). The correlation coefficients of the cultivars classified as the group that showed negative, high correlations (Fig. 4) dropped rapidly at five hours before flower opening. At five and four hours before flower opening, the correlation coefficients of the group with negative, high correlations remained at approximately −0.7. Although the obtained negative correlation was weak, the correlation coefficients of 'Takanari' and 'Koshihikari' dropped at 5 hr before flower opening. We did not calculate the correlation coefficients between flower opening time and solar radiation in 'Xiaomazhan' because the flowers of this cultivar opened before 1000.

2.2.3 Evaluation of the effects of solar radiation, on cultivar's, air temperature, vapor-pressure deficit, and wind speed on the flower opening time

The multiple correlation coefficients determined by generalized linear model increased after 0400 and peaked between 0700 and 0800 (Fig.6) and then slightly decreased after 0800. The highest multiple correlation coefficient (adjusted R^2 = 0.849, p < 0.001) was obtained for the 0800–0900 period (Table 3). During this period, three factors (cultivar type, solar radiation, and air temperature) were significant (p < 0.05). The standardized partial regression coefficient for air temperature was −0.131, indicating that higher air temperature during this period resulted in an earlier flower opening time. Similarly, the standardized partial regression coefficient of solar radiation was −0.002, indicating that higher solar radiation during this period also resulted in an earlier flower opening time. The standardized partial regression coefficients for the other two weather factors (vapor-pressure deficit and wind speed) were not significant except for wind speed during the 0500–0600 period. The estimated contributions (hr) of each cultivar were −2.16, −0.89, −0.37, −0.01, 0, 0.41, and 0.437 for 'Xiaomazhan', 'IR72', 'Takanari', 'Fujihikari', 'Asahi', 'Koshihikari', and 'Milky Queen' respectively.

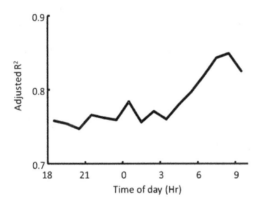

Fig. 6. The multiple correlation coefficients (adjusted R^2) obtained from multiple regression analysis using general linear models. The correlations for four weather factors (air temperature, solar radiation, vapor-pressure deficit, and wind speed) and the flower opening time for the 1-hr periods based on Japan Standard Time were analyzed.

Period	Multiple-correlation coefficient	F values in generalized linear models				
		Cultivar	Air temperature	Solar radiation	Vapor-pressure deficit	Wind speed
0500–0600	0.797***	12.44***	1.22*	1.43*	0.36 ns	1.88**
0600–0700	0.819***	12.77***	1.40*	3.94*	0.11 ns	0.33 ns
0700–0800	0.843***	12.36***	0.73 ns	3.62***	0.21 ns	0.01 ns
0800–0900	0.849***	11.84***	1.02*	2.03**	0.17 ns	0.03 ns
0900–1000	0.825***	12.12***	1.14*	0.58 ns	0.14 ns	0.00 ns

Period	Standardized partial regression coefficient			
	Air temperature	Solar radiation	Vapor-pressure deficit	Wind speed
0500–0600	−0.108	−0.035	0.042	1.124
0600–0700	−0.129	−0.007	0.026	0.601
0700–0800	−0.102	−0.003	0.029	0.045
0800–0900	−0.131	−0.002	0.027	0.059
0900–1000	−0.147	−0.001	0.024	0.008

ns (not significant). *, **, and *** (significant at $p < 0.05$, $p < 0.01$, and $p < 0.001$) respectively.

Table 3. Results of multiple-regression analysis using general linear models for the correlations between five hourly-averaged weather factors (air temperature, solar radiation, vapor-pressure deficit, and wind speed) and the flower opening time for the 1-hr periods based on Japan Standard Time and the significance of the rice cultivar.

The contributions of the four weather factors (air temperature, solar radiation, vapor-pressure deficit, and wind speed) were estimated for each cultivar (Table 4.). The contributions of solar radiation and air temperature were higher than those of vapor-pressure deficit and wind speed. Among cultivars, the variation in solar radiation was larger than that in air temperature. The contribution of solar radiation in 'Fujihikari' was 45.2 min, whereas its contribution in 'Xiaomazhan' was 3.9 min. The contributions of solar radiation in the cultivars classified as the group that showed negative, high correlations (Fig. 4.) were relatively high.

Cultivar	Solar radiation	Air temperature	Vapor-pressure deficit	Wind speed
'Fujihikari'	45.2	48.6	13.7	3.6
'Koshihikari'	28.9	39.4	15.3	4.0
'Xiaomazhan'	3.9	50.1	12.6	0.8
'Milky Queen'	28.1	26.0	15.3	3.4
'IR72'	37.1	48.0	14.4	1.4
'Takanari'	5.0	20.3	10.9	1.6
'Asahi'	33.1	45.4	11.5	1.2

Table 4. Estimated contributions (expressed as min) of the four weather factors (air temperature, solar radiation, vapor-pressure deficit, and wind speed) using the results of general linear models (Table 3.).

2.2.4 Observation of a diurnal pattern in multiple correlation coefficients in each cultivar

Two distinctive peaks were observed in the diurnal change of multiple correlation coefficients in each cultivar (Fig. 7). One peak was observed immediately after sunset (2000–2300) and the other was observed immediately after sunrise (0500–0700). In 'Fujihikari', 'IR72', and 'Milky Queen', highest multiple correlation coefficients of 0.816, 0.559, and 0.571, respectively were obtained for the 0600–0700 period. In 'Xiaomazhan', the highest multiple correlation coefficient (0.720) was obtained for the 0500–0600 period. In these four cultivars, the peak immediately after sunrise was higher than that immediately after sunset. In 'Fujihikari' and 'Koshihikari', highest multiple correlation coefficients of 0.570 and 0.484, respectively were obtained for the 2100–2200 period. The highest multiple correlation coefficient (0.716) in 'Asahi' was obtained for the 2200–2300 period. In these three cultivars, the peak immediately after sunrise was lower than that immediately after sunset.

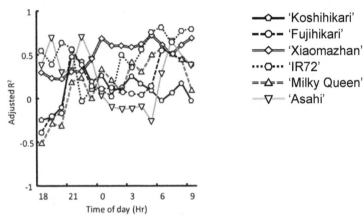

Fig. 7. Multiple correlation coefficients (Adjusted R^2) for each hour between 1800 on the day before flowering and 0900 on the flowering day for each cultivar.

2.3 Discussion

Higher solar radiation after sunrise resulted in an earlier flower opening time because the six cultivars excluding 'Takanari' showed negative correlations between the flower opening time and mean hourly solar radiation for every hour between 0500 and 1000 (Fig. 4). High solar radiation in the morning is related to an early flower opening time in several rice cultivars (Kobayasi et al., 2010). Similarly, solar radiation from 0400 to 0800 influenced the flower opening time in 'Koshihikari', but not in EG0 (Nakagawa & Nagata, 2007). In this experiment, cultivar differences in the sensitivity of the flower opening time to solar radiation were observed. The cultivars examined in this experiment were classified into two groups: cultivars with negative, high correlations between solar radiation and the flower opening time and those with relatively weak correlations. However, some of our results were inconsistent in that 'Koshihikari' showed a high response to solar radiation in previous studies (Kobayasi et al., 2010; Nakagawa & Nagata, 2007), but not in this experiment. In 2010, record heat and sunshine occurred, and maximum air temperature was higher than 34°C for 23 days, while, in normal years, the maximum air temperature is rarely higher than 34°C. The mechanism of solar radiation and air temperature in determining the flower opening time is not unclear. Air temperature and solar radiation may synergistically affect the flower opening time by altering panicle tissue temperature.

The correlation coefficients between the flower opening time and hourly solar radiation dropped rapidly at five hours before flower opening for each hour over seven 1-hr periods based on the mean flower opening time in the six cultivars except 'Xiaomazhan' whose flowers opened before 1000 (Fig. 5). The sensitivity of rice flowers to solar radiation during flower opening may increase before flower opening. Pollen grains continue to develop until flowering occurs, and they accumulate with starch 1 day before anthesis (Koike & Satake, 1987). After the end of starch engorgement in the pollen grains, the starch is rapidly digested at the end of the grain opposite to the germ poles 3–4 hr before flower opening; and more than 70% of pollen grains become sugar-type grains by the time of anther dehiscence (Koike & Satake, 1987). These findings suggest that development in sugar-type grains and the digestion of starch in the pollen grains are related to the beginning of flower opening. The digestion of starch in the pollen grains starts immediately after sunrise except in 'Xiaomazhan'. These results suggest that an increase in solar radiation triggers flower opening processes. However, the mechanism(s) by which solar radiation triggers starch digestion remains unclear. Pollen grains in anthers receive little solar radiation because they are covered by the palea and lemma of a rice flower. Solar radiation may increase anther temperature and promote starch digestion.

We found the highest multiple correlation coefficient (adjusted $R^2 = 0.849$, $p < 0.001$) for the 0800–0900 period (Table 3). During this period, three factors (cultivar, solar radiation, and air temperature) were significant ($p < 0.05$). The standardized partial regression coefficient for air temperature was −0.131, indicating that higher air temperature during this period resulted in an earlier flower opening time. Similarly, the standardized partial regression coefficient of solar radiation was −0.002, indicating that higher solar radiation during this period also resulted in an earlier flower opening time. The standardized partial regression coefficients for the contributions of vapor-pressure deficit and wind speed to the flower opening time were relatively small. These results indicate that it is necessary to consider two weather factors (air temperature and solar radiation) when analyzing the effects of weather on the flower opening time and, the contributions of vapor-pressure deficit and wind speed to the flower opening time were small. We estimated the contributions (hr) to the flower opening time. The contributions of 'Xiaomazhan', 'IR72', 'Takanari', 'Fujihikari', 'Asahi', 'Koshihikari', and 'Milky Queen' were −2.16, −0.89, −0.37, −0.01, 0, 0.41, and 0.437, respectively. Three indica cultivars ('Xiaomazhan', 'IR72', 'Takanari') showed early morning flowering, and the contribution to the flower opening time advancement was the highest in 'Xiaomazhan'. This result agreed with the previous results that 'Xiaomazhan' showed the earliest flower opening time in Japan (Kobayasi et al., 2010) and Jiangsu Province, China (Zhao et al., 2010). On the other hand, 'Milky Queen' did not show early morning flowering in this experiment. This result suggests that weather factors are also important in determining the flower opening time, and the consideration of all weather effects is required in breeding rice cultivars with early morning flowering.

Two distinctive peaks were observed in the diurnal change of multiple correlation coefficients in each cultivar (Fig. 7). One peak was immediately after sunset (2000–2300), and the other was immediately after sunrise (0500–0700). This suggests that other aspects of light conditions such as light cycle may influence the flower opening time in addition to the amount of solar radiation. The flowers of rice plants grown in a chamber tend to open 1–2 hr later than those grown outdoors (Imaki et al., 1982). Artificial dark or light treatments have been reported to affect the flower opening time (Nishiyama & Blanco, 1981). The effects of the diurnal cycle of light and temperature on the flower opening time in *P. nil* flowers have been studied experimentally (Kaihara & Takimoto, 1979, 1980, 1981a, 1981b, 1983). *P. nil* flowers subjected to various photoperiods bloomed approximately 10 hr after light-off when the light period was 10 hr or longer and approximately 20 hr after light-on when the light period was shorter (Kaihara & Takimoto, 1979). The higher air temperature during the dark period resulted in a later flower opening time with the temperature during the last half of the dark period having a stronger effect than that during the first half (Kaihara & Takimoto, 1980). At the lower temperature, the flower opening time is probably determined by the time of the latest preceding light-off (or light-on) signal (Kaihara & Takimoto, 1981a). Rice plants grown in a glasshouse or growth chamber have been reported to open flowers 1–2 hr later than those grown outdoors (Imaki et al., 1982).

In *P. nil* flowers, Kaihara & Takimoto (1981b) found that petals of the buds are the sites of photo- and thermo-perception; flower opening is caused mainly by the epinasty of petal midribs. On the other hand, we do not know the sites of photo- and thermo-perception in rice. Rice flowers lack petals and sepals. Rice lodicules, which are considered to be organs homologous to petals, expand when a rice flower opens. However, lodicules are covered with a palea and lemma and receive a low level solar radiation. Furthermore, plant growth regulators affect the flower opening time. Among plant growth regulators, abscisic acid promotes the flower opening time in *P. nil* (Kaihara & Takimoto, 1983). In rice, methyl jasmonate affects the flower opening time (Kobayasi & Atsuta, 2010; Zeng et al., 1999). Methyl jasmonate is important for tapetum, stamen, and pollen development in rice (Hirano et al., 2008). Some of plant growth regulators in anthers or lodicules may be triggers for flower opening by increasing solar radiation and air temperature. Advanced study of the sites of photo- and thermo-perception and of the mechanism for flower opening time determination, with consideration of plant growth regulators is needed. Under field conditions, where light intensity varies, it may be difficult to separate the effects of light intensity and the cycle of light and dark. Future research is necessary to study the effect of the light cycle on the flower opening time using growth chambers or supplementary light under field conditions to separate this factor from the effects of light intensity.

3. Experiment 2: Effects of solar radiation on rice panicle temperature

Global warming could increase the probability of heat-induced sterility in rice, which is most sensitive during anthesis (Matsui et al., 1997a). In addition, an elevated CO_2 concentration raises canopy temperature through stomatal closure, thereby exacerbating heat-stress-induced sterility in rice (Matsui et al., 1997b). The mechanism of heat-stress-induced sterility is explained by the inhibition of anther dehiscence when flowering under high temperature (Satake & Yoshida, 1978). Atmospheric temperature is usually employed to estimate temperature-dependent stress, but panicle temperature is obviously the major determinant of heat-induced sterility (Zhao et al. 2010). The results in section 2 suggest that panicle temperature also affects the flower opening time, which is strongly related with heat-stress-induced sterility in rice. Together with other microclimates and cultivar-related factors such as transpirational conductance and panicle shape, solar radiation can affect panicle temperature and exacerbate or mitigate temperature-dependent stresses. However, it is difficult to predict panicle temperature change under field conditions because panicles have shapes and characteristics that differ from those leaves on transpiration (Ishihara et al., 1990), which generally leads to a sizeable difference between panicle temperature and leaf/air temperatures (Nishiyama 1981).

Yoshimoto et al. (2005) developed a heat balance model to simulate panicle temperature and its transpiration. However, the model needs many parameters of microclimates. We developed a simple and convenient empirical model to estimate the contribution of solar radiation to panicle temperature using generalized linear models in Experiment 2. The contribution of solar radiation for panicle temperature was estimated using this model.

3.1 Materials and methods

The study was conducted at the same location in Experiment 1. We used six rice cultivars in this study. The nursery growing and transplanting methods used were similar to those in Experiment 1. Table 6 shows the seeding dates and dates for measuring panicle temperature. The planting density was 22.2 hills m^{-2}. The area of each plot was 12.8 m^2. A basal dressing of 4.0 g m^{-2} of N, 4.9 g m^{-2} of P$_2$O$_5$, and 4.3 g m^{-2} of K$_2$O was applied. Top dressing was not used. Culture methods such as irrigation and pesticide used followed the standard local practices for rice production in Shimane Prefecture.

Cultivar	Seeding dates	Measurement dates
'Fujihikari'	April 20	July 17, 18
'Koshihikari'	May 18, June 20	August 13, 15, September 6
'Xiaomazhan'	April 20	July 29
'Milky Queen'	May 18, June 20	August 13, 15, September 6
'IR72'	May 18	September 6
'Asahi'	May 18	September 6

Table 5. Seeding dates and measurement dates for six *O. sativa* cultivars in Experiment 2.

3.1.1 Measurements of panicle temperature and weather

We measured panicle and leaf temperatures every hour between 0600 and 1200 on the flower opening days (Table 5), using infrared thermography (Neo Thermo TVS-700, Nippon Avionics co. LTD., Tokyo, Japan) with a fixed emissivity factor of 1. Weather factors (solar radiation, air temperature, relative humidity, and wind speed) were measured every 5 min using a wireless weather station (Wireless Vantage Pro, Davis Instruments, Hayward, CA, USA) located at Shimane University. Measurement details and calculating weather factors are the same in Experiment 1.

3.1.2 Statistical analysis

As there are inherent relationships among weather factors, relatively high correlations may exist among air temperature, solar radiation, vapor-pressure deficit, and wind speed. Therefore, to evaluate their individual effects as well as the cultivar effects on panicle temperature, we analyzed our data by means of generalized linear models and multiple regression procedures using SPSS (Version 14J for Windows, SPSS Japan Inc., Tokyo, Japan). In this analysis, air temperature, solar radiation, and vapor-pressure deficit values were used in one hour intervals between 6000 and 1200 based on Japan Standard Time. The relative contribution of each weather component to panicle temperature was determined using their standardized partial regression coefficients, and the overall strength of the relationships was quantified using the multiple correlation coefficients.

3.2 Results

The weather in 2011 was variable, in contrast with the record hot weather in 2010 (Fig. 9). Air temperature and solar radiation in mid-July to mid-August were high. In 2011, two strong typhoons affected Japan in late July and late August; as a result, air temperature and solar radiation in late July and late August dropped rapidly.

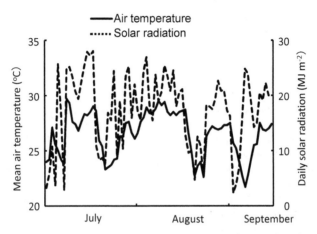

Fig. 8. Daily mean air temperatures and total solar radiation during the observation period.

The weather during the observation hours (0600–1200) differed every on the measuring days (Table 6). On July 17, air temperature was higher than 30°C, and solar radiation was high (13.5 MJ m^{-2}). On the other hand, solar radiation was low (5.0 MJ m^{-2}) on July 18. On September 6, air temperature was below 22°C, but solar radiation was high (12.1 MJ m^{-2}).

Measurement date	Air temperature (°C)	Solar radiation (MJ m^{-2})	Vapor-pressure deficit (hPa)	Wind speed (m s^{-1})
July 17	30.2	13.5	9.6	2.5
July 18	29.2	5.0	8.8	3.3
July 29	28.4	11.5	7.7	5.9
August 13	29.5	12.5	9.0	2.4
August 15	29.2	7.9	8.3	3.8
September 6	22.4	12.1	4.8	2.5

Table 6. Mean air temperature, total solar radiation, mean vapor-pressure deficit, and wind speed between 0600 and 1200 on measuring days in Experiment 2.

3.2.1 Meteorological variables, panicle and leaf temperatures

The air temperature on August 13 was similar to that on August 15; on the other hand, the solar radiation was approximately 58% higher than that on August 15 (Table 6). Solar radiation on August 15 increased between 0600 and 0800, but it decreased between 0800 and 0900 (Fig. 9). The level of solar radiation between 0800 and 1200 on August 15 was lower than that on August 13.

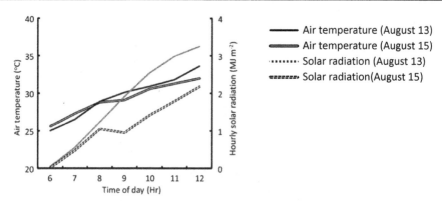

Fig. 9. Solar radiation and air temperature between 0600 and 1200 on August 13 and 15.

Panicle and leaf temperatures in both cultivars were as high as the air temperature between 0600 and 0700. As solar radiation increased, both panicle and leaf temperatures increased above the air temperature, and the differences between the panicle and air temperatures and between the leaf and air temperatures increased. In both cultivars and on both days, the panicle temperature was higher than the leaf temperature by approximately 1°C. On August 15, the increases in panicle and leaf temperatures in both cultivars were small between 0800 and 1200, probably because of lower solar radiation. The differences in panicle and leaf temperatures between 'Milky Queen' and 'Koshihikari' were also small.

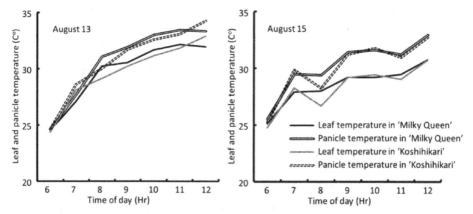

Fig. 10. Leaf temperature and panicle temperature in two rice cultivars ('Milky Queen' and 'Koshihikari') between 0600 and 1200 on August 13 (left figure) and 15 (right figure).

The air temperature on September 6 was lower than that on August 13 and 15; on the other hand, solar radiation on September 6 was as high as that on August 13 (Table 6). Solar radiation on September 6 increased linearly between 0600 and 1200 just as on August 13. However, the air temperature between 0600 and 1200 was less than 30°C, although air temperature increased linearly at a rate of 1.5°C hr^{-1}; this rate was higher than that on August 13 (1.4°C hr^{-1}).

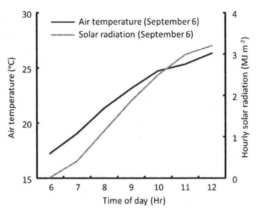

Fig. 11. Solar radiation and air temperature between 0600 and 1200 on September 6.

Panicle and leaf temperatures in the four cultivars were slightly lower than the air temperature at 0600. As solar radiation increased, panicle and leaf temperatures increased rapidly. In all cultivars, the panicle temperature was higher than leaf temperature by approximately 1–2°C. However, panicle and leaf temperatures between 0600 and 1200 were less than 30°C although they increased at the rate of 1.8°C hr^{-1}, which was a higher rate than that on August 13 (1.4°C hr^{-1}). At 1100, panicle and leaf temperatures in 'Asahi' and 'Koshihikari' dropped, but the reason for this decrease was unclear.

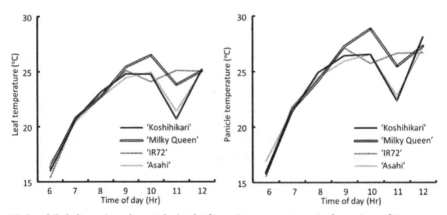

Fig. 12. Leaf (left figure) and panicle (right figure) temperatures in four rice cultivars ('Koshihikari', 'Milky Queen', 'IR72'and 'Asahi') between 0600 and 1200 on September 6.

3.2.2 Evaluation of the effects of solar radiation, cultivar, air temperature, vapor-pressure deficit, and wind speed on panicle and leaf temperatures

The multiple correlation coefficient determined by the general linear model for leaf temperature was 0.911 ($p < 0.001$), and three weather parameters (air temperature, solar radiation, and vapor-pressure deficit) significantly influenced leaf temperature (Table 7). A cultivar effect was not observed. The standardized partial regression coefficient for air temperature was 1.580; on the other hand, the standardized partial regression coefficient for solar radiation was small (0.180). The multiple correlation coefficient determined by the general linear model for panicle temperature was 0.891 ($p < 0.001$), and three weather factors (air temperature, solar radiation, and vapor-pressure deficit) also significantly influenced on panicle temperature (Table 7). A cultivar effect was not detected. The standardized partial regression coefficient for air temperature was 1.430; on the other hand, the standardized partial regression coefficient for solar radiation (0.225) was slightly larger than that of the generalized linear model for leaf temperature.

Dependent variables	Multiple correlation coefficient	F values in the generalized linear models				
		Cultivar	Air temperature	Solar radiation	Vapor-pressure deficit	Wind speed
Leaf temperature	0.911***	0.42 ns	84.91***	6.96*	11.30**	2.92 ns
Panicle temperature	0.891***	1.21 ns	85.21***	13.32*	10.82*	0.06 ns

Dependent variables	Standardized partial regression coefficient			
	Air temperature	Solar radiation	Vapor-pressure deficit	Wind speed
Leaf temperature	1.580	0.180	−0.702	−0.072
Panicle temperature	1.430	0.225	−0.620	−0.010

ns (not significant). *, **, and *** (significant at $p < 0.05$, $p < 0.01$, and $p < 0.001$) respectively.

Table 7. Results of multiple regression analysis using general linear models for the correlations between four weather factors (air temperature, solar radiation, vapor-pressure deficit, and wind speed) and panicle and leaf temperatures for every hour during 0600 and 1200.

3.3 Discussion

The direct effects of solar radiation on panicle and leaf temperatures were smaller than those of air temperature and vapor-pressure deficit because the standardized partial regression coefficient for air temperature was much larger than that for solar radiation (Table 7.). Higher solar radiation on August 13 resulted in higher panicle and leaf temperatures (Fig. 10.), but air temperature was also affected by solar radiation; higher air temperature also resulted in higher panicle and leaf temperatures. Yoshimoto et al. (2007) estimated the panicle temperature in New South Wales, Australia and in Jiangsu Province, China, using their heat balance model to simulate panicle temperature and its transpiration. In New South Wales, solar radiation was higher and exceeded 1000 W m^{-2} at midday with higher air temperatures of nearly 40°C; on the other hand, solar radiation is approximately 800 W m^{-2} at midday with modest air temperatures of nearly 35°C in Jiangsu Province, China. However, the estimated panicle temperature was lower in New South Wales than in Jiangsu Province due to low relative humidity of <50% and high wind speeds (2-6 m s^{-2}). Based on these results, we concluded that the effect of solar radiation on panicle temperature is indirect and through increasing air temperature.

Cultivar effects on panicle and leaf temperatures were not observed. However, differences in cultivar may exist in the ability to reduce panicle temperature through panicle transpiration. The change in the reflectance of the surface of the leaves and panicles may mitigate high solar radiation. Purple panicles with low reflectance and white and hairy panicles with high reflectance may respond differently to solar radiation. Research that includes tissue reflectance and conductance for panicle transpiration is needed to estimate panicle temperature accurately. An attempt should be made to conduct another experiment under a wider range of weather factors under field conditions.

4. Conclusion

1. Higher solar radiation after sunrise resulted in an earlier flower opening time (Fig. 4). The cultivars examined in this experiment were classified into two groups: cultivars with negative, high correlations between solar radiation and the flower opening time and those with relatively weak correlations.
2. The correlation coefficients between the flower opening time and hourly solar radiation on the mean flower opening time dropped rapidly at 5 hr before flower opening (Fig. 5). This result suggests that the sensitivity of rice flowers to solar radiation in flower opening increases before flower opening.
3. We obtained the highest multiple correlation coefficient (adjusted $R^2 = 0.849$, $p < 0.001$) for the 0800-0900 period (Table 3). During this period, three factors (cultivar, solar radiation, and air temperature) were significant ($p < 0.05$). The standardized partial regression coefficient for air temperature was −0.131, and that a higher air temperature during this period resulted in an earlier flower opening time. Similarly, the standardized partial regression coefficient of solar radiation was −0.002, and that higher solar radiation during this period also resulted in an earlier flower opening time.
4. Two distinctive peaks, one immediately after sunset and one immediately after sunrise were observed in the diurnal change of multiple correlation coefficients (Fig. 7). This result suggests that other aspects of solar radiation such as the light cycle may influence flower opening time in addition to the amount of solar radiation.
5. The direct effects of solar radiation on panicle and leaf temperatures were smaller than those of air temperature and vapor-pressure deficit.

5. Acknowledgments

We thank Dr. Fumihiko Adachi (Shimane University) for scientific support in obtaining weather data from the Shimane University Web site (http://www.ipc.shimane-u.ac.jp/weather/station/index.html). and Ms. Yuka Omura for technical support in obtaining the data regarding flower opening time.

6. References

Buck, A.L. (1981). New equations for computing vapor pressure and enhancement factor, *Journal of Applied Meteorology*, Vol. 20, No. 12, (Debember 1981), pp. 1527-1532, ISSN 0894-8763

Hirano, K.; Aya, K.; Hobo, T.; Sakakibara, H.; Kojima, M.; Shim, R.A.; Hasegawa, Y.; Ueguchi-Tanaka, M. & Matsuoka, M. (2008). Comprehensive transcriptome analysis of phytohormone biosynthesis and signaling genes in microspore/pollen and tapetum of rice, *Plant & Cell Physiology*, Vol. 49, No. 10, (October 2008), pp. 1429-1450, ISSN 0032-0781

Horie, T.; Matsui, T.; Nakagawa, H. & Omasa, K. (1996). Effects of elevated CO_2 and global climate change on rice yield in Japan, In : *Climate Change and Plants in East Asia*, K. Omasa ; K. Kai ; H. Taoda ; Z. Uchijima and M. Yoshino (Eds.), pp. 39-56, Springer-Verlag, ISBN 4-431-70176-1, Tokyo, Japan

Hoshikawa, K. (1989). *The Growing Rice Plant–An Anatomical Monograph*, Nosan Gyoson Bunka Kyokai, ISBN 978-4-540-88113-8, Tokyo, Japan

Imaki, T.; Jyokei, K. & Hara, K. (1982). Flower opening under the controlled environments in rice plants, *Bulletin of the Faculty of Agriculture, Shimane University*, Vol. 16, (December, 1982), pp 1-7, ISSN 0370-940X*

Imaki, T.; Jyokei, K. & Yamada, I. (1983). Sterility caused by high temperature at flowering in rice plants, *Bulletin of the Faculty of Agriculture, Shimane University*, Vol. 17, (December, 1983), pp 1-7, ISSN 0370-940X *

Imaki, T.; Tokunaga, S. & Obara, N. (1987). High temperature sterility of rice spikelets at flowering in relation to flowering time, *Japanese Journal of Crop Science*, Vol. 56, Extra issue 2, (October 1987), pp. 209-210, ISSN 0011-1848**

Ishihara, K.; Kiyota, E. & Imaizumi, N. (1990). Transpiration and Photosynthesis characteristics of the panicle in comparison with the flag leaf in the rice plant, *Japanese Journal of Crop Science*, Vol. 59, No. 2, (September 1990), pp. 321-326, ISSN 0011-1848***

Ishimaru, T.; Hirabayashi, H.; Ida, M.; Takai, T.; San-Oh, Y.A.; Yoshinaga, S.; Ando, I.; Ogawa, T. & Kondo, M. (2010). A genetic resource for early-morning flowering trait of wild rice *Oryza officinalis* to mitigate high temperature-induced spikelet sterility at anthesis, *Annals of Botany*, Vol. 106, No. 3, (September 2010), pp. 515-520, ISSN 0305-7364

Jagadish, S.V.K.; Craufurd, P.Q. & Wheeler, T.R. (2007). High temperature stress and spikelet fertility in rice (*Oryza sativa* L.), *Journal of Experimental Botany*, Vol. 58, No. 7, (May 2007), pp. 1627-1635, ISSN 0022-0957

Jagadish, S.V.K.; Craufurd, P.Q. & Wheeler, T.R. (2008). Phenotyping parents of mapping populations of rice for heat tolerance during anthesis, *Crop Science*, Vol. 48, No.3, (May 2008), pp. 1140-1146, ISSN 0011-183X

Kaihara, S. & Takimoto, A. (1979). Environmental factors controlling the time of flower-opening in *Pharbitis nil*, *Plant & Cell Physiology*, Vol. 20, No. 8, (December 1979), pp. 1659-1666, ISSN 0032-0781

Kaihara, S. & Takimoto, A. (1980). Studies on the light controlling the time of flower-opening in *Pharbitis nil*, *Plant & Cell Physiology*, Vol. 21, No. 1, (February 1980), pp. 21-26, ISSN 0032-0781

Kaihara, S. & Takimoto A. (1981a). Effects of light and temperature on flower-opening of *Pharbitis nil*, *Plant & Cell Physiology*, Vol. 22, No. 2, (April 1981), pp. 215-221, ISSN 0032-0781

Kaihara, S. & Takimoto A. (1981b). Physical basis of flower-opening in *Pharbitis nil*, *Plant & Cell Physiology*, Vol. 22, No. 2, (April 1981), pp. 307-310, ISSN 0032-0781

Kaihara, S. & Takimoto A. (1983). Effects of plant growth regulators on flower-opening of *Pharbitis nil*, *Plant & Cell Physiology*, Vol. 24, No. 3, (April 1983), pp. 309-316, ISSN 0032-0781

Kim, H.Y.; Horie, T.; Nakagawa, H. & Wada, K. (1996). Effect of elevated CO_2 concentration and high temperature on growth and yield of rice. II. The effect on yield and its components of Akihikari rice, *Japanese Journal of Crop Science*, Vol. 65, No. 4, (December 1996), pp. 644-651, ISSN 0011-1848***

Kobayasi, K.; Matsui, T.; Yoshimoto, M. and Hasegawa, T. (2010). Effects of temperature, solar radiation, and vapor-pressure deficit on flower opening time in rice, *Plant Production Science*, Vol. 13, No. 1, (January 2010), pp. 21-28, ISSN 1343-943X

Kobayasi, K. & Atsuta, Y. (2010). Sterility and poor pollination due to early flower opening induced by methyl jasmonate, *Plant Production Science*, Vol. 13, No. 1, (January 2010), pp. 29-36, ISSN 1343-943X

Koike, S. & Satake, T. (1987). Sterility caused by cooling treatment at the flowering stage in rice plants. II. The abnormal digestion of starch in pollen grain and metabolic changes in anthers following cooling treatment, *Japanese Journal of Crop Science*, Vol. 56, No. 4, (December 1987), pp. 666-672, ISSN 0011-1848

Matsui, T.; Omasa, K. and Horie, T. (1997a). High temperature-induced spikelet sterility of japonica rice at flowering in relation to air temperature, humidity and wind velocity conditions, *Japanese Journal of Crop Science*, Vol. 66, No. 3, (September 1997), pp. 449-455, ISSN 0011-1848

Matsui, T.; Namuco, O.S.; Ziska, L. H. and Horie, T. (1997b). Effects of high temperature and CO_2 concentration on spikelet sterility in indica rice, *Field Crops Research*, Vol. 51, No. 3, (April 1997), pp. 213-219. ISSN 0378-4290

Matsui, T.; Kobayasi, K.; Yoshimoto, M. & Hasegawa, T. (2007). Stability of rice pollination in the field under hot and dry conditions in the Riverina Region of New South Wales, Australia, *Plant Production Science*, Vol. 10, No. 1, (January 2007), pp. 57-63, ISSN 1343-943X

Nakagawa, H. & Nagata, A. (2007). Internal and environmental factors affecting the time of flower-opening in rice, *Japanese Journal of Crop Science*, Vol. 76, Extra issue 2, (September 2007), pp. 280-281, ISSN 0011-1848**

Nishiyama, I. & Blanco, L. (1980). Avoidance of high temperature sterility by flower opening in the early morning, *Japan Agricultural Research Quarterly*, Vol. 14, No. 2, (April 1980), pp. 116-117, ISSN 0021-3551

Nishiyama, I. (1981). Temperature inside the flower of rice plants, *Japanese Journal of Crop Science*, Vol. 50, No. 1, (March 1981), pp. 54-58, ISSN 0011-1848

Nishiyama, I. & Blanco, L. (1981). Artificial control of flower opening time during the day in rice plants. I. Preliminary experiments, *Japanese Journal of Crop Science*, Vol. 50, No. 1, (March 1981), pp. 59-66, ISSN 0011-1848

Nishiyama, I. & Satake, T. (1981). High temperature damages in rice plants, *Japanese Journal of Tropical Agriculture*, Vol. 25, No.1, (March 1981), pp. 14-19, ISSN 0021-5260**

Satake, T. & Yoshida, S. (1978). High temperature-induced sterility in indica rices at flowering, *Japanese Journal of Crop Science*, Vol. 47, No. 1, (March 1978), pp. 6-17, ISSN 0011-1848

Tsuboi, Y. (1961). Ecological studies on rice plants with regard to damages caused by wind, *Bulletin of the National Institute of Agricutural Sciences*, Series A, No. 8, (March, 1961). pp. 1-156, ISSN 0077-4820*

Wang, C.; Yang, J.; Wa, J. and Cai, Q. (2004). Influence of high and low temperature stress on fertility and yield of rice (*Oryza sativa* L.): Case study with the Yangtze River rice cropping region in China, *Abstract of World Rice Research Conference 2004*, Tsukuba, Japan, (November 2004), pp 97

Yoshimoto, M.; Oue, H.; Takahashi, H. & Kobayashi, K. (2005). The effects of FACE (Free-Air CO_2 Enrichment) on temperatures and transpiration of rice panicles at flowering stage, *Journal of Agricultural Meteorology*, Vol. 60, No. 5, (December 2005), pp. 597-600, ISSN 0021-8588

Yoshimoto, M.; Matsui, T.; Kobayasi, K.; Nakagawa, H.; Fukuoka, M. & Hasegawa, T. (2007). Micrometeorological effects on heat induced spikelet sterility of rice estimated by energy balance model, *Japanese Journal of Crop Science*, Vol. 76, Extra issue 2, (September 2007), pp. 162-163, ISSN 0011-1848**

Zeng, X.C.; Zhou, X.; Zhang, W.; Murofushi, N.; Kitahara, T. & Kamuro, Y. (1999). Opening of rice floret in rapid response to methyl jasmonate, *Journal of Plant Growth Regulation*, Vol. 18, No. 4, (December 1999), pp. 153-158, ISSN 0721-7595

Zhao, L.; Kobayasi, K.; Hasegawa, T.; Wang, C.L.; Yoshimoto, M.; Wan, J. & Matsui, T. (2010). Traits responsible for variation in pollination and seed set among six cultivars grown in a miniature paddy field with free air at a hot, humid spot in China, *Agriculture, Ecosystem and Environment*, Vol. 139, No.1-2, (October 2010), pp. 110-115, ISSN 0167-8809

* In Japanese with English summary.
** In Japanese.
*** In Japanese with English abstract.

Permissions

The contributors of this book come from diverse backgrounds, making this book a truly international effort. This book will bring forth new frontiers with its revolutionizing research information and detailed analysis of the nascent developments around the world.

We would like to thank E. B. Babatunde, for lending his expertise to make the book truly unique. He has played a crucial role in the development of this book. Without his invaluable contribution this book wouldn't have been possible. He has made vital efforts to compile up to date information on the varied aspects of this subject to make this book a valuable addition to the collection of many professionals and students.

This book was conceptualized with the vision of imparting up-to-date information and advanced data in this field. To ensure the same, a matchless editorial board was set up. Every individual on the board went through rigorous rounds of assessment to prove their worth. After which they invested a large part of their time researching and compiling the most relevant data for our readers. Conferences and sessions were held from time to time between the editorial board and the contributing authors to present the data in the most comprehensible form. The editorial team has worked tirelessly to provide valuable and valid information to help people across the globe.

Every chapter published in this book has been scrutinized by our experts. Their significance has been extensively debated. The topics covered herein carry significant findings which will fuel the growth of the discipline. They may even be implemented as practical applications or may be referred to as a beginning point for another development. Chapters in this book were first published by InTech; hereby published with permission under the Creative Commons Attribution License or equivalent.

The editorial board has been involved in producing this book since its inception. They have spent rigorous hours researching and exploring the diverse topics which have resulted in the successful publishing of this book. They have passed on their knowledge of decades through this book. To expedite this challenging task, the publisher supported the team at every step. A small team of assistant editors was also appointed to further simplify the editing procedure and attain best results for the readers.

Our editorial team has been hand-picked from every corner of the world. Their multi-ethnicity adds dynamic inputs to the discussions which result in innovative outcomes. These outcomes are then further discussed with the researchers and contributors who give their valuable feedback and opinion regarding the same. The feedback is then collaborated with the researches and they are edited in a comprehensive manner to aid the understanding of the subject.

Apart from the editorial board, the designing team has also invested a significant amount of their time in understanding the subject and creating the most relevant covers. They scrutinized every image to scout for the most suitable representation of the subject and create an appropriate cover for the book.

The publishing team has been involved in this book since its early stages. They were actively engaged in every process, be it collecting the data, connecting with the contributors or procuring relevant information. The team has been an ardent support to the editorial, designing and production team. Their endless efforts to recruit the best for this project, has resulted in the accomplishment of this book. They are a veteran in the field of academics and their pool of knowledge is as vast as their experience in printing. Their expertise and guidance has proved useful at every step. Their uncompromising quality standards have made this book an exceptional effort. Their encouragement from time to time has been an inspiration for everyone.

The publisher and the editorial board hope that this book will prove to be a valuable piece of knowledge for researchers, students, practitioners and scholars across the globe.

List of Contributors

E. B. Babatunde
Covenant University, Ota, Ogun State, Nigeria

E. O. Falayi
Department of Physics, Tai Solarin University of Education, Ijebu-Ode, Nigeria

A. B. Rabiu
Department of Physics, Federal University of Technology, Akure, Nigeria

Goro Yamanaka, Hiroshi Ishizaki, Hiroyuki Tsujino, Hideyuki Nakano and Mikitoshi Hirabara
Meteorological Research Institute, Japan Meteorological Agency, Japan

Edgar G. Pavia
Centro de Investigación Científica y de Educación Superior de Ensenada, Mexico

Kalju Eerme
Tartu Observatory, Estonia

Hu Bo
State Key Laboratory of Atmospheric Boundary Layer Physics and Atmospheric Chemistry (LAPC), Institute of Atmospheric Physics, Chinese Academy of Sciences, Beijing, China

Esperanza Carrasco and Alberto Carramiñana
Instituto Nacional de Astrofísica, Óptica y Electrónica, Puebla, México

Remy Avila
Centro de Física Aplicada y Tecnología Avanzada, Universidad Nacional Autónoma de México, Santiago de Querétaro, México

Leonardo J. Sánchez and Irene Cruz-González
Instituto de Astronomía, Universidad Nacional, Autónoma de México, México D.F., México

Isabel Tamara Pedron
Paraná Western State University, Brazil

E. B. Babatunde
Covenant University, Ota, Ogun State, Nigeria

Carlos Campillo, Rafael Fortes and Maria del Henar Prieto
Centro de Investigación finca la Orden-Valdesequera, Spain

M. Azizul Moqsud
Kyushu University, Japan

Amy L. Norris and Thomas H. Kunz
Boston University, USA

Kazuhiro Kobayasi
Shimane University, Japan

Printed in the USA
CPSIA information can be obtained
at www.ICGtesting.com
JSHW011449221024
72173JS00004B/1009

9 781632 394132